Wonders of Numbers

DR. GOOGOL PRESENTS

Wonders of Numbers

Adventures in Mathematics, Mind, and Meaning

Clifford A. Pickover

OXFORD
UNIVERSITY PRESS
2001

OXFORD
UNIVERSITY PRESS

Oxford New York

Athens   Auckland   Bangkok   Bogotá   Buenos Aires   Calcutta
Cape Town   Chennai   Dar es Salaam   Delhi   Florence   Hong Kong   Istanbul
Karachi   Kuala Lumpur   Madrid   Melbourne   Mexico City   Mumbai   Nairobi
Paris   São Paulo   Shanghai   Singapore   Taipei   Tokyo   Toronto   Warsaw
and associated companies in
Berlin   Ibadan

Copyright © 2001 by Clifford A. Pickover

Published by Oxford University Press, Inc.
198 Madison Avenue, New York, New York 10016

Library of Congress Cataloging-in-Publication Data
Pickover, Clifford A.
Wonders of numbers: adventures in mathematics, mind, and meaning /
by Clifford A. Pickover.
p.   cm.
Includes bibliographical references and index.
At head of title: Dr. Googol presents.
ISBN 0-19-513342-0 (alk. paper)
1. Mathematical recreations. 2. Number theory. I. Title.
II. Title: Dr. Googol presents.
QA95.P53 2000
793.7'4-dc21   99-27044

1 3 5 7 9 8 6 4 2
Printed in the United States of America
on acid-free paper

This book is dedicated not to a person but rather to an amusing mathematical wonder: the Apocalyptic Magic Square—a rather bizarre six-by-six magic square in which all of its entries are prime numbers (divisible only by themselves and 1), and each row, column, and diagonal sum to 666, the Number of the Beast.

THE APOCALYPTIC MAGIC SQUARE

3	107	5	131	109	311
7	331	193	11	83	41
103	53	71	89	151	199
113	61	97	197	167	31
367	13	173	59	17	37
73	101	127	179	139	47

❀ For additional wondrous features of this square, see Chapter 101.

We are in the position of a little child entering a huge library
whose walls are covered to the ceiling with books in many
different tongues. The child does not understand the
languages in which they are written. He notes a
definite plan in the arrangement of books,
a mysterious order which he does not
comprehend, but only
dimly suspects.
—*Albert Einstein*

Amusement
is one of humankind's
strongest motivating forces.
Although mathematicians sometimes
belittle a colleague's work by calling it
"recreational" mathematics, much serious
mathematics has come out of recreational problems,
which test mathematical logic and reveal mathematical truths.
—*Ivars Peterson,* Islands of Truth

The mathematician's job is to transport us to new seas,
while deepening the waters
and lengthening
horizons.
—*Dr. Francis O. Googol*

Acknowledgments

ACKNOWLEDGMENTS FROM CLIFFORD A. PICKOVER

Legendary mathematician Dr. Francis O. Googol currently resides on a small island off the coast of Sri Lanka. Because he desires privacy to continue his research, he has allowed my name to appear on this book's title page. In the past, I have frequently collaborated with Dr. Googol and edited his work. You can reach Dr. Googol by writing to me, and you can read more about the extraordinary life of Dr. Googol in the "Word from the Publisher" that follows this section. Dr. Googol admits to pillaging a few of my older papers, books, lectures, and patents for ideas, but he has brought them up to date with reader comments and startlingly fresh insight and presentation.

ACKNOWLEDGMENTS FROM DR. FRANCIS GOOGOL

Martin Gardner and Ian Stewart, two scintillating stars in the universe of recreational mathematics and mathematics education, are always a source of inspiration. Martin Gardner, a mathematician, journalist, humorist, rationalist, and prolific author, has long stunned the world by giving countless people an incentive to study and become fascinated by mathematics.

Many other individuals have provided intellectual stimulation over the years: Arthur C. Clarke, J. Clint Sprott, Ivars Peterson, Paul Hoffman, Theoni Pappas, Douglas Hofstader, Charles Ashbacher, Dorian Devins, Rudy Rucker, John Conway, Jack Cohen, and Isaac and Janet Asimov.

Dr. Googol thanks Brian Mansfield for his creative advice and encouragement. Aside from drawing the various number mazes, Brian also created all of the cartoon representations of Dr. Googol from rare photographs in Googol's private archives. Dr. Googol also thanks Kevin Brown, Olivier Gerard, Dennis Gordon, Robert E. Stong, and Carl Speare for further advice and encouragement.

He also owes a special debt of gratitude to Dr. John J. O'Connor and Professor Edmund F. Robertson (School of Mathematics and Statistics, University of St. Andrews, Scotland) for their wonderful "MacTutor History of Mathematics Archive," http://www-history.mcs.st-andrews.ac.uk/history/index.html. This web page allows users to access biographical data of more than 1300 mathematicians, and Dr. Googol used this wonderful archive extensively for background information for Chapters 29, 33, and 38.

A Word from the Publisher about Dr. Googol

Francis Googol's date of birth is unknown. According to court records, he was born in London, England, and has held various "jobs" including mathematician, world explorer, and inventor. A prolific author of over 300 publications, Googol achieved his greatest fame with his book *Number Madness*, in which he argued that Neanderthals invented a primitive form of calculus. He also conducted pioneering studies of parabolas and statistics and was knighted in 1998. Dr. Googol is a practical scientist, always testing his theories using apparatuses of his own design.

Today, Dr. Googol has an obsessive predilection for quantifying anything that he views—from the curves of women's bodies to the number of brush strokes used to paint his portrait. It is rumored that he even published anonymously a paper in *Nature* on the length of rope necessary for breaking a criminal's neck without decapitation. In short, Googol is obsessed with the idea that anything can be counted, correlated, and understood as some sort of pattern. Clements Markham (former president of the Geographical Society) once remarked, "His mind is mathematical and statistical with little or no imagination."

When asked his advice on life, Googol responded: "Travel and do mathematics."

❀ ❀ ❀

Francis Googol, great-great-great-grandson of Charles Darwin, was born to a family of bankers and gunsmiths of the Quaker faith. His family life was happy. Googol's mother, Violetta, lived to 91, and most of her children lived to their 90s or late 80s. Perhaps the longevity of his ancestors accounts for Googol's very long life.

When Francis Googol was born, 13-year-old sister Elizabeth asked to be his primary caretaker. She placed Googol's cot in her room and began teaching him numbers, which he could point to and recognize before he could speak. He would cry if the numbers were removed from sight.

As an adult, Googol became bored by life in England and felt the urge to explore the world. "I craved travel," he said, "as I did all adventure." For the next decade, he embarked on a shattering odyssey of self-discovery; in fact, his biography reads more like Pirsig's *Zen and the Art of Motorcycle Maintenance* or Simon's *Jupiter's Travels* than like the life story of a mathematical genius. Googol suddenly moved like a roller coaster over some of the world's most mysterious physical and psychological terrain: studies of the female monkeys at Kathmandu, camel rides through Egyptian desserts, death-defying escapes in the jungles of Tanzania. . . . Anyone who hears about Googol's journeys is enthralled by Googol's descriptions of the exotic places and people, by his ability to adjust to adversity, by his humor and incisiveness, but above all by the realization that to understand his world, he had to make himself vulnerable to it so that it could change him.

Preface

One Fish, Two Fish, and Beyond . . .

> The trouble with integers is that we have examined only the small ones.
> Maybe all the exciting stuff happens at really big numbers, ones we can't
> get our hands on or even begin to think about in any very definite way.
> So maybe all the action is really inaccessible and we're just fiddling
> around. Our brains have evolved to get us out of the rain, find where the
> berries are, and keep us from getting killed. Our brains did not evolve to
> help us grasp really large numbers or to look at things in a hundred
> thousand dimensions.
>
> —*Ronald Graham*

> Mathematics, rightly viewed, possesses not only truth, but supreme
> beauty—a beauty cold and austere, like that of sculpture.
> —*Bertrand Russell,* Mysticism and Logic, 1918

> The primary source of all mathematics is the integers.
> —*Herman Minkowski*

Dr. Googol loves numbers. Whole numbers. Big ones like **1,000,000**. And little ones like 2 or 3. In this book, you will see integers more often than fractions like ½, trigonometric functions like "sine," or complicated, long-winded numbers like $\pi = 3.1415926$. . . . He cares mainly about the integers.

Dr. Googol, world-famous explorer and brilliant mathematician, knows that his obsession with integers sounds silly to many of you, but integers are a great way to transcend space and time. Contemplating the wondrous relationships among these numbers stretches the imagination, and the usefulness of these numbers allows us to build spaceships and investigate the very fabric of our universe. Numbers will be our first means of communication with intelligent alien races.

Ancient people, like the Greeks, had a deep fascination with numbers. Could it be that in difficult times numbers were the only constant thing in an ever-shifting world? To the Pythagoreans, an ancient Greek sect, numbers were tangible, immutable, comfortable, eternal—more reliable then friends, less threatening than Zeus.

The mysterious, odd, and *fun* puzzles in this book should cause even the most left-brained readers to fall in love with numbers. The quirky and exclusive surveys on mathematicians' lives, scandals, and passions will entertain people at all levels of mathematical sophistication. In fact, this book focuses on creativity, discovery, and challenge. *Parts 1* and *4* are especially tuned for amusing classroom explorations and experiments by beginners. *Part 2* is for classroom debate and for causing arguments around the dinner table or on the Internet. *Part 3* contains problems that sometimes require a little bit more mathematical manipulation.

When Dr. Googol talks to students about the strange numbers in this book, they are always fascinated to learn that it is possible for them to break numerical world records and make new discoveries with a personal computer. Most of the ideas can be explored with just a pencil and paper!

Number theory—the study of properties of the integers—is an ancient discipline. Much mysticism accompanied early treatises; for example, Pythagoreans explained many events in the universe in terms of whole numbers. Only a few hundred years ago courses in numerology—the study of mystical and religious properties of numbers—were required for all college students, and even today such numbers as 13, 7, and 666 conjure up emotional reactions in many people. Today, integer arithmetic is important in a wide spectrum of human activities and has repeatedly played a crucial role in the evolution of the natural sciences. (For a description of the use of number theory in communications, computer science, cryptography, physics, biology, and art, see Manfred Schroeder's *Number Theory in Science and Communication*.)

One of the abiding sins of mathematicians is an obsession with completeness—an urge to go back to first principles to explain their works. As a result, readers must often wade through pages of background before getting to the essential ingredients. To avoid this problem, each chapter in this book is less than 5 pages in length. Want to know about undulating numbers? Turn to Chapter 52, and in a few pages you'll have a quick challenge. Interested in Fibonacci numbers? Turn to Chapter 71 for the same. Want a ranking of the 8 most influential female mathematicians? Turn to Chapter 33. Want a list of the Unabomber's 10 most mathematical technical papers? Turn to Chapter 40. Want to know why Roman numerals aren't used anymore? Turn to Chapter 2. What are the latest practical applications of fractal geometry? Turn to the "Further Exploring" section of Chapter 54. Why was the first woman mathematician murdered? Turn to Chapter 29. You'll quickly get the essence of surveys, problems, games, and questions!

One advantage of this format is that you can jump right in to experiment and have fun, without having to sort through a lot of detritus. The book is not intended for mathematicians looking for formal mathematical explanations. Of course, this approach has some disadvantages. In just a few pages, Dr. Googol can't go into any depth on a subject. You won't find much historical context or extended discussion. That's okay. He provides lots of extra material in the "Further Exploring" and "Further Reading" sections.

To some extent, the choice of topics for inclusion in this book is arbitrary, although they give a nice introduction to some common and unusual problems in number theory and recreational mathematics. They are also problems that Dr.

Googol has researched himself and on which he has received mail from readers. Many questions are representative of a wider class of problems of interest to mathematicians today. Some information is repeated so that you can quickly dive into a chapter picked at random. The chapters vary in difficulty, so you are free to browse.

⊛ ⊕ ⊛

Why care about integers? The brilliant mathematician Paul Erdös (discussed in detail in Chapter 46) was fascinated by number theory and the notion that he could pose problems, using integers, that were often simple to state but notoriously difficult to solve. Erdös believed that if one can state a problem in mathematics that is unsolved and over 100 years old, it is a problem in number theory. There is a harmony in the universe that can be expressed by whole numbers. Numerical patterns describe the arrangement of florets in a daisy, the reproduction of rabbits, the orbit of the planets, the harmonies of music, and the relationships between elements in the periodic table. Leopold Kronecker (1823–1891), a German algebraist and number theorist, once said, "The integers came from God and all else was man-made." His implication was that the primary source of all mathematics is the integers. Since the time of Pythagoras, the role of integer ratios in musical scales has been widely appreciated.

More important, integers have been crucial in the evolution of humanity's scientific understanding. For example, in the 18th century, French chemist Antoine Lavoisier discovered that chemical compounds are composed of fixed proportions of elements corresponding to the ratios of small integers. This was very strong evidence for the existence of atoms. In 1925, certain integer relations between the wavelengths of spectral lines emitted by excited atoms gave early clues to the structure of atoms. The near-integer ratios of atomic weights was evidence that the atomic nucleus is made up of an integer number of similar nucleons (protons and neutrons). The deviations from integer ratios led to the discovery of elemental isotopes (variants with nearly identical chemical behavior but with different radioactive properties). Small divergences in pure isotopes' atomic weights from exact integers confirmed Einstein's famous equation $E = mc^2$ and also the possibility of atomic bombs. Integers are everywhere in atomic physics. Integer relations are fundamental strands in the mathematical weave—or, as German mathematician Carl Friedrich Gauss said, "Mathematics is the queen of sciences—and number theory is the queen of mathematics."

Prepare yourself for a strange journey as *Wonders of Numbers* unlocks the doors of your imagination. The thought-provoking mysteries, puzzles, and problems range from the most beautiful formula of Ramanujan (India's most famous mathematician) to the Leviathan number, a number so big that it makes a trillion pale in comparison. Each chapter is a world of paradox and mystery. Grab a pencil. Do not fear. Some of the topics in the book may appear to be curiosities, with little practical application or purpose. However, Dr. Googol has found these experiments to be useful and educational—as have the many students, educators, and scientists who have written to him during his long lifetime. Throughout history, experiments, ideas, and conclusions originating in the play of the mind have found striking and unexpected practical applications. In order to encourage your involvement, Dr. Googol provides computational hints.

As this book goes to press, Oxford University Press is delighted to announce a web site (www.oup-usa.org/sc/0195133420) that contains a smorgasbord of computer program listings provided by the author. Readers have often requested online code that they can study and with which they may easily experiment. We hope the code clarifies some of the concepts discussed in the book. Code is available for the following:

- ⊙ Chapter 2. Why Don't We Use Roman Numerals Anymore (BASIC program to generate Roman numerals when you type in any number)

- ⊙ Chapter 16. Jerusalem Overdrive (C program to scan for Latin Squares)

- ⊙ Chapter 17. The Pipes of Papua (Pseudocode for creating Papua rhythms)

- ⊙ Chapter 22. Klingon Paths (C and BASIC code to generate and explore Klingon paths)

- ⊙ Chapter 49. Hailstone Numbers (BASIC code for computing hailstone numbers and path lengths)

- ⊙ Chapter 50. The Spring of Khosrow Carpet (BASIC code for Persian carpet designs)

- ⊙ Chapter 51. The Omega Prism (BASIC code for finding the number of intersected tiles)

- ⊙ Chapter 53. Alien Snow: A Tour of Checkerboard Worlds (C code for exploring alien snow)

- ⊙ Chapter 54. Beauty, Symmetry, and Pascal's Triangle (BASIC code for computing and drawing Pascal's Triangle)

- ⊙ Chapter 56. Dr. Googol's Prime Plaid (BASIC code for exploring prime numbers and plaids)

- ⊙ Chapter 62. Triangular Numbers (BASIC code for computing triangular numbers)

- ⊙ Chapter 63. Hexagonal Cats (BASIC code for computing polygonal numbers)

- ⊙ Chapter 64. The *X-Files* Number (BASIC code for computing *X-Files* "End-of-the-World" Numbers)

- ⊙ Chapter 66. The Hunt for Elusive Squarions (BASIC code for generating pair square numbers)

- ⊙ Chapter 68. Pentagonal Pie (BASIC code for computing Catalan numbers)

- ⊙ Chapter 71. Mr. Fibonacci's Neighborhood (BASIC code for computing Fibonacci numbers)

⊙ Chapter 73. The Wonderful Emirp, 1597 (REXX code for computing prime Fibonacci numbers)

⊙ Chapter 83. The Leviathan Number (C and BASIC code for comparing Stirling and factorial values)

⊙ Chapter 85. The Aliens in *Independence Day* (C and BASIC code for computing number and sex of humans)

⊙ Chapter 88. The Latest Gossip on Narcissistic Numbers (BASIC code for searching for all cubical narcissistic numbers. Also, C code for factorion searches)

⊙ Chapter 89. The *abcdefgh* problem (REXX code for finding solutions to the *abcdefgh* problem)

⊙ Chapter 94. Perfect, Amicable, and Sublime Numbers (BASIC code for finding perfect and amicable numbers)

⊙ Chapter 96. Cards, Frogs, and Fractal Sequences (REXX code for computing fractal signature sequences. Also, BASIC code to compute Batrachions)

⊙ Chapter 99. Everything You Wanted to Know about Triangles but Were Afraid to Ask (BASIC code for generating Pythagorean triangles and for computing side lengths of triangles that pray)

⊙ Chapter 100. Cavern Genesis as a Self-Organizing System (C code for exploring stalactite formation)

⊙ Chapter 123. Zen Archery (Java code for solving Zen problems)

For many of you, seeing computer code will clarify concepts in ways mere words cannot.

Contents

⊙

PART I
FUN PUZZLES
AND QUICK THOUGHTS

☉

PART II
QUIRKY QUESTIONS, LISTS,
AND SURVEYS

⊙

PART III
FIENDISHLY DIFFICULT
DIGITAL DELIGHTS

⊙
PART IV
THE PERUVIAN COLLECTION

Part i

Fun Puzzles
and Quick
Thoughts

Your vision will become clear only when you can look
into your own heart. Who looks outside, dreams;
who looks inside, awakens.

—*Carl Jung*

Where there is an open mind, there will always be a frontier.
—*Charles Kettering*

Mathematics is the hammer that shatters the ice
of our unconscious.

—*Dr. Francis O. Googol*

Chapter 1

Attack of the Amateurs

Every productive research scientist cultivates and relies upon nonrational
processes to direct his or her own creative thinking. Watson and Crick
used visualization to conceive the DNA molecule's configuration.
Einstein used visualization to imagine riding on a light beam.
Mathematician Ramanujan usually saw a vision of his family Goddess
Narnagiri whenever he conceived of a new mathematical formula. The
heart of good science is the harmonious integration of good luck in making uncommonly made observations, nonrational processes that are only
poorly suggested by the words "creativity" and "intuition."
—*John Waters,* Skeptical Inquirer

Amazingly, lack of formal education can be an advantage. We get stuck in
our old ways. Sometimes, progress is made when someone from the outside looks at mathematics with new eyes.
—*Doris Schattschneider,* Los Angeles Times

Are you a mathematical amateur? Do not fret. Many amazing mathematical findings have been made by amateurs, from homemakers to lawyers. These amateurs
developed new ways to look at problems that stumped the experts.

Have any of you seen the movie *Good Will Hunting,* in which 20-year-old
Will Hunting survives in his rough, working-class South Boston neighborhood?
Like his friends, Hunting does menial jobs between stints at the local bar and
run-ins with the law. He's never been to college, except to scrub floors as a jani-
tor at MIT. Yet he can summon obscure historical references from his photographic memory and almost instantly solve math problems that frustrate the
most brilliant professors.

This is not as far-fetched as it sounds! Although you might think that new
mathematical discoveries can only be made by professors with years of training,

beginners have also made substantial contributions. Here are some of Dr. Googol's favorite examples:

☉ In the 1970s, Marjorie Rice, a San Diego housewife and mother of 5, was working at her kitchen table when she discovered numerous new geometrical patterns that professors had thought were impossible. Rice had no training beyond high school, but by 1976 she had discovered 58 special kinds of pentagonal tiles, most of them previously unknown. Her most advanced diploma was a 1939 high school degree for which she had taken only one general math course. The moral to the story? It's never too late to enter fields and make new discoveries. Another moral: Never underestimate your mother!

☉ In 1998, college student Roland Clarkson discovered the largest prime number known at the time. (A prime number, like 13, is evenly divisible only by 1 and itself.) The number was so large that it could fill several books. In fact, some of the largest prime numbers these days are found by college students using a network of cooperating personal computers and software downloadable from the Internet. (See "Further Exploring" for Chapter 56 to view the latest prime number records.)

☉ In the early 1600s, Pierre de Fermat, a French lawyer, made brilliant discoveries in number theory. Although he was an "amateur" mathematician, he created mathematical puzzles such as Fermat's Last Theorem, which was not solved until 1994. Fermat was no ordinary lawyer indeed. He is considered, along with Blaise Pascal, as the founder of probability theory. As the coinventor of analytic geometry along with René Descartes, he is considered one of the first modern mathematicians.

☉ In the mid-1990s, Texas banker Andrew Beal posed a perplexing mathematical problem and offered $5,000 for its solution. The value of the prize increases by $5,000 per year up to $50,000 until it is solved. In particular, Beal was curious about the equation $A^x + B^y = C^z$. The 6 letters represent integers, with x, y, and z greater than 2. (Fermat's Last Theorem involves the special case in which the exponents x, y, and z are the same.) Oddly enough, Beal noticed, when a solution of this general equation existed, then A, B, and C have a common factor. For example, in the equation $3^6 + 18^3 = 3^8$, the numbers 3, 18, and 3 all have the factor 3. Using computers at his bank, Beal checked equations with exponents up to 100 but could not discover a solution that didn't involve a common factor. He wondered if this is always true. R. Daniel Mauldin of the University of North Texas commented in the December 1997 *Notices of the American Mathematical Society*, "It is remarkable that occasionally someone working in isolation, and with no connections to the mathematical community, formulates a problem so close to current research activity."

☉ In 1998, 17-year-old Colin Percival calculated the five trillionth binary digit of pi. (Pi is the ratio of a circle's circumference to its diameter, and its digits

1.1 In 1998, 17-year-old Colin Percival calculated the five trillionth binary digit of pi. His accomplishment is significant not only because it was a record-breaker but because, for the first time ever, the calculations were distributed among 25 computers around the world. (Photo by Marianne Meadahl.)

go on forever. Binary numbers are defined in Chapter 21's "Further Exploring" section.) In 1999, computer scientist Yasumasa Kanada and his coworkers at the University of Tokyo Information Technology Center computed pi to 206,158,430,000 decimal digits. Percival (Figure 1.1) discovered that pi's five trillionth *bit*, or binary digit, is a 0. His accomplishment is significant not only because it was a record-breaker but because, for the first time ever, the calculations were distributed among 25 computers around the world. In all, the project, dubbed PiHex, took 5 months of real time to complete and a year and a half of computer time. Percival, who graduated from high school in June 1998, had been attending Simon Fraser University in Canada concurrently since he was 13.

⊙ In 1998, self-taught inventor Harlan Brothers and meteorologist John Knox developed an improved way of calculating a fundamental constant, *e* (often rounded to 2.718). Studies of exponential growth—from bacterial colonies to interest rates—rely on *e*, which can't be expressed as a fraction and can only be approximated using computers. Knox comments, "What we've done is bring mathematics back to the people" by demonstrating that amateurs can find more accurate ways of calculating fundamental mathematical constants. (Incidentally, *e* is known to more than 50 million decimal places.)

⊙ In 1998, Dame Kathleen Ollerenshaw and David Brée made important discoveries regarding a certain class of magic squares—number arrays whose rows, columns, and diagonals sum to the same number. Although their particular discovery had eluded mathematicians for centuries, neither discoverer was a typical mathematician. Ollerenshaw spent much of her professional life as a high-level administrator for several English universities. Brée has held university positions in business studies, psychology, and artificial intelligence. Even more remarkable is the fact that Ollerenshaw was 85 when she and Brée proved the conjectures she had earlier made. (For more information, see Ian Stewart, "Most-perfect magic squares." *Scientific American.* November, 281 (5): 122–123, 1999)

Hundreds of years ago, most mathematical discoveries were made by lawyers, military officers, secretaries, and other "amateurs" with an interest in mathematics. After all, back then, very few people could make a living doing pure mathematics. Modern-day French mathematician Olivier Gerard wrote to Dr. Googol:

> I believe that amateurs will continue to make contributions to science and mathematics. Computers and networks allow amateurs to work as efficiently as professionals and to cooperate with one another. When one considers the time wasted by many professionals in grant writing and for other paperwork justifying their activity, the amateurs may even have a slight edge in certain cases. However, the amateurs often lack the valuable experience of teaching or having a mentor.

This is not to say that amateurs can make progress in the most obscure areas in mathematics. Consider, for example, the strange list in Chapter 42 that includes the 10 most difficult-to-understand areas of mathematics, as voted on by mathematicians. It would be nearly impossible for most people on Earth to understand these areas, let alone make contributions in them. Nevertheless, the mathematical ocean is wide and accommodating to new swimmers. Wonderful mathematical patterns, from intricately detailed fractals to visually-pleasing tilings, are ripe for study by beginners. In fact, the late-1970s discovery of the Mandelbrot set—an intricate mathematical shape that the *Guinness Book of World Records* called "the most complicated object in mathematics"—could have been made and graphically rendered by anyone with a high school math education (Figure 1.2). In cases such as this, the computer is a magnificent tool that allows amateurs to make new discoveries that border between art and science. Of course, the high schooler may not understand why the Mandelbrot set is so complicated or why it is mathematically significant. A fully informed interpretation of these discoveries may require a trained mind; however, exciting exploration is often possible without erudition.

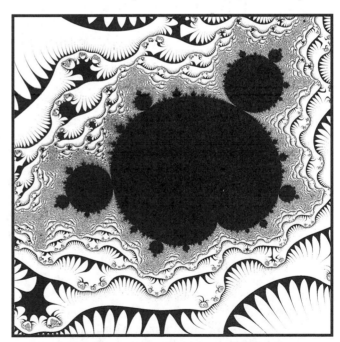

1.2 The Mandelbrot set is described in the 1991 *Guinness Book of World Records* **as the most complicated object in mathematics. The book states, "a mathematical description of the shape's outline would require an infinity of information and yet the pattern can be generated from a few lines of computer code."**

Chapter 2

Why Don't We Use Roman Numerals Anymore?

Rarely do I solve problems through a rationally deductive process. Instead I value a free association of ideas, a jumble of three or four ideas bouncing around in my mind. As the urge for resolution increases, the bouncing around stops, and I settle on just one idea or strategy.
—*Heinz Pagels,* Dreams of Reason

Science and art are similar. New scientific theories do not automatically result from tedious data collection. To conceive a hypothesis is as creative an act as writing a poem. When a hypothesis elegantly explains an aspect of reality more clearly than ever before, there is cause for great wonder and aesthetic pleasure.
—*Lucio Miele,* Skeptical Inquirer

Dr. Googol was walking through the ruins of the Roman Coliseum, daydreaming about his favorite of all things-numbers. Suddenly, he was accosted by a small boy.

"Sir," said the boy, "why don't we use Roman numerals anymore?"

Dr. Googol took a step back. "Are you talking to me?"

"You are the famous Dr. Googol?"

"Ah, yes," said Dr. Googol, "I can answer your question, but before I tell you, you must solve a small puzzle with Roman numerals. I don't think this puzzle dates back to Roman times, but it looks so simple that it could well be quite ancient." Dr. Googol drew the Roman numerals I, II, and III on 6 columns as schematically illustrated in the aerial view in Figure 2.1.

Dr. Googol took a pad of paper from his pocket and started drawing. "Given the 6 columns (represented by circles I, II, and III), is it possible to connect circle I to I, II to II, and III to III, with lines that do not cross or go outside

Si vis feire utrum mulier tua fit cafta

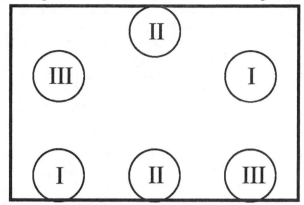

2.1 The Coliseum puzzle.

the surrounding frame? Your lines must be along the floor. They may be curvy, but they cannot touch or cross one another. You can't draw lines through the columns."

The boy studied the figure for several minutes. "Sir, surely this puzzle is impossible to solve."

"It is possible, but I find most people who can't solve the puzzle can solve it if they put it away for a day and then look at it again."

"Wait!" the boy said. "Before attempting your problem, try mine." He handed Dr. Googol a card:

$$XI + I = X$$

The boy looked deeply into Dr. Googol's eyes. "Without using a pencil, how would you make this equation true?"

As Dr. Googol and the boy pondered the puzzles, Dr. Googol also began to tell the boy why Roman numerals survived for so many centuries but eventually were discarded like old shoes.

❁ ❁ ❁

Today we rarely use Roman numerals except on monuments and special documents—and for dates at the end of movie credits to make it difficult to determine when a movie was actually made. You also sometimes see Roman numerals on clock faces, which, incidentally, almost always show four as **IIII** instead of

the traditional **IV**. (*Ever wonder why?* See the "Further Exploring" section.) We are familiar with Roman numerals because they were the only ones used in Europe for a thousand years. The Roman number system was based on similar ones used by the Etruscans, with the letters **I**, **V**, **X**, **L**, and **C** being based on the Etruscan originals. The Roman number system was useful because it expressed all numbers from 1 to 1,000,000 with a total of 7 symbols: **I** for 1, **V** for 5, **X** for 10, **L** for 50, **C** for 100, **D** for 500, and **M** for 1,000. Roman numerals are read from left to right. The symbols representing the largest quantities are placed at the left. Immediately to the right are the symbols representing the next largest quantities, and so on. The symbols are usually added together. For example, **LX** = 60, and **MMCIII** = 2103. **M̄** represents 1,000,000—a small bar placed over the numeral multiplies the numeral by 1,000. Using an infinite number of bars, Romans could have represented the numbers from 1 to infinity! In practice, however, 2 bars were the most ever used.

Numerals are written symbols for numbers. The earliest numerals were simply groups of vertical or horizontal lines, each line corresponding to the number 1. Today, the Arabic system of number notation is used in most parts of the world. This system was first developed by the Hindus and was used in India by the 3rd century B.C. At that time, the numerals 1, 4, and 6 were written as they are today. The Hindu numeral system was probably introduced into the Arab world about the 7th or 8th century A.D. The first recorded use of the system in Europe was in A.D. 976.

Most of Europe switched from Roman to Arabic numerals in the Middle Ages, partly due to Leonardo Fibonacci's 13th-century book *Liber Abaci*, in which he extols the virtues of the Hindu-Arabic numeral system. (This is the same beloved Mr. Fibonacci discussed by Dr. Googol in Chapter 71.) Islamic thinking wasn't far away from the European minds of the Middle Ages. After all, the Muslims had ruled Sicily, Spain, and North Africa, and when the Europeans finally kicked them out, the Muslims left behind their important mathematical legacy. Many of us forget that Islam was a more powerful culture—and more scientifically advanced—than European civilizations in the centuries after the Western Roman Empire fell. Baghdad was an incredible center of learning.

This isn't to say Roman numerals disappeared entirely in the Middle Ages. Many accountants still used them because additional and subtraction can be easy with Roman numerals. For example, if you want to subtract 15 from 67, in the Arabic system you subtract 5 from 7, and 1 from 6. But in the Roman system, you'd simply erase an **X** and a **V** from **LXVII** to get **LII**. It's subtraction by erasing.

However, Arabic numerals hold greater power. Because we switched from the Roman to the Arabic system, humankind can now formulate exotic theories about space and time, contemplate gravitational wave theory, and explore the stars. Arabic numerals are superior to Roman numerals because Arabic numerals have a "place" system in which the value of a numeral is determined by its position. A 1 can mean one, ten, one hundred, or one thousand, depending on its position in a numerical string. This is one reason why it's so much easier to write 1998 than **MCMXCVIII**—one thousand (**M**) plus one hundred less than a thousand (**CM**) plus ten less than a hundred (**XC**) plus five (**V**) plus one plus one plus one (**III**). Try doing arithmetic with this Roman monstrosity. On the other hand, positional notation greatly simplifies all forms of written numerical calculation.

Around A.D. 200, the Hindus, possibly with Arab help, also invented 0, the greatest of all mathematical inventions. (The Babylonians had a special symbol for the "absence" of a number around 300 B.C., but it wasn't a true zero symbol because they didn't use it consistently. Nor did they think of this "absence of a number" as a kind of number, anymore than we think that the "absence of an ear" is a kind of ear.) The number 0 makes it possible to differentiate between 11, 101, and 1,001 without the use of additional symbols, and all numbers can be expressed in terms of 10 symbols, the numerals from 1 to 9 plus 0. During the Middle Ages, the calculational demands of capitalism broke down any remaining resistance to the "infidel symbol" 0 and ensured that by the early 17th century Hindu numerals reigned supreme. Even during Roman times, Roman numerals were used more to *record* numbers, while most calculations were done using the abacus and piling up stones.

⊛ ⊛ ⊛

How far back in time do numerals go? Imagine yourself transported back to the year 20,000 B.C. You are 40 kilometers from the Spanish Mediterranean at the cave of La Pileta. You shine your flashlight on the wall and see parallel marks, groups of 5, 6, or more numbers (Figure 2.2). Clusters of lines are connected across the top with another line, like a comb, or crossed through in a way that reminds you of the modern way of checking things in groups of 5. Were

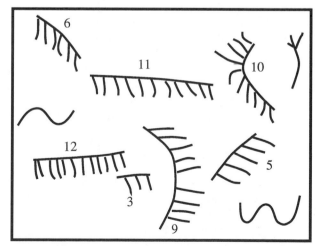

2.2 Designs on the wall of the Number Cave. Some researchers believe the markings represent numbers. If you were to explore the cave and consider the teeth of the "combs" as units, you could read all numbers up to 14. In one area of the cave, the numbers 9, 10, 11, and 12 appear close together. Could it be that the artist was counting something, recording data, or experimenting with mathematics?

the cave people counting something? You can visit the "Number Cave" today, but modern archeologists are not sure of the markings' significance. Nevertheless, the discovery of the Number Cave certainly contradicts old-fashioned notions that cave people of this period made guttural noises and were only concerned with feeding and breeding. If the people who drew these designs mastered numbers, they had intellects beyond the minimal demands of hunting. Also remember that if we were to still regard Mayan friezes and decorated pyramids as merely art, we'd be wrong. Luckily, mathematically minded scholars studied them and discovered their numerical significance.

The earliest forms of number notation that used straight lines for grouping 1s were inconvenient when dealing with large numbers. By 3400 B.C. in Egypt, and 3000 B.C. in Mesopotamia, a special symbol was adopted for the number 10. The addition of this second number symbol made it possible to express the number 11 with 2 symbols instead of 11, and the number 99 with 18 symbols instead of 99.

In Babylonian cuneiform notation, the numeral used for 1 was also used for 60 and for powers of 60; the value of the numeral was indicated by its context. The Egyptian hieroglyphic system evolved special symbols (resembling ropes, lotus plants, etc.) for 10, 100, 1000, and 10,000. The ancient Greeks had 2 systems of numerals. The earlier of these was based on the initial letters of the names of numbers: The number 5 was indicated by the letter *pi*; 10 by the letter *delta*; 100 by the antique form of the letter *H*; 1000 by the letter *chi*; and 10,000 by the letter *mu*. The second system, introduced in the 3rd century B.C., used all the letters of the Greek alphabet plus 3 letters borrowed from the Phoenician alphabet as number symbols. The first 9 letters of the alphabet were used for the numbers 1 to 9, the second 9 letters for the tens from 10 to 90, and the last 9 letters for the hundreds from 100 to 900. Thousands were indicated by placing a bar to the left of the appropriate numeral, and tens of thousands by placing the appropriate letter over the letter *M*. This more advanced Greek system had the advantage that large numbers could be expressed with a minimum of symbols, but it had the disadvantage of requiring the user to memorize a total of 27 symbols.

✻ See the "Further Exploring" section for discussions of the puzzles.

▪ See [www.oup-usa.org/sc/0195133420] for computer code that generates Roman numerals.

Chapter 3

In a Casino

> The heavens call to you and circle about you, displaying to you their
> eternal splendors, and your eye gazes only to earth.
>
> —*Dante*

Some individuals have extraordinary memories when it comes to memorizing
cards in a standard playing-card deck. For example, Dominic O'Brien from
Great Britain memorized, on a single sighting, a random sequence of 40 separate
decks of cards (2,080 cards in all) that had been shuffled together, with only one
mistake! The fastest time on record for memorizing a single deck of shuffled
cards is 42 seconds.

Dr. Googol was interested in similar feats of mental agility and was attending
a card-memorization contest at the largest casino in the world—the Foxwoods
Resort Casino in Ledyard, Connecticut. One of the casino's employees, dressed
as a Roman gladiator, came to him and slammed a deck of cards (Figure 3.1) on
the table.

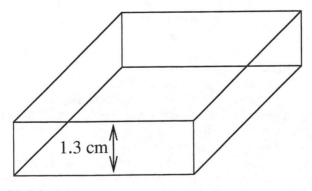

1.3 cm

3.1 A deck of cards.

"My good man," Dr. Googol said, "I personally don't have such a good memory."

"Don't worry," the huge man said with a grin. "This tests another kind of card ability. If a pack of playing cards measures 1.3 centimeters when viewed sideways, what would the measurement be if all the Kings were removed?"

The gladiator handed Dr. Googol a ruler in case Dr. Googol needed it.

⊛ ⊛ ⊛

Can you help Dr. Googol? Hurry! The casino employee will give him $1,000 if you can solve this problem within a minute.

❀ For a solution, see "Further Exploring."

Chapter 4

The Ultimate Bible Code

> The aim of science is not to find the "meaning" of the world.
> The world has no meaning. It simply *is*.
> —*John Bainville, "Beauty, Charm and Strangeness:*
> *Science as Metaphor," Science 281, 1998*

Dr. Googol was visiting Martin Gardner, the planet's foremost mathematical puzzle expert and an all-around wonderful human. It was nearly dusk when Dr. Googol followed Gardner around his North Carolina mansion filled with all manner of mathematical oddities—from glass models of Klein bottles (objects with just 1 surface) to strange tiles arranged in attractive shapes to metallic fractal sculptures of unimaginable complexity.

"Dr. Googol, let me show you something." Martin Gardner withdrew an ancient King James Bible from a bookshelf and drew a box around the first 3 verses of Genesis.

> 1 In the beginning God created the heaven and the Earth.
> 2 And the earth was without form, and void; and darkness was upon the face of the deep. And the Spirit of God moved upon the face of the waters.
> 3 And God said, Let there be light: and there was light.

Gardner pointed to the Bible. "Select any of the 10 words in the first verse: *In the beginning God created the heaven and the Earth.*"

"Got it," Dr. Googol said.

"Count the number of letters in the chosen word and call this number n_1. Then go to the word that is n_1 words ahead. (For example, if you picked the first *the*, go to *created*.) Now count the number of letters in this new word—call it n_2—then jump ahead another n_2 words. Continue until your chain of words enters the third verse of Genesis."

Dr. Googol nodded. "Okay, I am in the third verse."

"On what word does your count end?"

"God!"

"Dr. Googol, consider my next question carefully. Your **soul** may depend on it. Does your answer prove that God exists and that the Bible is a reflection of **ultimate reality**?"

✿ For the mind-boggling answer, see "Further Exploring." Your view of reality will change as you embark on this shattering odyssey of self-discovery.

Chapter 5

How Much Blood?

> Why does there seem to be something inhuman about regarding human beings like roses and refusing to make any distinction between the inside of their bodies and the outside?
> —*Yukio Mishima*

Dr. Googol was lying in a hospital room, receiving a blood transfusion to rid him of a parasite he had recently picked up while exploring the Congo.

He began to wonder . . . What is the volume of human blood on Earth today? In other words, if all approximately 6 billion people from every country on Earth were drained of their blood by some terrible vampire machine, what size container would the machine require to store the blood? The answer to this is quite surprising. Think about it before reading further.

The average adult male has about 6 quarts of blood, but a large part of the Earth's human population is women and children, so let's assume that each person has an average of a gallon of blood. This gives 6 billion gallons of blood in the world. Given that there are 7.48 gallons per cubic foot, this gives us roughly

> ♦ **800,000,000 cubic feet of human blood** ♦

in the world. The cube root of this value indicates that all the blood in the world would fit in a cube about 927 feet on a side. To give you a feel for this figure, the length of each side of the base of the Great Pyramid in Egypt is 755 feet. The length of the famous British passenger ship SS *Queen Mary* was close to 1,000 feet. The height of the Empire State Building, with antenna, is 1,400 feet. This means that a box with a side as long as the SS *Queen Mary* could contain the blood of every man, woman, and child living on Earth today. Most people would guess that a much bigger container would be needed.

John Paulos, in his remarkable book *Innumeracy*, discusses blood volumes as well as other interesting fluid volumes, such as the volume of water rained down upon the Earth during the Flood in the book of Genesis. Considering the biblical statement "All the high hills that were under the whole heaven were covered," Paulos computed that half a billion cubic miles of water had to have covered the Earth. Since it rained for 40 days and 40 nights (960 hours), the rain must have fallen at a rate of at least 15 feet per hour. Paulos remarks that this is "certainly enough to sink any aircraft carrier, much less an ark with thousands of animals on board."

❀ If all this talk about blood hasn't disturbed you too much, see "Further Exploring" for additional bloody challenges.

Chapter 6

Where Are the Ants?

The ants and their semifluid secretions teach us that pattern, pattern, pattern is the foundational element by which the creatures of the physical world reveal a perfect working model of the divine ideal.

—*Don DeLillo,* Ratner's Star

As a child, Dr. Googol had an "ant farm" consisting of sand sandwiched between 2 plates of glass separated by several millimeters. When ants were added to the enclosure, they would soon tunnel into the sand, creating a maze of intricate paths and chambers. Since the space between the glass plates was very thin, confining the ants to a 2-dimensional world, it was always easy to observe the ants and their constructions. Every day, Dr. Googol added a little food and water to the enclosure.

As an adult, Dr. Googol brought an ant farm, schematically illustrated in Figure 6.1, to his students. It had 3 chambers marked *A, B,* and *C.* Dr. Googol added 25 ants to the upper area on top above the soil. He then covered the glass with a dark cloth and waited 25 minutes.

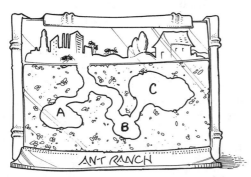

6.1 An ant farm. After the ants randomly walk for a few hours, where do you expect the ants most likely to be: in chamber A, B, or C? (Drawing by April Pedersen.)

Dr. Googol looked at his class of attentive students. "Assuming that the ants wander around randomly, can any of you tell me in which chamber reside the most ants? How would your answer change if there were an additional tunnel connecting chamber *C* to *A*?"

One of the students raised his hand. "And what do we get if we give you the correct answer?"

"A box of delicious chocolate-covered ants."

"Not very appetizing," said a girl with a pierced tongue.

Dr. Googol nodded. "Okay, to the students who get this correct and can explain their reasoning, I will give free copies of Dr. Cliff Pickover's phenomenal blockbuster *Time: A Traveler's Guide.*"

"All right!" the students screamed. With this special incentive, the students became excited and tried their best to predict the chamber holding the most ants. What is your prediction?

❁ For the solution, see "Further Exploring."

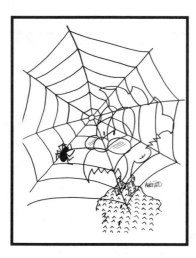

Chapter 7

Spidery Math

The structures with which mathematics deals are more like lace, the leaves
of trees and the play of the light and shadow on a human face than they
are like buildings and machines, the least of their representatives.
                                                                    —*Scott Buchanan*

Dr. Googol has always been interested in spiderwebs, and he continually searches for beautiful specimens throughout the world. Spiderwebs come in all shapes, sizes, and orientations. The largest of all webs are the aerial ones spun by tropical orb weavers of the genus *Nephila*—they can grow up to 18 feet in circumference!

Spiders sometimes make mistakes. Researchers have found that spiders under the influence of mind-altering drugs spin abnormal webs. Marijuana, for example, causes spiders to leave large spaces between the framework threads and inner spirals. Spiders on benzedrine produce an erratic, seemingly unfinished web, and caffeine leads to haphazardly spun threads.

How does all this relate to a fascinating mathematical puzzle? One day while walking through the woods, Dr. Googol came upon a huge orb web more than a foot in diameter. As the sun reflected from its shiny surfaces, he developed this brain boggler.

Consider a spider hallucinating under the influence of some drug. While spinning the web, the spider leaves certain gaps in it. In Figure 7.1, there are three gaps. Dr. Googol calls this simple web a (2, 2) web because it is made from 2 radial lines and 2 circular lines.

At each node (intersection) in the web, the spider constructs a little number that indicates the number of other nodes along the same radial line and circular line he would get to before being stopped by something—either a gap or an outer edge. In Figure 7.2 the spider has marked the top node 4, because as he slides down radially, he gets to 1 node before the gap, and as he slides circularly, he hits 3 other nodes—1 as he heads counterclockwise, and 2 in the clockwise direction.

Figure 7.3 shows a (4, 3) web. The wife of the spider who spun it has come home, devoured her husband (as is the custom of some female spiders), and repaired the web. She has left his numbers in place as a reminder not to become romantically involved with addicted spiders. Can you determine where the gaps in the web would have been located?

Finally, "spider numbers" are defined as the sum of the numbers at each node in a web. For example, the (2, 2) web in Figure 7.2 has a spider number of 44. Using just 4 gaps, what are the smallest and largest spider numbers you can produce for a (2, 2) web and a (4, 3) web?

❋ For solutions to this spidery problem, see "Further Exploring."

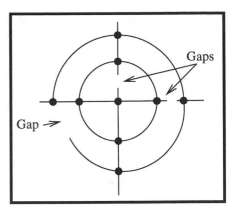

7.1 Spiderweb showing 3 gaps.

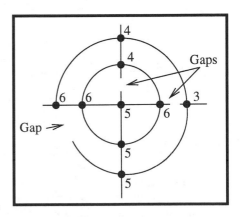

7.2 Numbering the web.

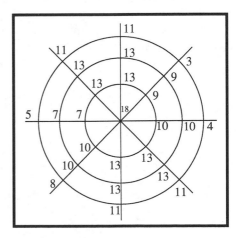

7.3 A (4,3) web.

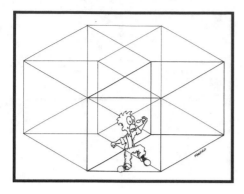

Chapter 8

Lost in Hyperspace

Imagination is more important than knowledge.

—*Albert Einstein*

Dr. Googol has invented numerous problems for the *Star Trek* scriptwriters. Many involve mathematical problems that test their understanding of space, time, and higher dimensions. Here's his favorite puzzle.

Two starships, the *Enterprise* and the *Excelsior*, start at opposite ends of a circular track (Figure 8.1). When Captain Kirk says "go," the ships start to travel in opposite directions with constant speed. (In other words, one ship goes clockwise, the other counterclockwise.)

From its departure point to the first time they cross paths, the *Enterprise* travels 800 light-years. And from the first time they cross to the second time they cross, *Excelsior* travels 200 light-years. With so little information, is it possible to determine the length of the track? Would your answer change if the track were another closed curve, but not a circle?

8.1 The starships *Enterprise* and *Excelsior*, before they start their journeys to where no man has gone before.

✤ For a wonderful solution, see "Further Exploring."

Chapter 9

Along Came a Spider

It's the sides of the mountain which sustain life, not the top.
Here's where things grow.
—*Robert Pirsig,* Zen and the Art of Motorcycle Maintenance

Dr. Googol was in a Peruvian rain forest, 15 miles south of the beautiful Lake Titicaca, when he dreamed up this tortuous brain boggler. A month later, while in Virginia, Dr. Googol gave this puzzle to all CIA employees to help them improve their analytical skills.

Three spiders named Mr. Eight, Mr. Nine, and Mr. Ten are crawling on a Peruvian jungle floor. One spider has 8 legs; one spider has 9 legs; one spider has 10 legs. All of them are usually quite happy and enjoy the diversity of animals with whom they share the jungle. Today, however, the hot weather is giving them bad tempers.

"I think it is interesting," says Mr. Ten, "that none of us have the same number of legs that our names would suggest."

"Who the heck cares?" replies the spider with 9 legs.

How many legs does Mr. Nine have? Amazingly, it is possible to determine the answer, despite the little information given.

❁ ❁ ❁

Now for the second part of the puzzle. The same 3 spiders have built 3 webs. One web holds just flies, the other just mosquitoes, and the third both flies and mosquitoes. They label their 3 webs "flies," "mosquitoes," and "flies and mosquitoes." All 3 labels are incorrect. The insects are wrapped up tightly in web strands. How many insects does a spider have to unwrap to correctly label the webs?

Please try to solve at least one of these tantalizing problems. If too difficult, draw diagrams and think about them with some friends. If you are a teacher, have students work on the puzzles in teams. Whatever you do, don't skip this problem and go to the next one. If you take this lazy approach, a live, 2-dimensional spider will emerge from the tiny web, which the publisher's overworked typesetter has with luck placed right here: ▨

🕷 For a solution, see "Further Exploring."

Chapter 10

Numbers beyond imagination

The study of the infinite is much more than a dry, academic game. The intellectual pursuit of the Absolute Infinite is a form of the soul's quest for God. Whether or not the goal is ever reached, an awareness of the process brings enlightenment.
—*Rudy Rucker,* Infinity and the Mind

For a human, there are gigaplex possible thoughts. [A gigaplex is the number written as 1 followed by a billion zeros.]
—*Rudy Rucker,* Infinity and the Mind

Dr. Googol sat on a sandy beach, typing on his notebook computer while downloading the results of his Big Number Contest via a satellite link to the Internet.

A few minutes ago, he had asked his fellow Web-heads to construct an expression for a very large number using only the following 8 symbols:

$$1\ 2\ 3\ 4\ (\ )\ .\ -$$

Each digit could be used only once.

Within a half hour, a teenager in Florida came up with $4^3 - 12 = 52$. (The expression 4^3 denotes exponentiation and is simply $4 \times 4 \times 4$.)

"Not bad for a start," Dr. Googol typed on his notebook computer. "Can anyone come up with a solution greater than 52?"

Dr. Googol got up, stretched, and wiggled his toes in the sand. By the time he got back to his computer a gentleman from North Carolina had come up with 31^{42}. This huge number had 63 digits.

"You can do better," Dr. Googol typed as his pulse rose with exponentially increasing anticipation.

From various locations around the country came the reply 3^{421}. It had 201 digits!

"Very good," he said, shaking with pleasure.

A woman from New York exclaimed, "I take the prize with $.1^{-432}$. It has 433 digits!"

"Excellent," he yelled aloud, although no one could hear him but the seagulls. A nearby bird quickly took to the sky. He typed back to the woman, "Good work. You recalled that a number raised to a negative power is simply 1 over the number raised to the positive value of the power. You also realized that to determine the number of digits in a number you simply take the log of the number and add 1. This means that $.1^{-432} = 1/.1^{432} = 10^{432}$. The log of 10^{432} is 432, and the number of digits is 433."

Dr. Googol wondered: *Is it possible to beat the woman's fantastic 433-digit answer?*

❀ For the world-record holder and more information on numbers too large to contemplate, see "Further Exploring."

Chapter 11

Cupid's Arrow

The mathematician may be compared to a designer of garments who is
utterly oblivious of the creatures whom his garments may fit.
To be sure, his art originated in the necessity for clothing such creatures,
but this was long ago; to this day a shape will occasionally appear which
will fit into the garment as if the garment had been made for it.
Then there is no end of surprise and delight!

—*Tobias Dantzig*

It is Valentine's Day 2000. Dr. Googol is ambling along the Tiber River, watching the beautiful passersby and enjoying the crisp weather, when a sudden wrenching pain in his right atrium interrupts his stroll. As he clutches at his heart and falls to the ground, he has a vision of a peculiar man with wings and a bow who lands nearby.

"Just trying out a new arrow my uncle Divisio, God of Arithmetic, gave me," the man says. Reaching toward Dr. Googol, the man pulls an arrow studded with 5 disks out of Dr. Googol's chest (Figure 11.1). "Not like the old one, this," he continues, running his hand lovingly over the disks. "You get to choose who you want as your sweetheart if you can solve the puzzle."

"Use the numbers 1 through 9," the man tells Dr. Googol, "placing 1 digit in each of the circles according to the following rule: Each pair of digits connected by a line must make a 2-digit number that is evenly divisible by either 7 or 13. For

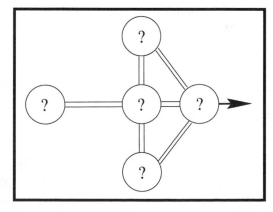

11.1. Cupid's arrow.

example, 7 and 8 connected by a line would be appropriate because the number 78 is divisible by 13. You can consider the 2 digits in either order, and no digit can be used more than once."

"For every solution you find," the winged man adds before flying off, "you win someone's heart. If you can find a solution in which lines connect the top and bottom disks to the base at left as well, you will *always* be lucky in love. There are at least 5 hearts out there for you. Can you win the others?"

❀ For a solution, see "Further Exploring."

[Editor's note: Dr. Googol shortly woke up from his fainting spell. Physicians pronounced his heart normal. His "heart pain" was diagnosed as severe indigestion resulting from a recently eaten wasabi-pepperoni pizza.]

Chapter 12

Poseidon Arrays

Truly the gods have not from the beginning revealed all things to mortals,
but by long seeking, mortals make progress in discovery.
—*Xenophanes of Colophon*

Poseidon arrays are those in which successive rows are equal to the first row multiplied by consecutive numbers. That's a mouthful! An example will help clarify this. The following pattern

1	1	1
2	2	2
3	3	3

is such an array because the second row is twice the first, and the third row is 3 times the first. Dr. Googol began to wonder if there were similar Poseidon arrays where each digit is used only once. After much thought, he discovered

1	9	2
3	8	4
5	7	6

Notice that 384 is twice the number in the first row, and that 576 is 3 times the number in the first row. Are there other ways of arranging the numbers to produce the same result, using each digit only once and the same rules? Remember, the second row must be twice the first. The third row must be 3 times the first row.

❋ For a solution and additional speculation, see "Further Exploring."

Chapter 13

Scales of Justice

The popular image of mathematics as a collection of precise facts,
linked together by well-defined logical paths, is revealed to be false.
There is randomness and hence uncertainty in mathematics, just
as there is in physics.
—*Paul Davis,* The Mind of God

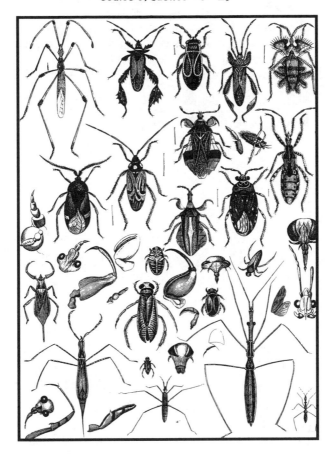

Dr. Googol was trapped in the laboratory of a crazy Egyptian entomologist. All manner of beetles and bugs flew within jars, climbed the walls, and dangled from the ceiling.

"This is sick," Dr. Googol screamed.

"Sick?" the scientist said. "I'll show you sick."

He went to a piece of paper on the table where he had cutouts of his favorite insects. He placed the cutouts on schematic drawings of scales. For example, on the first scale 2 ants were in one pan and exactly balanced a grasshopper and wasp in the other pan:

Ant Ant	Grasshopper Wasp
Ant Cockroach	Grasshopper
Cockroach	?

"The first 2 sets of scales are in balance," he said while popping a few ants into his mouth as a snack. "I want you to assign values to the insects' weights and tell me which insect or sets of insects replaces the empty side of the third scale in order to balance it. Each insect species is of a different weight. Assume that the cockroach is balanced by some collection of insects."

Can you help Dr. Googol solve this puzzle and win his release? What strategy did you use?

❀ For a solution, see "Further Exploring."

Chapter 14

Mystery Squares

> He calmly rode on, leaving it to his horse's discretion to go which way it
> pleased, firmly believing that in this consisted the very essence of adventures.
> —*Cervantes,* Don Quixote

Dr. Googol has placed the numbers 1, 2, 3, and 4 at the corners of a square. Can you try to arrange 5, 6, 7, 8, 9, 10, 11, and 12 along the sides of the square so that the numbers along each side all add up to the same number? (If you don't at least try to solve this intriguing enigma, Dr. Googol may visit you at home—not entirely pleasant, since Dr. Googol doesn't stop talking and posing problems.)

Below is an example where the sums are all *unequal.* For instance, the top row adds up to 18, and the left column adds up to 16. (Notice the 1, 2, 3, and 4 at the corners.)

1	7	8	2
6			9
5			10
4	11	12	3

How many solutions can you find in which the numbers along each side add up to the same sum? Remember, the numbers 1, 2, 3, and 4 are to remain fixed in place.

✸ For a solution, see "Further Exploring."

Chapter 15

Quincunx

We think of the number "five" as applying to appropriate groups of any entities whatsoever—to five fishes, five children, five apples, five days. . . . We are merely thinking of those relationships between those two groups which are entirely independent of the individual essences of any of the members of either group. This is a very remarkable feat of abstraction; and it must have taken ages for the human race to rise to it.
—*Alfred North Whitehead*

Applications, computers, and mathematics form a tightly coupled system yielding results never before possible and ideas never before imagined.
—*Lynn Arthur Steen*

The enormous usefulness of mathematics in natural sciences is something bordering on the mysterious, and there is no rational explanation for it. It is not at all natural that "laws of nature" exist, much less that man is able to discover them. The miracle of the appropriateness of the language of mathematics for the formulation of the laws of physics is a wonderful gift which we neither understand nor deserve.
—*Eugene P. Wigner,* "The Unreasonable Effectiveness of Mathematics in the Natural Sciences"

Five is Dr. Googol's favorite number, and 5-fold symmetry is his favorite symmetry. Would you care for a barrage of mathematical trivia befitting only the most ardent mathophiles?

⊙ Not only is 5 the hypotenuse of the smallest *Pythagorean triangle*, but it is also the smallest automorphic number. Let me explain. A Pythagorean triangle is a right-angled triangle with integral sides. For example, the smallest Pythagorean triangle has side lengths 3, 4, and 5. An automorphic number *n*, when multiplied by itself, leads to a product whose rightmost digits are *n*. Not counting the trivial case of the number 1, 5 and 6 are the smallest automorphic numbers because 5 × 5 = 25 and 6 × 6 = 36. Examining a larger number, the square of 25 is 625. Note that 25 appears as the final 2 digits of 625.

⊙ Five is probably the only odd *untouchable number*. (The legendary and bizarre mathematician Paul Erdös called a number "untouchable" if it is never the sum of the proper divisors of any other number. The sequence of untouchable numbers starts 2, 5, 52, 88, 96, 120. A "divisor" of a number *N* is a number *d* which divides *N*; it's also called a factor. A "proper divisor" is simply a divisor of a number *N* excluding *N* itself.)

⊙ Also, there are 5 Platonic solids. (The 5 Platonic solids are the tetrahedron, cube, octahedron, dodecahedron, and icosahedron. All the faces of a Platonic solid must be congruent regular polygons.)

⊙ The word *quincunx* is the name for the pattern

on a die, and it involves both 5 and 1. It's also the name for a particular type of 5-domed cathedral, like St. Mark's Cathedral in Venice. (Certain Khmer temples in Southeast Asia also use this configuration.)

Dr. John Lienhard of the University of Houston points out to Dr. Googol that most 19th-century forts were square or pentagonal (Figure 15.1), with "bastions" on each corner that gave the old forts the shape of great stone "snowflakes." (Bastions are spade-shaped widenings of the corners that let defenders fire parallel to the walls.)

Fort Sumter was 5-sided and sat on the tip of an island in Charleston Bay. (The first engagement of the Civil War took place at Fort Sumter, and in a few years most of the fort was reduced to brick rubble.) Water came right up to 4 of its walls. Only the fifth wall needed the protection of bastions. Dr. Lienhard suspects that 5 was a typical solution to the problem of placing bastions close enough together without increasing the costs of construction and manning the walls.

15.1 A typical early 19th-century pentagonal fortification. (From the 1832 *Edinburgh Encyclopedia*.)

Five occurs in the symmetry of several creatures in science-fiction literature. For example, Naomi Mitchison's *Memoirs of a Spacewoman* describes "Radiates," intelligent 5-armed creatures resembling starfish (Figure 15.2). They live in villages composed of long, low buildings decorated with fungi that grow in spiral patterns. Radiates don't think in terms of dualities, having instead a 5-valued system of logic.

Five-fold symmetrical organs are sometimes described in science-fiction stories. For example, the Old Ones in H. P. Lovecraft's *At the Mountains of Madness* are incredibly tough and durable creatures, having characteristics of both plants and animals. They also possess an extraordinary array of senses to help them survive. Hairlike projections and eyes

15.2 A Radiate from Naomi Mitchison's novel *Memoirs of a Spacewoman*. (Drawing by Michelle Sullivan.)

on stalks at the top of their heads permit vision. The colorful, prismatic hairs seem to supplement the vision of the eyes, and in the absence of visible light, the species is able to "see" using the hairs. Their complex nervous system and 5-lobed brains process senses other than the human ones of sight, smell, hearing, touch, and taste. When the Old Ones open their eyes and fully retract their eyelids, virtually the entire surface of the eye is apparent.

The number 5 is also remarkable for its appearance in Earthly biology and in art. Five-fold symmetry in biology is fairly common, as evidenced by a variety of animal species such as the starfish and other invertebrates. Five-fold symmetry

15.3 Several terra-cotta inlays from the smaller dome chamber of the Masjid-i-Jami in Isfahan (A.D. 1088).

15.4 The 5-pointed Star of Bethelehem.

15.5. Badge from the Leicester family.

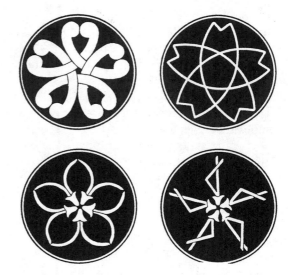

15.6. Several Japanese crests exhibiting 5-fold symmetry.

also appears in mathematics; for example, in numerous uniform polyhedra. Five-fold symmetry is relatively rare, however, in the art forms produced by humans. Perhaps partly because pentagonal motifs do not tightly pack on the plane, they are much rarer than other symmetries in historic and artistic ornament. Nevertheless, there are occasional interesting examples of pentagonal ornaments in artistic symbols and designs. The oldest and most important examples of 5-fold symmetry and odd-number symmetry are the 5-pointed star and triangle, first used in cave paintings and in the Near East since about 6000 B.C. Since then they have been used in sacred symbols by the Celts, Hindus, Jews, and Moslems. Later (circa 10th century A.D.) the 5-pointed star was adopted by medieval craftspeople such as stonecutters and carpenters. In the 12th century, it was adopted by magicians and alchemists.

❁ ❁ ❁

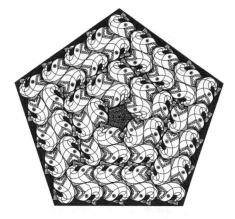 

15.7 *Pentagon with Fishes,* by Peter Raedschelders.

15.8 *Tropical Fishes,* by Peter Raedschelders.

To begin this picture essay, Dr. Googol invites you to consider some of the Persian designs and motifs with pentagonal symmetry. Over the centuries, Persia (Iran) has been periodically invaded, and elements of the invading cultures were incorporated into the native artistic traditions. Much of Persian art contains highly symmetrical designs. Examples of symmetrical ornaments appear on silk weaves, printed fabrics, carpets, ceramics, stone, and calligraphy. Occasionally, we find a 5-fold symmetrical design in Persian ornament. Figure 15.3 shows terra-cotta inlays from the smaller dome chamber of the Masjid-i-Jami in Isfahan (A.D. 1088).

Religious symbols sometimes contain pentagonal symmetry; an example, shown in Figure 15.4, is the 5-pointed Star of Bethelehem. Various symmetrical designs have also appeared in heraldic shapes. In the Middle Ages these designs on badges, coats of arms, and helmets generally indicated genealogy or family name. Figure 15.5 shows a badge from the Leicester family. The Japanese also had similar family symbols for the expression of heraldry. The family symbol, or *mon*, was known in Japan as early as A.D. 900 and reached its highest development during feudal times. Figure 15.6 shows several Japanese crests containing 5-fold symmetry. These kinds of crests are found on many household articles, including clothing.

Symmetrical ornaments, such as those in this chapter, have persisted from ancient to modern times. The different kinds of symmetry have been most fully explored in Arabic and Moorish design. The later Islamic artists were forbidden by religion to represent the human form, so they naturally turned to elaborate geometric themes. To explore the full range of symmetry in historic ornament, you may wish to study the work of Ernst Gombrich, who discusses the psychology of decorative art and presents several additional examples of 5-fold symmetry.

Finally, Belgian artist Peter Raedschelders frequently uses 5-fold symmetry in his art, and several of his recent works are presented here (Figures 15.7–15.10).

One of his passions is to determine mathematically interesting ways to pack regular pentagons with fish and snakes (Figures 15.7–15.9). He enjoys the challenge because other artists often shy away from the difficult packing of a pentagon. Notice that the snakes are moving along a strangely shaped single surface. Figure 15.10 illustrates a train that is able to ride along the various seemingly planar surfaces of this weird star. Hop on, and take a long, exciting ride!

15.9 *Five Snakes,* **by Peter Raedschelders.**

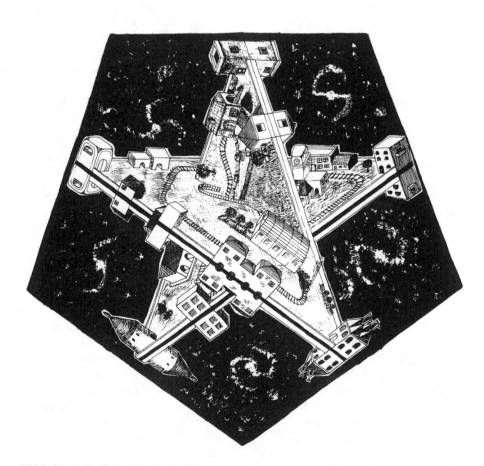

15.10 *Train,* **by Peter Raedschelders.**

Chapter 16

Jerusalem Overdrive

Who carved the nucleus, before it fell, into six horns of ice?
—*Johannes Kepler*

Dr. Googol was in Jerusalem, overseeing the construction of a new multidenominational religious center that would house prayer rooms for the 3 major religions: Judaism (✡), Christianity (✟), and Islam (☾). To make it more difficult for terrorists to bomb any single religious group, and to minimize religious conflicts, the architect is to design the center as a 3-by-3 matrix of prayer rooms so that (when viewed from above) each row and column contains only 1 prayer room of a particular religious denomination. An aerial view of the religious center looks like a tic-tac-toe board in which you are not permitted to have 2 of the same religions in any row or column. Is this possible?

The following is an arrangement prior to your attempt to minimize conflict:

 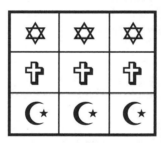 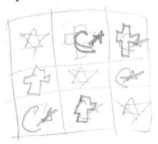

For a second problem, consider that you must place the prayer rooms so that each row and column contains exactly *2* religions. Is this possible?

You can design a computer program to solve this problem by representing the 3 religions as red, green, and amber squares in a 3-by-3 checkerboard. The program uses 3 squares of each color. Have the computer randomly pick combinations, and display them as fast as it can, until a solution is found. The rapidly changing random checkerboard is fascinating to watch, and there are quite a lot of different possible arrangements. In fact, for a 3-by-3 checkerboard there are 1,680 distinct patterns. If it took your computer 1 second to compute and display each 3-by-3 random pattern, how long would it take, on average, to solve the problem and display a winning solution? (There is more than 1 winning solution.)

❁ For a solution, and more on religious patterns and magic squares, see "Further Exploring."

Chapter 17

The Pipes of Papua

In Samoa, when elementary schools were first established, the natives developed an absolute craze for arithmetical calculations. They laid aside their weapons and were to be seen going about armed with slate and pencil, setting sums and problems to one another and to European visitors. The Honourable Frederick Walpole declares that his visit to the beautiful island was positively embittered by ceaseless multiplication and division.

—*T. Briffault*

I like that abstract image of life as something like an efficient factory machine, probably because actual life, up close and personal, seems so messy and strange. It's nice to be able to pull away every once in awhile and say, "There's a pattern there after all! I'm not sure what it means, but by God, I see it!"

—*Stephen King,* Four past Midnight

Late last autumn, while enjoying the brisk New England air, Dr. Googol took a walk with Omar Khayyam, his octogenarian friend. Omar whispered a tale about his buddies who had once explored Papua New Guinea in the southwestern Pacific Ocean. Dr. Googol should tell you right up front that he can never be certain as to the accuracy of Omar's tales. During the past 10 years his stories have evolved into highly embellished tales, composed of myth and truth, perhaps more of the former than the latter, depending on his mood. Whatever the case, Dr. Googol recounts his colorful story here and lets you decide about the authenticity of Omar's old recollections.

Omar's friends were camping on a riverbank when they heard strange flutes or wooden pipes. There was a certain rhythm to the pipes, but the tones never quite repeated themselves. Occasionally a drum seemed to beat the same rhythm. A few men explored the surrounding bush but, even after much searching, never succeeded in locating the source of the sounds. Sometimes the sounds seemed to come from the north, at other times from the east.

The sounds emanated from a 2-tone pipe. Edward Fitzgerald was one of the explorers on the journey, and he was sufficiently interested in this peculiar phenomenon to record it in his tattered notebook, using 👍 and 👎 to represent the long and short tones he heard. Luckily the pipe sounds were slow enough to allow the explorer to accurately record the rhythmic pattern. The first few entries were:

Then the player would pause for a minute and then start again. On the next line of the notebook were the drawings:

The notebook contained several pages of these symbols. By midnight, the pages of the notebook were exhausted.

Years later, Omar came into possession of the notebook from Fitzgerald, who croaked, "It's the strangest thing ye ever heard. It ain't exactly irregular and it ain't exactly regular, either." Omar, who has some mathematical training, spent many days examining the pages of 👍 and 👎 symbols. His conclusion was startling.

Dr. Googol and Omar continued their walk in the cool night air. Suddenly, Omar stopped dead in the middle of the sidewalk under an amber streetlight. He looked Dr. Googol in the eye. "You might not believe this, but that strange pattern of 👍 and 👎 symbols turned out to be a well-known, exotic pattern of binary numbers called the *Morse-Thue* sequence—it's visually represented with a string of 0s and 1s." Omar went on to explain that the sequence is named in honor of the Norwegian mathematician Axel Thue (1863–1922) (pronounced "tew") and Marston Morse of Princeton (1892–1977). Thue introduced the sequence as an example of a nonperiodic, recursively computable string of symbols—a phrase that should become clear to you in the following discussion. Morse did further research on the sequence in the 1920s.

There are many ways to generate the Morse-Thue sequence. One way is to start with a 0 and then repeatedly do the following replacements: 0 → 01 and 1 → 10. In other words, whenever you see a 0 you replace it with 01. Whenever you see a 1 you replace it with 10. Starting with a single 0, we get the following successive "generations":

0
0 1
0 1 1 0
0 1 1 0 1 0 0 1
0 1 1 0 1 0 0 1 1 0 0 1 0 1 1 0

Try generating this with a pencil and paper. You begin with 0, and replace it with 01. Now you have a sequence of two digits. Replace the 0 with 01 and the 1 with 10. This produces the sequence 0110. The next binary pattern is 01101001. Notice that 0110 is symmetrical, a palindrome, but the next pattern, 01101001, is not. But hold on! The very next pattern, 0110100110010110, is a palindrome again. Does this pattern continue to hold for alternate sequences? The mysteries of this remarkable sequence have only begun.

Notice that the fourth line of the sequence can translate into the ☙ ☙ ☙ ☙ ☙ ☙ ☙ ☙ sounds in Omar's story if you let ☙ represent 0 and ☙ represent 1. Amazing!

You can generate the pipe sequence in another way: each generation is obtained from the preceding one by appending its complement. This means that if you see a 0110 you append to it a 1001. There is yet a third way to generate the sequence. Start with the numbers 0, 1, 2, 3, . . . and write them in binary notation: 0, 1, 10, 11, 100, 101, 110, 111, ... (Binary numbers are explained in detail in the "Further Exploring" for Chapter 21. Hop there now if you need background information.) Now calculate the sum of the digits modulo 2 for each binary number. That is, divide the number by 2 and use the remainder. For example, the binary number 11 becomes 2 when the digits are summed, which is represented as 0 in the final sequence. This yields the sequence 0, 1, 1, 0, 1, 0, 0, 1, . . . , which is the same sequence yielded by the other methods!

Let Dr. Googol tell you why this sequence is so fascinating. For one, it is *self-similar*. This means you can take pieces of the sequence and generate the entire infinite sequence! For example, retaining every other term of the infinite sequence reproduces the sequence. Try it. Similarly, retaining every other pair also reproduces the sequence. In other words, you take the first 2 numbers, skip the next 2 numbers, etc. Also, the sequence does not have any periodicities, as would a repetitious sequence such as 00, 11, 00, 11. However, although aperiodic, the sequence is anything but random. It has strong short-range and long-range structures. For example, there can never be more than 2 adjacent terms that are identical. One method for finding patterns in a sequence, the Fourier spectrum, shows pronounced peaks when used to analyze the sequence. Using this mathematical method, you can make a graph showing the frequencies in the data plotted versus position in the sequence, with the more intense frequency components shown in the third dimension, or more simply as a darker point on a 2-dimensional graph.

The sequence grows very quickly. The following is the sequence for the eighth generation.

```
0110100110010110100101100110100110010110011010010110100110
0101101001011001101001011010011001011001101001100101101001
0110011010011001011001101001011010011001011001101001100101
1010010110011010010110100110010110100101100110100110010110
0110100101101001100 10110
```

```
0110100110010110100101100110100110010110011010010
1101001100101101001011001101001011010011001011001
1010011001011010010110011010011001011001101001011
0100110010110011010011001011010010110011010010110
1001100101101001011001101001100101100110100101101
0011001011010010110011010010110100110010110011010
0110010110100101100110100101101001100101101001011
0011010011001011001101001011010011001011001101 00
1100101101001011001101001100101100110100101101001
1001011010010110011010011001011001101001011001011 
0010110100101100110100110010110011010010110100110
0101100110100110010110100101100110100101101001100
1011010010110100101100110010110011010010110100110
0110100101100110100101101001011001101001100101101
0010110100110010110100101100110100101101001100101
1100110100110010110100101100110100101101001100101
1010010110011010011001011001101001011010011001011
0100101100110100101101001100101100110100110010110
1001011001101001100101100110100101101001100101100
1101001100101101001011001101001011010011001011010
0101100110100110010110011010010110100110010110100
1011001101001011010011001010110011010011001011010
0110011010010110100110010110100101100110100110010
1100110100101101001100101101001101001100101101001 0
1001101001011010011001011001010110100110010110 1001
0011010010110100110010110011010011001011010010110
0110100101101001100101101001011001101001100101100
1101001011010011001011010010110011010011001011010
0101100110100101101001100101101001011001101001011
1010010110100110010110100101100110100101101001100 10
0101100110100110010110100101011001101001100101100 11
0100101101001100101100110100110010110100101100110
1001011010011001011010010110011010011001011001101
0010110100110010110011010011001011010010110011010
0110010110011010010110100110010110100101100110100
1011010011001011001101001100101101001011001101001
1001011001101001011001101001100101100110100110101
0010110100110010110011010011001011010010110011010
0101100110100101101001100101101001011001101001011
0100110010110011010011001011010010110011010010110
1001100101101001011001101001100101100110100101101
0011001011001101001100101101001011001101001100010
1100110100101101001100101101001011001101001011010
0110010110011010011001011010010110011010010110010 1
```

Table 17.1 A Morse–Thue sequence for the 11th generation.

Table 17.1 shows the sequence for the eleventh generation. Sometimes certain patterns emerge when a sequence is stacked up on itself in this manner. Can you see any patterns here? Another way to represent the Morse-Thue sequence is to redraw it as a "bar code" of sorts, placing vertical lines wherever a 1 occurs and skipping a space wherever a 0 occurs. To make the positions of "11" entries clear to the human eye, wherever two 1s appear consecutively, try joining them by short ladder-like steps. Dr. Googol also likes to draw the Morse-Thue sequence with botanical shapes. Here the 1s are replaced by flowers and the 0s by spaces:

0110100110010110100101100110100110010110011010010110100
1100101101001011001101001011010011001011001101001100101
1010010110

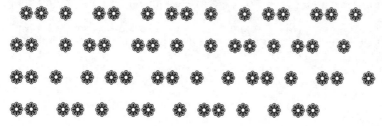

The diagram looks even better when tall trees are used. Can you arrange the rows and columns in a way that better reveals the sequence's patterns? What would it be like to walk through this strangely spaced forest? Imagine holding the hand of someone you love as you explore an infinite Morse-Thue forest that stretches for as far as your eye can see.

❀ For more on the musical qualities of these patterns, see "Further Exploring."

▥ For computer hints, see [www.oup-usa.org/sc/0195133420].

Chapter 18

The Fractal Society

I believe that scientific knowledge has fractal properties, that no matter how much we learn, whatever is left, however small it may seem, is just as infinitely complex as the whole was to start with. That, I think, is the secret of the Universe.

—*Isaac Asimov,* I, Asimov.

God gave us the darkness so we could see the stars.

—*Johnny Cash,* "Farmer's Almanac"

Dr. Googol belongs to a group of mathematicians who meet each month in a secret club. Status in their Fractal Society is based on the prowess with which an

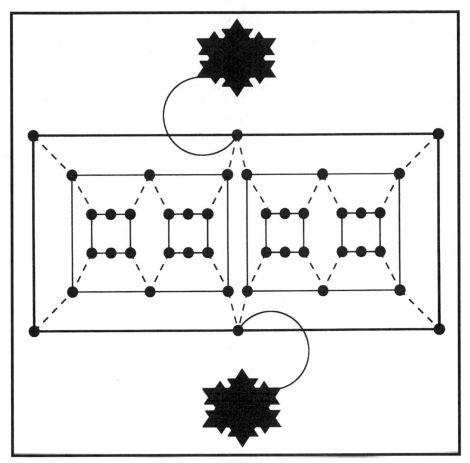

18.1 Fractal Fantasies playing board (degree 2).

individual plays mathematical games and proves mathematical theorems. The center of such activity is a building called the Imaginarium, which is shaped like a Mandelbrot set. There are various pleasurable rewards bestowed upon club members in proportion to the novelty of theorems they solve. Dr. Googol's favorite society game is called Fractal Fantasies.

The playing board for the Fractal Fantasies is a fractal nesting of interconnected rectangles. (Figure 18.1). Dr. Googol is so enthralled with this game that he has cut the design into the roofing slabs of his home and the surface of his kitchen table. The board for Fractal Fantasies contains rectangles within rectangles interconnected with dashed lines as shown in Figure 18.1. There are always two rectangles within the rectangles that encompass them. The degree of nesting can be varied. Beginners play with only a few nested rectangles, while grand masters play with many recursively positioned rectangles. Tournaments last for days, with breaks only for eating and sleeping. The playing board illustrated in Figure 18.1 is called a "degree 2" board, because it has two different sizes of rectangles within the large bounding rectangle. Beginners usually start with a degree 1 board, and grand masters have been known to use a degree 20 board. One player uses white playing pieces (like stones); the other uses black. Each player starts with a number of pieces equal to half the number of vertices (dots) on the board minus 2. For the board here, each player gets 19 stones. With alternate moves, the players begin by placing a stone at points on the black dots that are empty. As they place stones, each player attempts to form a row of 3 stones along any 1 of the horizontal sides of any rectangle. This 3-in-a-row assembly of stones is called a Googol. When all the stones have been placed, players take turns moving a piece to a neighboring vacant space along one of the dashed or straight connecting lines. When a player succeeds in forming a Googol (either during the alternate placement of pieces at the beginning of the game, or during alternate moves along lines to adjacent empty points), then the player captures any 1 of the opponent's pieces on the board and removes it from the board. These removed stones may be kept in star-shaped receptacles represented by the black stars at the top and bottom of the board in Figure 18.1 (In some versions of the game, an opposing stone cannot be taken from an opposing Googol.) A player loses when he or she no longer has any pieces or cannot make a move.

Mathematicians and philosophers will no doubt spend many years pondering a range of questions, particularly for boards with higher nesting. Computer programmers will design programs allowing the board to be magnified in different areas, permitting convenient playing at different size scales. They'll all wish they had fractal consciousnesses allowing the contemplation of all levels of the game simultaneously.

Many of Dr. Googol's dearest friends have spent years of their lives pondering the following questions relating to Fractal Fantasies. No one has succeeded in answering these questions for games with degree higher than 2. Various centers have been established and funded in order to answer the following research questions:

1. What is the maximum number of pieces that can be on the board without *any* forming a row?

2. Is there a best opening move?

3. If the large bounding square has a side 1 foot in length, and each successive generation of square has a length 1/6 of the previous, what is the total length of lines on the board?

4. If a spider were to start anywhere on the board and walk to cover all the lines, what would be the shortest possible route on the board?

5. How many positions are possible after 1 move by each player?

6. How large would a degree 100 board have to be in order for the smallest squares to be seen? How many playing pieces would be used? What length of time would be required to play such a bizarre game?

❀ For reader comments, see "Further Exploring."

Chapter 19

The Triangle Cycle

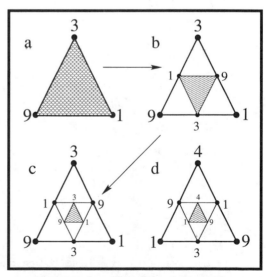

The mathematical rules of the universe are visible to men
in the form of beauty.

—*John Michel*

Dr. Googol spends long hours contemplating a number puzzle called the Triangle Cycle. The puzzle begins with that simple shape of geometry—the triangle—and soon becomes fiendishly complex. Place a single digit at each corner of a triangle so that the lines that connect adjacent digits create 2-digit numbers that are multiples of either 7 or 13. (The 2-digit number needs to be a multiple in only 1 direction.) For example, a line connecting 1 and 9 is valid because you can read it as 19 or 91, and 91 equals 13 × 7. We can make a triangle starting with these two digits by putting a 3 in the third corner, as seen in Figure 19.1a. One of these lines connects 1 and 3, forming the number 13 (13 times 1), while the other connects 3 and 9, forming 39 (13 times 3).

But the puzzle is far from over. Draw a new triangle inside the first, with its corners cutting the sides of the larger triangle in half. Now pick 3 more numbers for the new corners. Be careful: you've actually created 4 triangles, and each has to obey the rules outlined above.

19.1 Playing board for the Triangle Cycle game.
(a) A starting position. (b) Triangles within triangles. (c) A cycle 1 solution. (d) A cycle 2 solution

The easiest solution uses the same 3 numbers (see Figure 19.1b). Let's first take a look at the 6 lines that make up the outer triangle. In clockwise order starting from the top corner, they form the following valid numbers: 39, 91, 13, 39, 91, and 13. On the inner triangle, counterclockwise from the top left corner, the numbers are 13, 39, and 91.

You can draw a third triangle inside the second that is a copy of the first, as shown in Figure 19.1c—and a fourth and a fifth and so on until infinity. The triangles flip up and down ($\triangle \triangledown \triangle \triangledown \ldots$) forever. Dr. Googol likes to call this a cycle 1 solution because it can repeat the same triangle forever. A cycle 2 solution, on the other hand, flips back and forth between two different triangles; Figure 19.1d shows one example.

A cycle 4 solution, as you might expect, uses 4 different triangles. Can you figure 1 out? Can you find higher cycles?

❀ For a solution and additional challenges, see "Further Exploring."

Chapter 20

iQ-Block

> Even if the rules of nature are finite, like those of chess, might not science
> still prove to be an infinitely rich, rewarding game?
> —*John Horgan,* Scientific American 267(6), *1992.*

An interesting example of cultural contamination occurred in a secluded West African valley when Dr. Googol left behind a mathematical puzzle called IQ-Block (manufactured by Hercules, designated as Item No. P991A, UK Registered No. 2013287, and made in Hong Kong). The puzzle, schematically illustrated in Figure 20.1, consists of 10 brightly colored polygonal pieces of plastic. The 10 pieces fit together to form a square. Only 1 piece is shaped like a rectangle. The others are more complex. One is shaped like a Z. The remainder are L-shaped. To play IQ-Block, first choose a shape you like, place it in the upper left, and do not change its position as you try to place the other 9 blocks into the remaining space on the square playing board. The manufacturer boasts, "There are more than 60 different kinds of arrangements" of pieces that will fill in the square playing board. The company also states, "It is an incredible game. Join us in challenging your IQ."

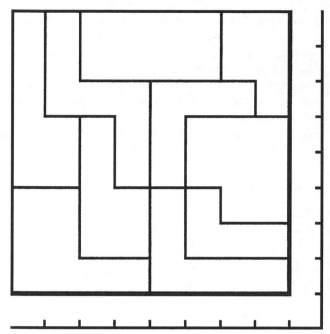

20.1. IQ-Block. Scramble the puzzle pieces and see if your friends can put them together again in a square. Is there more than 1 way to do this?

After the natives found IQ-Block protruding from the jungle floor in the Congo, they quickly translated the playing manual and passed out copies to all members of their local clan. They began to study the game and hold tournaments. During these tournaments, masters attempted to form as many possible different arrangements of pieces within the bounding square as possible before a 10-minute alarm rang.

Here are some challenges and digressions:

⊙ The manufacturer of IQ-Block boasts that there are over 60 different ways of placing the pieces together to form a square. Is this correct? Just how many different arrangements are there? Some Congo philosophers argued that there are only 10 different unique arrangements, while others asserted that there are over 1000 ways of solving the puzzle! Who is closer to the truth?

⊙ On a particularly frigid evening, in a fit of frustration, a master at the game swallowed a polygonal playing piece to prevent his opponent from finding any solutions. Their glistening eyes locked in open warfare. Suddenly a blush of pleasure rose to his opponent's cheeks—and she then created a square of slightly smaller dimensions. Can you create a square after removing a piece, using all the remaining pieces?

⊙ After giving this puzzle to 5 friends, Dr. Googol found that none could create a square by arranging the 10 pieces. So, even if there are "60 different ways" of solving this, you should not despair if the task seems too difficult. Try IQ-Block on your friends, colleagues, children, and students to see if any can even find a *single* way of arranging the pieces to form a square. Dr. Googol looks forward to hearing from you regarding this intriguing puzzle.

❀ For a solution and additional challenges, see "Further Exploring."

Chapter 21

Riffraff

The ratio of the height of the Sears Building in Chicago
to the height of the Woolworth Building in New York is the same to
four significant digits (1.816 vs. 1816) as the ratio of the mass of a
proton to the mass of an electron.
—*John Paulos,* Innumeracy

After weeks of searching, Dr. Googol has finally found a beautiful, spacious, low-rent apartment with French doors and a southern exposure. But on his first night, he discovers a slight disadvantage: all the apartments surrounding his are filled with musicians who practice only after the sun goes down. Not only can't Dr. Googol sleep, but each musician plays the most unmelodious pattern. Above him is the maniacal mathematical trumpeter Fermats Navarro. Every night he plays the same thing. He starts with a long note (shown on the next page as a ☺), then plays a long note followed by a short blast (shown on the next page as a ☹), and then plays a longer phrase of long and short blasts, continuing through the night, each phrase longer than the last:

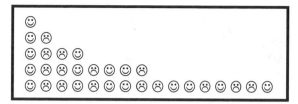

Below Dr. Googol is the similarly maniacal trombone player Curtis Euler. He jams his fist into the mouth of the trombone to create normal and muted sounds symbolized by ☐ (when the trombone's mouth is open) and ☐ (when his fist closes the opening). The odious melody grows ever longer:

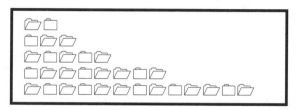

The apartment to the east of Dr. Googol's is occupied by violinist Itzhak Pythagoras. He plays his seemingly random riff of short and long notes over and over again:

$$⊖⊗⊖⊗⊖⊗⊖⊖⊖⊖⊗⊖⊗⊖⊖⊖⊖⊗⊖⊗⊖⊖⊖⊖⊗⊖⊖⊖$$
$$⊖⊗⊖⊗⊖⊖⊖⊖⊖⊗⊖⊖⊖⊗⊖⊗⊖ \ldots$$

This can be represented as a string of 0s (long notes) and 1s (short notes):

01101010001010001010001000001010000010001010 . . .

On the west side is the great saxophonist Hank Möbius. He plays a run of 77 notes, then a run of 49, then one of 36 (♫ ♫ ♫ ♫ ♫ ♫ ♫ ♫ ♫ ♫ ♫ ♫ ♫ ♫ ♫ ♫ ♫ ♫), one of 18 (♫ ♫ ♫ ♫ ♫ ♫ ♫ ♫ ♫), and finally one of 8 (♫ ♫ ♫ ♫).

After a week without sleep, Dr. Googol goes to all his neighbors and asks them if they could play during the day. They all give him the same response: "If you can figure out the pattern in my playing, I'll stop playing at night."

Can you help Dr. Googol with his very difficult problem? If you are a teacher, have your students work in teams.

❉ For solutions to these difficult problems, and for more odd and challenging number sequences, see "Further Exploring."

Chapter 22

Klingon Paths

The advancement and perfection of mathematics are intimately connected with the prosperity of the State.

—*Napoleon*

Dr. Googol was watching *Star Trek* on television when he invented this gruesome puzzle. This grid of numbers is Klingon City, and it's a tough place to live. Each Klingon inhabiting this world carries a bomb worn at the hip as a testament to his courage. As a Klingon walks through the grid of squares, the first time he comes in contact with a number, his bomb receives a signal; if the bomb

6	8	18	15	24	20	2	20
6	2	15	2	17	15	3	7
0	11	18	16	20	15	1	11
6	2	6	13	4	17	20	16
5	12	7	2	3	5	18	23
7	13	3	2	2	11	4	23
16	23	10	2	4	12	5	10
17	12	10	1	13	12	6	20

is exposed to that number a second time, the bomb explodes and the Klingon dies. Klingons, being brave warriors, never show fear—in fact, they love the brutal challenge of the game.

A Klingon can walk on any square in Klingon City and can move horizontally or vertically but not diagonally. What is the longest path the Klingon can take without dying? Remember, the Klingon must wander around while trying to avoid numbers encountered previously—otherwise the Klingon explodes.

❀ For the solution and additional challenges, see "Further Exploring."
▪ For a computer program to study this class of puzzles, see [www.oup-usa.org/sc/0195133420].

Chapter 23

Ouroboros Autophagy

> Blindness to the aesthetic element in mathematics is widespread and can account for a feeling that mathematics is dry as dust, as exciting as a telephone book. . . . Contrariwise, appreciation of this element makes the subject live in a wonderful manner and burn as no other creation of the human mind seems to do.
> —*Philip J. Davis and Reuben Hersh,* The Mathematical Experience

Ouroboros, the mythical serpent always seen chewing or swallowing its own tail, is a symbol of growth, destruction, and the cyclic nature of the universe. Our Ouroboros is made up of 13 sections, each of which houses a number (Figure 23.1). Wrapped inside the outer serpent, which contains the numbers 0 through 12, are 4 generations of circular serpents, each also marked by 13 sections. These sections, though, do not have their numbers yet. You must use the numbers in the first (outer) serpent to find the numbers in the second, the numbers in the second to find those in the third, and so on. Here's how it works: The number you put in the first section of the second serpent will indicate the total number of 0s to be found among the second serpent's sections. The section below the 1 of the first serpent indicates the total number of 1s in the second serpent. The section below the 2 indicates the total number of 2s in the second serpent, and so on. For example, the 3 in the section below the 0 of the first serpent would

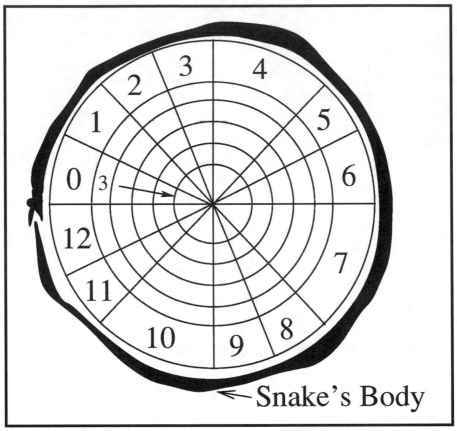

23.1 Each Ouroboros is made of 13 sections containing a number. (The concentric snakes get smaller and smaller as they progress toward the middle of the figure.)

indicate that there must be exactly 3 zeros in the second serpent (but there's no 3 in that position in the real answer).

Use the second serpent to find the numbers in the third. The number in the section under our fictional 3 would indicate how many 3s appear in the third serpent. When you've found all the numbers in the third serpent, use them to figure out the number in the fourth, then use the fourth serpent to solve the fifth. Eventually Ouroboros will begin to cycle with the same 2 sets of numbers. How many serpents does it take before it begins to cycle?

What would happen if each serpent were made up of 10 sections, the first with the numbers 0, 1, 2, 3, 4, 5, 6, 7, 8, 9? What if each serpent were made of 20 sections, the first with the numbers 0 to 19? A hundred sections, numbers 0 to 99? How about 0 to 5? What if the outer serpent's numbers were 1, 2, 2, 3, 3, 3, 4, 4, 4, 4? Can you think of any other Ouroboros numbers that need more serpents before they cycle (or don't cycle)?

❀ See "Further Exploring" for more ophidian delights.

Chapter 24

Interview with a Number

> The natural numbers came from God and all else was man-made.
> —*Leopold Kronecker*

If we are to believe bestselling novelist Anne Rice, vampires resemble humans in many respects but live secret lives hidden among the rest of us mortals. There are also vampires in the world of mathematics, numbers that look like normal figures but bear a disguised difference. They are actually the products of 2 progenitor numbers that when multiplied survive, scrambled together, in the vampire number. Consider 1 such case: $27 \times 81 = 2{,}187$. Another vampire number is 1,435, which is the product of 35 and 41.

Dr. Googol defines true vampires, such as the 2 previous examples, as meeting 3 rules. They have an even number of digits. Each of the progenitor numbers contains half the digits of the vampire. And finally, a true vampire is not created simply by adding 0s to the ends of the numbers, as in

$$270{,}000 \times 810{,}000 = 218{,}700{,}000{,}000$$

True vampires would never be so obvious.

Vampire numbers secretly inhabit our number system, but most have been undetected so far. When Dr. Googol grabbed his silver mirror and wooden stake and began his search for them, he found, in addition to the 2 listed above, 5 other 4-digit vampire numbers. Can you find others? Can you find any vampire numbers with more digits in them?

✸ See "Further Exploring" for a solution.

Chapter 25

The Dream-Worms of Atlantis

There is no such thing as a problem without a gift for you in its hands.
You seek problems because you need their gifts.
—*Richard Bach*, Illusions

On the Pacific Ocean floor lives a group of mathematician mermaids who spend their days in contemplation of the following game called Ocean Dreams. The mermaids use bits of coral and trained marine worms, but we dry-landers can play with a pencil and paper on graph paper.

First make a 10-by-10 array of dots (Figure 25.1). Each worm consists of 5 connected lines that stretch over 6 dots. One of the dots at the end of the worm's

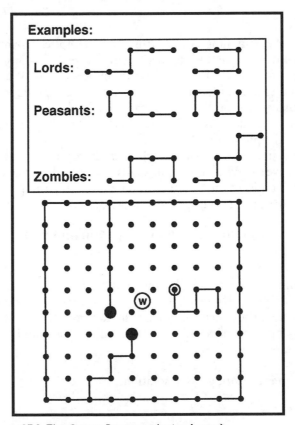

25.1 The Ocean Dreams playing board.

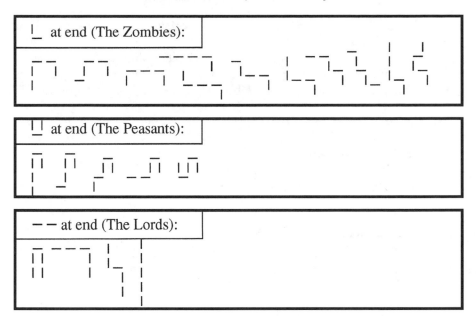

25.2 Several worm contortion patterns.

body is circled to indicate its head. The worms can contort their bodies at right angles to form different shapes, but they can't reuse a grid point through which another part of their body passes. The allowed movements are up, down, right, and left. Figure 25.2 shows some, but not all, worm contortion patterns. The Zombies have a |_ at one end. The Peasants have a |_| at one end. The Lords have a _ _ at one end. (Hint: There is 1 worm missing from the Zombies collection and 1 worm missing from the Peasants collection. Can you complete these sets in Figure 25.2?)

In the game, each player, on his or her turn, has to position a worm on the grid points. In Figure 25.1, player 1 has first placed a Lord with a black head. Player 2 has placed a peasant (at right) with a white head. Player 1 next places a Zombie with a black head. The worms of two players cannot overlap or share the same dot. In other words, every dot in the grid can only be occupied once. The worms cannot lie across one another but can be tightly folded and intertwined. The game is over when no one can add another worm to the grid. To determine who wins, count the number of Lords, Peasants, and Zombies. Each Lord is worth 3 points, each Peasant 2 points, each Zombie 1 point. One player may use open circles on the ends of his or her worms to denote their heads, while the other can use closed circles.

To make matters more interesting, there are 2 final constraints. There is a whirlpool created by a strange undertow at the center of the grid—denoted by a *W* with a circle around it. As a result of the undertow, the worms are pulled toward it. Therefore, the head of each worm must be closer to the center of the grid than the tail of each worm. Additionally, the worms discharge toxins—so, to play it safe, their heads may not be on adjacent, orthogonal grid units. (This means that the heads may not be next to one another in the up, down, left, or right direction.)

Dr. Googol looks forward to hearing from readers who have played Ocean Dreams. What is the best strategy? Does the first player have an advantage? Does this change when there are 3 or more players competing? What happens with bigger boards?

Remember: Dr. Googol has not listed all of the possible worm contortion patterns; there are more Zombies and Peasants than shown here. How many unique worm contortion patterns are there?

✸ For a solution, see "Further Exploring."

Chapter 26

Satanic Cycles

I'm one of those people who believe that life is a series of cycles—wheels within wheels, some meshing with others, some spinning alone, but all of them performing some finite, repeating function.
　　　　　　　　　　　　　　　　　—*Stephen King,* Four past Midnight

Dr. Googol brought a large unicycle to his classroom and started riding up and down the aisles.

"Sir," said a student after several minutes, "why are you doing this?"

Dr. Googol hopped off the unicycle. "Why? I'll tell you why. Listen to my tale about a demon bicyclist riding through the burning depths of Hell."

Most of Dr. Googol's students sat in rapt attention, although a boy with

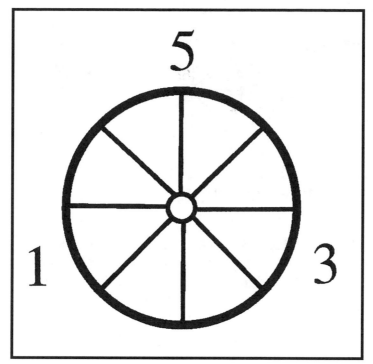

26.1 You can solve this wheel by inserting a *multiply* symbol between 5 and 3 to form 5 × 3 = 15.

spiked orange hair promptly got up and left the room. Dr. Googol did not seem to notice.

"In the crimson caves of Hell rides a bicyclist," Dr. Googol said. "He rides by the lost human souls and allows them to view his bicycle wheels for 1 minute. Surprisingly, each of his wheels has a mathematical formula that can be written out by starting at the correct number and following around the wheel's circular tire in a clockwise or counterclockwise direction until the formula is determined."

Dr. Googol went to the blackboard and drew the wheel in Figure 26.1. "For example, this bicycle wheel contains the formula 5 × 3 = 15. (You start at the 5 and proceed clockwise, inserting the appropriate mathematical symbols as needed.) If you are not able to determine the correct formula within 1 minute, you are relegated to the Stygian depths for all eternity. However, if you can correctly identify the formula before the bicyclist rides on, then you enter the empyrean realm of Paradise—not to mention getting an *A* in my class."

Dr. Googol held up a plaque inscribed with the bicycle wheels in Figure 26.2. Can you help his students identify the formulas they contain? Only the symbols +, -, ×, /, and = and exponentiation are permitted. You may use each of these symbols, at most, 2 times in your formulation. For example, 1 × 2 × 3 × 4

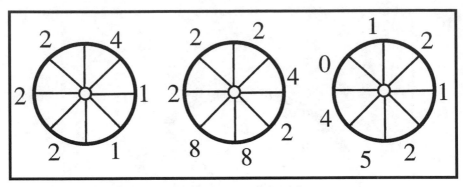

26.2. The eternal bicyclist. Can you solve these wheels?

would not be permitted because the multiplication symbol is used 3 times. Concatenation of digits to form multidigit numbers is allowed as often as needed. (You must proceed around the wheel back to the starting point, or beyond the starting point as in the 5 × 3 = 15 example). Can you solve Dr. Googol's wheels in Figure 26.2?

❀ See "Further Exploring" for solutions and further classroom experiments.

Chapter 27

Persistence

> Science is not about control. It is about cultivating a perpetual condition of wonder in the face of something that forever grows one step richer and subtler than our latest theory about it. It is about reverence, not mastery.
> —*Richard Power,* The Gold Bug Variations

Dr. Googol once lectured during a summer session at Harvard University. As he looked over his class of eager postdoctoral students, he smiled at Monica, his best pupil.

Dr. Googol started by drawing on the blackboard:

$$969, 486, 192, 18, 8$$

He turned back to the class. "Can anyone tell me how the following sequence arises?"

Monica instantly stuck up her hand. "Sir, in '969, 486, 192, 18, 8' each term is the product of the digits of the previous term."

"Monica, you are amazing. Now let me tell you about 969's persistence. The *persistence* of a number is the number of steps (4 in our example) before the number collapses to a single digit. Now, consider 2 mighty difficult questions:

"1. What is the smallest number with persistence 3?

"2. What is the smallest number with the persistence of 12? (Hint: This one is so difficult, don't even bother trying to solve it.)"

Dr. Googol looked at the befuddled students. Even Monica seemed distressed as she ran her shaking fingers through her dark hair.

Dr. Googol stared Monica straight in the eyes. "Monica, I will give a $100 bill to anyone who can answer question 1, and a $1,000 bill to anyone who can answer question 2. Take as long as you like to think about this extraordinarily delightful and devilish problem."

❦ For some commentary, see "Further Exploring."

Chapter 28

Hallucinogenic Highways

We live on a placid island of ignorance in the midst of black seas of infinity, and it is not meant that we should voyage far.
—*H. P. Lovecraft*, The Call of Cthulhu

👁 **Number Maze 1,
a visual intermission before
the next book part. . . .**

Dr. Googol was driving the interstate highways in Colorado when he developed this simple-looking but fiendishly difficult problem. Starting at the bottom

29.1 Hallucinogenic Highways. Can you reach the Finish with a sum of 84? (Drawing by Brian Mansfield.)

in Figure 28.1, you must find a path to the Finish in the Rocky Mountains by traveling the roads. Each time you pass a sign, you add the sign's number to a cumulative sum. Can you reach the Finish with a sum of 84?

If you are a teacher, have your students work in teams to solve this problem. What strategy did your students use? Have them develop their own highway puzzles using other operations including minus and multiply. For safety's sake, make sure students have not imbibed too many caffeinated beverages before embarking on these strange highways. (Your car is small, so you can avoid signs at intersections, like 27 and 7, if you turn sharply.)

❋ For a solution to this hellishly difficult maze, see "Further Reading."

Part ii

Quirky Questions, Lists, and Surveys

God exists since mathematics is consistent, and the devil exists
since we cannot prove the consistency.
—*Morris Kline,* Mathematical Thought
from Ancient to Modern Times

Goethe opposed the use of the microscope, since he believed that
what cannot be seen with the naked eye should not be seen, and that
what is hidden from us is hidden for a purpose. In this, Goethe was a
scandal among scientists, whose first, firm, and necessary principle is
that if something can be done, then it should be done.
—*John Bainville,* "Beauty, Charm and Strangeness: Science as
Metaphor," Science 281, *1998*

In addition to being fascinated with integers, Dr. Googol has always been
fond of making mathematical lists. He is always asking questions. What if this?
Who is that? Rank this! Why this? Many of the lists that follow have been constructed using information provided by exclusive surveys and discussions with
mathematicians around the world. The rankings are not the definitive word on
the subject. Rather, in the Talmudic tradition of presentation and analysis, the
lists are open for discussion. You will no doubt disagree with some of the rankings, but this makes for lively, philosophical debate. Enjoy! Dr. Googol welcomes your comments.

Chapter 29

Why Was the First Woman Mathematician Murdered?

The mathematical world of today owes Hypatia a great debt. . . . At the time of her death, she was the greatest mathematician then living in the Greco-Roman world, very likely the world as a whole."
—*M. Deakin,* American Mathematical Monthly, *1994*

Reserve your right to think, for even to think wrongly is better than not to think at all.

—*Hypatia*

The Pythagoreans of ancient Greece were fascinated by numbers, such as the triangular numbers mentioned in Chapter 62. In fact, the Pythagoreans worshiped numbers as gods. What ever became of Pythagorean thoughts and ideas once Pythagoras died? It turns out that Pythagoras's philosophy, modified by Plato, outlasted all other philosophies of ancient Greece. Even up to the 6th century A.D., the numerical gods were still worshiped, but during the Dark Ages their meaning was lost.

The Pythagoreans and their offshoot Platonists were the only ancient philosophical schools to allow women to share in the teaching and the only sects that produced outstanding woman philosophers. Unfortunately, one of their best, Hypatia of Alexandria (370–415), was martyred by being torn into shreds by a Christian mob—partly because she did not adhere to strict Christian principles. She considered herself a neo-Platonist, a pagan, and a follower of Pythagorean ideas. Interestingly, Hypatia is the first woman mathematician in the history of humanity of whom we have reasonably secure and detailed knowledge (Figure 29.1).

Hypatia was born in Alexandria during times of turbulent power struggles between Romans and militant Christians. Her father, Theon, was a respected mathematician and astronomer. When he recognized Hypatia's talents and desire to learn, he educated her even though most people of their era did not support the idea of educating women. Together they began writing books on Euclid's and Diophantus's works.

Hypatia was a respected, charismatic teacher, well liked by all her students. Because she was famous for being the greatest of problem solvers, mathematicians who had been stuck for months on particular problems would write to her seeking her advice. She was said to be physically attractive and determinedly celibate. When asked why she was obsessed with mathematics and would not marry, she replied that she was wedded to the truth.

Hypatia edited books on geometry, algebra, and astronomy. Her main focus was on Euclidean geometry and solving integer equations, and she authored a popular treatise on the conics of Apollonius. In one of her mathematical problems for her students, she asked them for the integer solution of the pair of simultaneous equations: $x - y = a$, $x^2 - y^2 = (x - y) + b$, where a and b are known. Can you find any integer values for x, y, a, and b that make both of these formulas true?

Aside from being a mathematician, Hypatia assisted in the design of astrolabes, mechanical devices that replicate the motion of the planets. She also helped design urinonmeters to measure the specific gravity of urine. These were of potential use in determining the proper dosages of diuretics used to treat illnesses.

29.1 Hypatia.

Sadly, we know more about Hypatia's death than about other significant events in her life. The Christians were her strongest philosophical rivals, and they officially discouraged her teachings, which were Pythagorean in nature with a religious dimension. Donning a philosopher's cloak and making her way through the city, she spoke publicly about the writings of Plato, Aristotle, and other philosophers to anyone who wished to hear.

On a warm March day in A.D. 414, after having engaged her students in a brilliant philosophical discussion, Hypatia guided her chariot confidently down the streets of Alexandria toward her home. She noticed a crowd in front of a church, and before she could turn her chariot away, two men pulled her out. "Kill the pagan!" they shouted. Like many victims of terrorists today, she may have been seized merely because she was a well-known figure and prominent on the other side of the religious divide.

The historian Edward Gibbon provided a sad account of her death:

> On a fatal day, in the holy season of Lent, Hypatia was torn from her chariot, stripped naked, dragged to the church, and inhumanely butchered by the hands of Peter the Reader and a troop of savage, merciless fanatics; her flesh was scraped from her bones with sharp oyster-shells, and her quivering limbs were delivered to the flames.

Her horrible death was recorded by 5th-century Christian historian Socrates Scholasticus. After reporting her murder to Rome, Orestes, a former student of Hypatia, requested an investigation. The investigation never took place, supposedly because of the lack of evidence and witnesses. Her murder adversely affected educational freedom for many years. Mathematics entered a period of stagnation, and it was not until after the Renaissance that another woman, Maria Agnesi, made her name as a mathematician.

Chapter 30

What if We Receive Messages from the Stars?

> We feel certain that the extraterrestrial message is a mathematical code of some kind. Probably a number code. Mathematics is the one language we might conceivably have in common with other forms of intelligent life in the universe. As I understand it, there is no reality more independent of our perception and more true to itself than mathematical reality.
> —*Don DeLillo,* Ratner's Star

What if spaceships from another world suddenly appeared in our skies? What if tomorrow morning you turned on your radio and heard a strange, pulsating tone, and what if you learned that the same thing was happening across our planet?

Dr. Googol often fantasizes that he is a handsome computer genius watching as a giant alien mothership arrives in Earth orbit and immediately begins to transmit a cyclic tone down to the nations of Earth. The world frantically tries to understand the aliens' intentions—until Dr. Googol deciphers the alien message: it's a countdown to weapon firing. The President of the United States attempts to reason with the creatures, who give Earthlings one choice: become their slaves, or die. The aliens demonstrate their massive orbital firepower by destroying large U.S. cities, and the military forces of many nations try to retaliate with little effect.

If we really ever do receive a message from the stars intended to be deciphered by us, just how will it be sent, and how difficult will it be to interpret? If we decided to reply, how would we send a message? One possibility is that we or aliens would use radio waves beamed into space at frequencies between 1 and 10,000 megahertz, because these frequencies travel relatively easily through space and through the atmosphere of planets like our own. The first part of the message would be easy to understand so it would attract attention, such as a series of pulses representing the numbers 1, 2, 3. . . . This could be followed by more intricate communications.

👽 👽 👽

Many science-fiction novels have dealt explicitly with alien signals and their decipherment. For example, in Buzz Aldrin and John Barnes's bestselling novel *Encounter with Tiber*, Earth's astronomers detect a signal from Alpha Centauri, the triple star of which the faintest component is the closest star to Earth, about 4.3 light-years away. Scientists first attempt to determine around which of the stars the alien transmitter is orbiting by analyzing the Doppler shift (change in frequency) occurring for waves coming from a moving object.

Bits and pieces of the signal seem to be strangely ordered, like a sequence of tones, 2 different pitches stuttering at an enormous rate. Unfortunately, the Earth's atmosphere is nearly opaque to radio at the transmission wavelength of 96 meters, because the signal cannot easily penetrate the ionosphere. Thus it is impossible to catch more than brief snatches of the message even with the most sensitive radio telescopes on the ground. Luckily, the scientists find a way to make use of a space station upon which they mount simple antennas to listen to the signal.

Despite their skepticism, the scientists continue to study the signal and discover it is a pattern of high tones, low tones, and silences. Assuming that the silences are spaces, and because the transmission comes as triple beeps, it seems likely that the message is in base 8.

Scientists call the high tones "beeps" (represented below by a 👽) and the low tones "honks" (represented by a 👽). A group of 3 beep-or-honk choices has 8 different arrangements:

The digits are likely to stand for the digits 0–7, which are the 8 digits for a base 8 system. The string of digits in the message could represent pictures or text.

The most common numbering system on Earth is base 10. In other words, we have 10 digits, 0 through 9. In our base 10 representation, each digit represents a power of 10. For example, the number 2,010 is $2 \times 10^3 + 0 \times 10^2 + 1 \times 10^1 + 0 \times 10^0$ where $10^3 = 1000$, $10^2 = 100$, $10^1 = 10$, and $10^0 = 1$. However, there's no reason to assume that aliens would use a base 10 number system, and it's unlikely that a message from the stars would arrive in base 10 numbers. On Earth, our mathematical calculations are based on 10 because we have 10 fingers. In fact, our language suggests the connection between fingers and our number system—we use the world *digit* to designate both a number and a finger. Because our base 10 system comes from our use of 10 fingers, what would a base 8 system tell us about the anatomy of aliens? Perhaps a base 8 system would denote an alien with a thumb and 3 fingers on each hand, or a creature with 8 tentacles, or a thumb and a finger on each of its 4 arms. An even wilder possibility is that the aliens have 3 heads and these are all the combinations of nodding and shaking that are possible! (Of course, it is possible that their number system tells us nothing about their anatomy. After all, what did the Babylonian's base 60 system tell us about their anatomy?)

As scientists study the message, they find it repeats every 11 hours and 20 minutes. Each group of 16,769,021 base 8 numbers takes about 2½ seconds to be received, so there are 16,384 such groups in all. What could it mean?

The first thing to check is the number 16,769,021. Does it have any unusual properties? It turns out that you can use a simple factoring program to determine it is equal to 4,093 times 4,097—2 prime numbers. Since a prime number isn't evenly divisible by another number (except itself and 1), an alien could transmit a gridlike pattern whose size is the product of two prime numbers. As a result, there are only a couple of possible arrangements for the numbers in the grid. (In fact, the pattern could be a photo consisting of an array of pixels as on your computer screen.) On the other hand, if the image were composed of, for example, 10,000,000 pixels with many factors, there would be a very large number of possible arrangements, such as $5 \times 200,000$, $10,000 \times 1,000$, and many others, and this would make the image difficult to decode.

👽 👽 👽

In *Encounter with Tiber*, it turns out that the 8 groups of honks and beeps represent 8 different intensity values in an image: 0 for black, 7 for white, and 1–6 for intermediate intensities. By representing these brightness values on a 4,093-by-4,097 grid, the astrophysicists determine that each transmission is a frame of a movie. When played sequentially on a computer, 8 creatures are seen waving as they climb into a spacecraft! Other more technical information follows including instructions on how to find an alien encyclopedia containing poems, paintings, music, literature, science, engineering, and jokes of a civilization centuries in advance of Earth's.

Would you like to view such an alien encyclopedia? In *Encounter with Tiber*, some people worry that humanity is not ready for advanced knowledge from the encyclopedia. "What if you'd given Napoleon the atomic bomb?" scientists and politicians ask. "What if the Civil War had been fought with airplanes dropping poison gas on cities?" Should the encyclopedia be made available to all the nations on Earth?

Do you think that communication with aliens would create widespread hysteria? The psychologist Carl Jung believed that contact with superior beings would be devastating and demoralizing to us because we'd find ourselves no more a match for them intellectually than our pets are for us. Such fears and jealousies might cause various extremists groups, such as the Ku Klux Klan, to try to kill the aliens.

👽 For further information on aliens, numbers, and messages from the stars, see "Further Exploring."

Chapter 31

A Ranking of the 5 Strangest Mathematicians Who Ever Lived

Erdös covered the floor with cereal. He couldn't close a window by himself. He woke you up at 4 A.M. shouting about number theory. Paul Erdös may have been the world's worst houseguest, but he was also the world's most generous and prolific mathematician.
—*Paul Hoffman,* The Man Who Loved Only Numbers

Most classmates regarded Ted Kaczynski as alien, or not at all.
—*Robert McFadden,* New York Times

Freedom means having power; not the power to control other people but the power to control the circumstance of one's own life.
—Unabomber Manifesto

There were five clear winners when respondents were asked to name the five strangest mathematicians who ever lived.

1. Paul Erdös (1913–1996) This legendary mathematician, one of the most prolific in history, was so devoted to math that he lived as a nomad with no home and no job. As discussed in fantastic detail in Chapter 46, sexual pleasure revolted him; even an accidental touch by anyone made him feel uncomfortable. During the last year of his life, at age 83, he continued churning out theorems and delivering lectures, defying conventional wisdom that mathematics is a young person's sport. On this subject, Erdös once said:

> The first sign of senility is when a man forgets his theorems. The second sign is when he forgets to zip up. The third sign is when he forgets to zip down.

Paul Hoffman, author of *The Man Who Loved Only Numbers*, notes:

> Erdös thought about more problems than any other mathematician in history and could recite the details of some 1,500 papers he had written. Fortified by coffee, Erdös did mathematics 19 hours a day, and when friends urged him to slow down, he always had the same response: "There'll be plenty of time to rest in the grave."

2. Srinivasa Ramanujan (1887–1920) Ramanujan, who started life as a clerk in the accounting department of the Madras post office, became India's greatest mathematical genius and one of the greatest 20th-century mathematicians. Ramanujan made substantial contributions to the analytical theory of numbers and worked on elliptic functions, continued fractions, and infinite series. He came from a poor family, and his mother took in boarders, which created a crowded home. Ramanujan was very shy and found it hard to speak. He excelled in math but usually failed all his other courses. When he was 13, he borrowed a high school student's math book and mastered it in a week. Because he was deprived of manuals that could teach him about rigorous proofs, Ramanujan developed rather strange methods to establish mathematical truths. Mathematician G. H. Hardy remarked:

> His ideas as to what constituted a mathematical proof were of the most shadowy description. All his results, new or old, right or wrong, had been arrived at by a process of mingled argument, intuition, and induction, of which he was entirely unable to give any coherent account.

Ramanujan, although self-taught in mathematics, was given a fellowship to the University of Madras in 1903, but the following year he lost it because he devoted all his time to mathematics and neglected his other subjects. Hardy, a professor at Trinity College, invited him to Cambridge on the basis of a now-historic letter Ramanujan wrote him, which contained some 100 theorems. A leading expert in calculus, Hardy found himself dealing with a collection of formulas completely unfamiliar to him. He said:

> These relations defeated me completely; I had never seen anything in the least like them before. A single look at them is enough to show that they could only be written down by a mathematician of the highest class.

Some years later, Ramanujan, weakened by his strict vegetarianism, became quite sick with tuberculosis. However, neither physicians nor his family could persuade him to stop his studies. He returned to India in February 1919 and died in April 1920 at the age of 32. During that period he wrote down about 600 theorems on loose sheets of paper. These were discovered only in 1976 by

Professor George Andrews of Pennsylvania State University, who termed them the Lost Notebook of Ramanujan. Many of Ramanujan's formulas came to occupy central places in modern theories of algebraic number theory, and today scholars wonder how he could envision such equations when he didn't have any of the supporting knowledge to understand them.

3. **Pythagoras** (580–500 B.C.) A Greek philosopher, Pythagoras was responsible for important developments in mathematics, astronomy, and the theory of music. Philosopher Bertrand Russell once wrote that Pythagoras was intellectually one of the most important men who ever lived, both when he was wise and when he was unwise. Pythagoras is also the most puzzling mathematician in history, because he founded a numerical religion whose main tenets were transmigration of souls and the sinfulness of eating beans, along with a host of other odd rules and regulations.

4. **Theodore Kaczynski** (b. 1942) Ted Kaczynski, also known as the Unabomber, was a mathematican who rose swiftly to academic heights even as he became an emotional cripple, loner, and murderer. Kaczynski's 25-year self-imposed exile in the Montana woods was particularly appropriate for this man who had always been alone. The May 26, 1996, *New York Times* noted that the cabin "suited this genius with gifts for solitude, perseverance, secrecy and meticulousness, for penetrating the mysteries of mathematics and the dangers of technology, but never love, never friendship." The remoteness of the cabin was probably as much a means of limiting others' access to him as it was a symbol of freedom. Before he became a hermit, Kaczynski wrote several notable papers on the mathematical properties of functions in circles and boundary functions. Although his IQ was measured as 170, he exhibited many odd characteristics: excessive (pathological) shyness, fascination with body sounds, a metronomic habit of rocking, and frequent concerns about germs, infections, and other health matters. His room at school stank of rotting food and was piled high with trash. After teaching for 2 years and publishing mathematical papers (Chapter 40) that impressed his peers and put him on a tenure track at one of the nation's most prestigious universities, he suddenly quit, spent nearly half his life in the woods, and killed three strangers and injured 22 others. Throughout his life, Kaczynski found it painful to make errors and corrected minor errors in others. He shut himself up in his bedroom for days at a time and seemed incapable of sympathy, human insight, and simple connections with people. Although Kaczynski does not have the eminence of Erdös, Ramanujan, or Pythagoras, his mathematics papers were sufficiently complex to warrant his inclusion on this brief list.

5. **John Nash** (1928–) This brilliant mathematician received the 1994 Nobel Prize in Economics. Nash's prize-winning work appeared almost half a century earlier in his slender 27-page Ph.D. thesis written at the age of 21.

In 1950, Princeton graduate student John Nash formulated a theorem that enabled the field of game theory to become an important influence in modern economics. Compulsively rational, he often turned life's decisions—whether to take the first elevator or wait for the next one, or whether to marry—into calculations of advantage and disadvantage, mathematical rules divorced from emotion. In 1958, *Fortune* singled Nash out for his achievements in game theory, algebraic geometry, and nonlinear theory, calling him the most brilliant of the younger generation of mathematicians. He seemed destined for continued achievements, but in 1959 he was institutionalized and diagnosed as schizophrenic. Brilliant when he was young, Nash slipped into and out of schizophrenia for decades, believing that aliens had made him emperor of Antarctica. At other times he believed himself to be a messianic figure. Princeton and its academic staff stood by Nash and kindly let him wander about the math department for almost thirty years. There he became a mute figure who scribbled bizarre equations on blackboards in the mathematics building and searched for secret messages in numbers. He believed that ordinary things—a telephone number, a colorful necktie, a dog racing across the grass, a Hebrew letter, a sentence in the newspaper—had hidden and important significance. Sadly, Nash's son was also schizophrenic, but he was sufficiently versed in math that Rutgers University granted him a Ph.D. John Nash once remarked: "I would not dare to say that there is a direct relation between mathematics and madness, but there is no doubt that great mathematicians suffer from maniacal characteristics, delirium, and symptoms of schizophrenia." The most famous biography on John Nash is Sylvia Nasar's *A Beautiful Mind*.

Chapter 32

Einstein, Ramanujan, Hawking

There is no branch of mathematics, however abstract, which may not someday be applied to the phenomena of the real world.
—*Nicolai Lobachevsky*

Dr. Googol surveyed many mathematicians on the following question:

Which of the following would have had the most profound effect on our world today?

1. Physicist **Albert Einstein** lived another 20 years with a clear mind.
2. Mathematician **Srinivasa Ramanujan** lived another 20 years with a clear mind.
3. Astronomer **Stephen Hawking** was not afflicted with Lou Gehrig's disease.

Several respondents suggested that Stephen Hawking's affliction forced him to concentrate on black hole theory and also increased public interest in these theories. Therefore, they thought that removal of his affliction would not have had a favorable effect on the world.

Some mathematicians believed that Albert Einstein could have made great contributions to the "theory of everything" if he had lived longer, but others suggested that Einstein (and Ramanujan) had reached their peak during their lives and would not have contributed significant additional information. For example, Einstein made early, important discoveries in the theory of Brownian noise as a model for microscopic phenomena, energy and charge as a quantified phenomena, light speed and mass as constraints on spacetime, and fundamental forces as a deformation of space. But all these achievements came from very peculiar analyses and interpretations of older works that were in place in the early 1900s. Toward the end of his life, he made little progress in synthesizing new theories.

Nevertheless, debate on Dr. Googol's questions still rages. Mathematician Charles Ashbacher suggests to Dr. Googol:

> There is no doubt in my mind that if Albert Einstein had lived another 20 years the world would be profoundly impacted. Einstein was not only the greatest physicist of the 20th century with obvious major accomplishments, but he was also very influential in other ways. It was the letter from Einstein to President Franklin Roosevelt that tilted the balance in favor of the Manhattan Project. He commanded so much respect in the world that it is possible he could have tempered some of the events of the world well into the 1980s. Any changes as a result of any new discoveries in physics would be icing on the cake.

Philosopher of science Dennis Gordon says:

> If Albert Einstein had been so fortunate to have had 20 additional years with a sound mind, perhaps he would have collaborated with a young and vigorous Stephen Hawking to either demonstrate or disprove the existence of the long-speculated gravitons.
>
> Given the same good fortune, maybe we would have gained some insight into Ramanujan's extraordinary intuition and thought processes. How was Ramanujan able to generate such astounding results when even he himself was often unable to offer proofs? I am reminded of the scene in the movie *Amadeus* in which Mozart is shown producing perfectly written symphonies on the first draft; maybe the thought processes of both geniuses were very similar. And, further, with 20 additional years, Ramanujan might have solved all of David Hilbert's famous 23 problems [discussed in Chapter 36] and then later humbled Hilbert with solutions to several more then-unknown problems.

Ramanujan, described in detail in Chapter 31, was a self-taught mathematical genius, who used his gut instinct to attack the frontiers of mathematical analysis of his time (modular functions, analytic number theory, partitions, iteration theory, transcendence properties). However, he sometimes advanced slowly and was unable to transfer his insights to others and other fields. Perhaps 20 years more of activity would not have changed his approach. On the other hand, respondents suggested that 20 years more would have been very significant for mathematicians like Archimedes, Roger Cotes, Niels Abel, Évariste Galois, Bernhard Riemann, Henri Poincaré, Jacques Herbrand, and Allan Turing, all of whom were extremely creative when they died.

Perhaps a greater effect would have resulted if Évariste Galois, the founder of modern algebra and group theory, had not been killed at age 21 in a gun duel. (Galois, a genius and child prodigy, is discussed in Chapter 36.) His mathematical ideas and innovative methods were too advanced for his time; therefore, few contemporaries could understand his insights. Many had trouble filling in the steps he saw as obvious. If Galois had lived longer and continued his work in group theory and algebraic equations, the world would have been affected immensely.

❀ ❀ ❀

Given all these thoughts, it is not clear which of the 3 situations would yield the biggest impact on humanity. Respondents generally thought that Hawking's affliction, while sad on a personal level, did not hinder his effect on math and science. A longer life for Einstein would have given the world a scientist hero for a longer period of time and therefore increased public interest in science. Perhaps a long life for Ramanujan would have had the greatest impact, especially considering his fantastic mathematical output and short life. This leads to other questions. What if Ramanujan had developed in a more nurturing environment? Although he would have been a better-trained mathematician, would he have become such a unique thinker? Could he have discovered so many wonderful formulas if he had been taught the rules of mathematics early on and pushed to publish his results with rigorous proofs? Perhaps his relative isolation and poverty enhanced the greatness of his mathematical thought. For Ramanjuan, equations were not just the means for proofs or calculations. The *beauty* of the equation was of paramount value. The intrinsic elegance of his formulas causes them to play key roles in the most unusual circumstances.

Ramanujan's most "beautiful" formula draws a shocking connection between an infinite series (at left) and a continued fraction (middle). It is wonderful that neither the series nor the chain fraction can be expressed through the famous numerical constants π and e, and their sum mysteriously equals $\sqrt{\pi e/2}$. Try to compute the value on the left side of the equals sign, for several terms, and see how it compares with the right side when substituting π = 3.141592 and e = 2.718282.

Ramanujan's Most Beautiful Formula

$$1 + \cfrac{1}{1\cdot3} + \cfrac{1}{1\cdot3\cdot5} + \cfrac{1}{1\cdot3\cdot5\cdot7} + \cfrac{1}{1\cdot3\cdot5\cdot7\cdot9} + \ldots + \cfrac{1}{1+\cfrac{1}{1+\cfrac{2}{1+\cfrac{3}{1+\cfrac{4}{1+\ldots}}}}} = \sqrt{\cfrac{\pi e}{2}}$$

(Please send the publisher a note of thanks for allowing Dr. Googol to insist on adding this typographical monstrosity to this book.)

Chapter 33

A Ranking of the 8 Most Influential Female Mathematicians

It is impossible to be a mathematician without being a poet in soul.
—*Sofia Kovalevskaya, quoted in* Agnessi to Zeno *by Sanderson Smith*

Despite horrible prejudice in earlier times, several women fought against the establishment and persevered in mathematics. Until the 20th century, very few women received much education, and the path to more advanced studies was usually blocked. Many of these women had to go against the wishes of their families if they wanted to learn. Some were even forced to assume false identities, study in terrible conditions, and work in intellectual isolation. Consequently, very few women contributed to mathematics. The following ranking of the 8 most influential female mathematicians was compiled through extensive research and by surveying mathematicians. These women did more than just influence the course of mathematics. They also affected people's perceptions of women's role in all intellectual endeavors.

Many of these women came from mathematical families. Emmy Noether, Hypatia, Maria Agnesi, and others never married, partly because it was not socially acceptable for women to pursue mathematical careers, and, therefore, men were not likely to wed brides with such controversial backgrounds. Russian mathematician Sofia Kovalevskaya was an exception to this rule; she arranged a marriage of convenience to a man who was agreeable to a platonic relationship. For Sofia and her husband, the marriage allowed them to escape their families and concentrate on their respective research. The marriage also allowed Sofia a greater freedom to travel because, at the time, it was more suitable for a married woman to travel around Europe than a single woman.

When Dr. Googol asked dozens of mathematicians, "Who were the most influential female mathematicians in history?" there were several clear favorites, listed below. Again, this list is not meant to be definitive; rather, it should stir debate and discussion. Much of the biographical information comes from Dr. John J. O'Connor and Professor Edmund F. Robertson's "MacTutor History of Mathematics Archive," http://www-history.mcs.st-andrews.ac.uk/history/index.html.

1. **Hypatia** (370–415) Hypatia, who is discussed more extensively in Chapter 29, was famous for giving the most popular discourses in Western civilization and for being the greatest of problem solvers. She was the first woman to make a significant contribution to the development of mathematics. The daughter of the mathematician Theon, she eventually became head of the Platonist school at Alexandria. She came to symbolize scientific ideas, which, unfortunately, the early Christians identified with paganism. She met her death

at the hands of a mob who dragged her from her chariot and peeled off her skin with oyster shells.

2. **Sofia Kovalevskaya** (1850–1891) Kovalevskaya made valuable contributions to the theory of differential equations and was the first woman to receive a doctorate in mathematics. (Note: In case you are confused when searching for information on S. K., both her first and last names seem to be spelled in various ways in English translation including Sonia, Sofya, and Sonya, and Kovalevsky, Kovalevski, Kovalevskia.) Like most other mathematical geniuses, Sofia fell in love with mathematics at a very young age. Sofia wrote in her autobiography: "The meaning of these concepts I naturally could not yet grasp, but they acted on my imagination, instilling in me a reverence for mathematics as an exalted and mysterious science which opens up to its initiates a new world of wonders, inaccessible to ordinary mortals." When Sofia was 11 years old, she hung calculus papers on all the walls of her bedroom. When learning mathematics from the family's tutor, she said, "I began to feel an attraction for my mathematics so intense that I started to neglect my other studies." Sofia's father decided to put a stop to her mathematics lessons, but she secretly read math books late at night. Sofia was forced to marry so that she could go abroad to pursue higher education. (Her father forbid her to study at a university, and Russian women were not permitted to live apart from their families without the written permission of their father or husband.)

At the age of 18, Sofia entered a sad and tense marriage with Vladimir Kovalevski, a young paleontologist. In 1869, Sofia went to Heidelberg to study mathematics but discovered that women could not go to the university. Eventually she persuaded the university authorities to let her attend lectures unofficially. Sofia immediately attracted the teachers' attention with her mathematical brilliance, and virtually all of her professors were delighted about their gifted student and spoke about her as an extraordinary phenomenon. (Spending the summer of 1869 in England, Sofia and her husband met Charles Darwin, Thomas Huxley, and George Eliot.)

In 1871, Sophia Kovalevskaya moved to Berlin to study with mathematician Karl Weierstrass, but again she was not allowed to attend courses at the university. Ironically, this actually helped Sofia, because it forced Weierstrass give her more personal attention. By the spring of 1874, Kovalevskaya had completed 3 papers, each of which Weierstrass deemed worthy of a doctorate. (The 3 papers were on partial differential equations, abelian integrals, and Saturn's rings.) In 1874, Kovalevskaya received her doctorate, summa cum laude, from Göttingen University. However, despite this doctorate and enthusiastic letters of recommendation from Weierstrass, Kovalevskaya was unable to obtain an academic position because she was a woman. Her crushing rejections resulted in a bitter 6-year period during which she did no research. In 1878, Kovalevskaya gave birth to a daughter, and then returned to her study of mathematics. In 1882, she began work on the refraction of light and wrote 3 articles on the topic.

She began to lecture on mathematics in Stockholm and was appointed to a professorship in June of that year. She taught courses on the latest topics in analysis, became an editor of the journal *Acta Mathematica*, interacted with all the famous mathematicians of Paris and Berlin, and organized many international conferences. In 1886, Kovalevskaya was awarded the Prix Bordin of the French Academy of Sciences for her paper on the solution on the rotation of a rigid body around a fixed point.

3. Emmy Amalie Noether (1882–1935) Noether was described by Albert Einstein as "the most significant creative mathematical genius thus far produced since the higher education of women began." She is best known for her contributions to abstract algebra and, in particular, for her study of "chain conditions on ideals of rings." Moreover, in 1915, she discovered a result in theoretical physics sometimes referred to as Noether's Theorem, which proves a relationship between symmetries in physics and conservation principles. This basic result in the general theory of relativity was praised by Einstein. Noether's work in the theory of invariants led to formulations for several concepts of Einstein's general theory of relativity. In 1933, despite her amazing accomplishments, the Nazis caused her dismissal from the University of Göttingen because she was Jewish. She later lectured at the Institute for Advanced Study in Princeton.

4. Sophie Germain (1776–1831) Germain made major contributions to number theory, acoustics, and elasticity. When she was 13, Sophie read a book about the death of Archimedes at the hands of a Roman soldier. She was so moved by this story that she decided to become a mathematician. (Legend had it that Archimedes was so engrossed in the study of a geometric figure in the sand that he failed to respond to the questioning of a Roman soldier. As a result he was speared to death. This sparked Sophie's interest; if someone could be so engrossed in a problem as to ignore a soldier and then die for it, the subject must be interesting!) Sophie's parents felt her interest in mathematics was inappropriate, so at night she secretly began studying the works of Isaac Newton and mathematician Leonhard Euler. Her parents responded by taking away her clothes once she was in bed and depriving her of heat and light so that she would be forced to stay in her bed at night instead of studying. This did not work as planned. Sophie would wrap herself in quilts and use candles she had hidden in order to study at night. Finally her parents realized that Sophie's passion for mathematics was "incurable," and they let her learn. Sophie obtained lecture notes for many courses from the Ecole Polytechnique. (Note the oddly similar situation of the young mathematician Mary Somerville, whose father took away her candles for studying and said, "We must put a stop to this, or we shall have Mary in a straitjacket one of these days.")

After reading Joseph-Louis Lagrange's lecture notes on analysis, Sophie used the pseudonym M. LeBlanc to submit a paper whose originality and insight made Lagrange search desperately for its author. When he discovered "M. LeBlanc" was a woman, his respect for her work remained, and he became her sponsor and mathematical counselor. Sophie proved that if x, y, and z are integers and if $x^5 + y^5 = z^5$, then either x, y, or z must be divisible by 5. "Germain's theorem" was an important step toward proving Fermat's Last Theorem for the case where n equals 5. This was to remain the most important result related to Fermat's Last Theorem from 1738 until the contributions of Ernst Eduard Kummer in 1840. (Fermat's Last Theorem says that if x, y, z, and n are positive integers, then $x^n + y^n = z^n$ cannot be solved for any n greater than 2.) Sophie also worked on theories of elasticity, publishing several memoirs on the subject. The most important of these deals with the "nature, bounds, and extent of elastic surfaces."

5. Maria Agnesi (1718-1799) Agnesi is noted for her work in differential calculus. When she was 7 years old, she mastered the Latin, Greek, and Hebrew languages, and at the age of 9 she published a Latin discourse defending higher education for women. During her teens, she privately studied the mathematics of Descartes, Newton, Leibniz, and Euler. She also tutored the family's younger children and was hostess at scientific and mathematical meetings arranged by her father. At the age of 20, she published *Propositiones Philosophicae*, a treatise on philosophy. For the next decade, she worked on her 2-volume mathematics book *Analytic Institutions for the Use of Italian Youth*, which was finally published in 1748. Volume 1 dealt with algebra and precalculus mathematics, and volume 2 discussed differential and integral calculus, infinite series, and differential equations. Her clearly written textbooks included a discussion of the cubic curve now know as the "witch of Agnesi." (The word *witch* is in fact a mistranslation of *versiera*, which can mean either "curve" or "witch.") Agnesi's book received immediate praise, and the Bologna Academy of Science elected her a member. In 1749, Pope Benedict XIV awarded her a gold medal, and the next year he appointed her to teach mathematics at the University of Bologna, an extremely rare position for a woman because very few women were allowed to even attend a university. However, she turned down the appointment, and, after the death of her father two years later, she stopped doing scientific work altogether. Agnesi devoted the last 47 years of her life to caring for sick and dying women.

6. Helena Raisowa (1917–1994) Raisowa grew up in Warsaw, at a time when the German invasion of Poland in 1939 made it very dangerous for her to pursue mathematics. Nevertheless, she persevered and studied for her master's degree. In 1944, when the Germans suppressed the Warsaw Uprising, Rasiowa's thesis burned together with her entire house. She survived with her mother in a cellar covered by ruins of the building. Her 1950 doctoral thesis (*Algebraic Treatment of the Functional Calculus of Lewis and Heyting*), presented to the University of Warsaw, was on algebra and logic. Rasiowa was promoted continually, reaching the rank of full professor in 1967. Her main research was in algebraic logic and the mathematical foundations of computer science. She always believed that there are deep relations among the methods of algebra, logic, and the foundations of computer science. In 1984, Rasiowa developed techniques that are now central to the study of artificial intelligence. Rasiowa wrote hundreds of papers and books and edited numerous journals.

7. Nina Karlovna Bari (1901–1961) Bari was an outstanding Russian mathematician, the first woman student at Moscow State University, and author of over 50 research articles and textbooks such as *Higher Algebra* (1932) and *The Theory of Series* (1936). She edited the complete works of mathematician Nikolai Luzin and was the editor of 2 important mathematics journals. She also translated Henri Lebesgue's famous book (on integration) into Russian. Her extensive research monograph on trigonometric series became a standard reference for mathematicians specializing in the theory of functions and trigonometric series.

8. **Grace Hopper** (1906–1992) Hopper taught mathematics at Vassar and in 1944 worked with mathematician Howard Aikin on the Harvard Mark I computer. At this time she coined the term *bug* for a computer fault. (The original "bug" was actually a gypsy moth that caused a hardware fault in the Mark I!) In 1949, Hopper designed improved computer compilers. She also helped develop Flow-Matic, the first English-language data-processing compiler. She retired from the Navy with the rank of commander in 1966, but she continued to help standardize the Navy's computer languages. In 1991, she was awarded the National Medal of Technology.

Some runners-up: **Julia Robinson** (1919–1982), who studied number theory and was the first woman mathematician to be elected to the National Academy of Sciences and first woman president of the American Mathematical Society; **Mary Cartwright** (1900–1998), who studied analytic function theory and was the first woman mathematician to be elected a Fellow of the Royal Society of England; **Sun-Yung Alice Chang** (b. 1948), who studies nonlinear partial differential equations and various problems in geometry; and **Karen Keskulla Uhlenbeck** (b. 1942), a leading expert on partial differential equations whose work has provided analytic aids for using instantons as an effective geometric tool. (Instantons are particle-like wave packets that occupy a small region of space and exist for a tiny instant.)

Chapter 34

A Ranking of the 5 Saddest Mathematical Scandals

> Contrary to popular belief, mathematics is a passionate subject.
> Mathematicians are driven by creative passions that are difficult to
> describe, but are no less forceful than those that compel a musician to
> compose or an artist to paint. The mathematician, the composer, the
> artist succumb to the same foibles as any human—love, hate, addictions,
> revenge, jealousies, desires for fame, and money.
> —*Theoni Pappas,* Mathematical Scandals

Here's a quick quiz on (and ranking of) quirky, sad mathematical scandals. Dr. Googol has never found a person who could identify all the people referred to below:

1. What brilliant, famous, and beautiful woman mathematician died in incredible pain because her mother withdrew all pain medication? (Hint: The woman is recognized for her contributions to computer programming. The mother wanted her daughter to die painfully so that her daughter's soul would be cleansed.)

2. Which brilliant mathematician was forced to become a human guinea pig and subjected to drug experiments to reverse his homosexuality? (Hint: He was a 1950s computer theorist whose mandatory drug therapy made him impotent and caused his breasts to enlarge. He also helped to break the codes of the German Enigma machines during World War II.)

3. What famous mathematician deliberately starved himself to death in 1978? (Hint: He was perhaps the most brilliant logician of the 1900s.)

4. Which innovative mathematician suffered from a series of nervous breakdowns over a period of 30 years and died in a mental institution? (Hint: He was one of the most brilliant mathematicians of the 19th century and an avid explorer of the infinite.)

5. What important 11th-century mathematician pretended he was insane so he would not be put to death? (Hint: He was born in Iraq and made contributions to optics.)

❀ For answers, see "Further Exploring." Do you know anyone who can identify all 5 people?

Chapter 35

The 10 Most important Unsolved Mathematical Problems

If we wish to make a new world we have the material ready. The first one, too, was made out of chaos.

—*Robert Quillen,* OMNI magazine

In this section, Dr. Googol presents a ranking of the 10 most famous and/or unsolved mathematical "problems" today, as voted on by other mathematicians. Notice that many of the items on the list have a "classic" flavor in the sense that most of these problems were posed before 1900. A few have their roots in the mathematics of ancient Greece. Also, many of these problems can be stated simply (at least to mathematicians), and solutions are likely to have a great importance for mathematics and its development in the next century. Some of these problems are excruciatingly difficult, so Dr. Googol advises you to skip the tongue-twisting mathematical jargon to get an overall feeling for the problems. The proof of the Riemann hypothesis was mentioned most often by the mathematicians Dr. Googol interviewed.

1. **Proof of the Riemann Hypothesis** The "zeta function" can be represented by a complicated-looking curve that is useful in number theory for investigating properties of primes. Written as $\zeta(x)$, the function was originally defined as the infinite sum $\zeta(x) = 1 + (1/2)^x + (1/3)^x + (1/4)^x + \ldots$ etc. When $x = 1$, this series has no finite sum. For values of x larger than 1, the series adds up to a finite number. If x is less than 1, the sum is again infinite. The actual zeta function, studied and discussed in the literature, is a more complicated function that is equivalent to this series for values of x greater than 1, but has finite values for any real or complex number, the real part of which is different from 1. (Complex numbers are of the form $a + bi$ where $i = \sqrt{-1}$ and a and b are real numbers). Here's the big question: For what values does this function equal 0? We know that the function equals 0 when x is $-2, -4, -6, \ldots$, and that the function has an infinite number of 0 values for the set of complex numbers, the real part of which is between 0 and 1—but we do not known exactly for what complex numbers these 0s occur. Mathematician Georg Bernhard Riemann (1826–1866) conjectured that these 0s probably occur for those complex numbers the real part of which equals 1/2. Although there is vast numerical evidence favoring this conjecture, it is still unproven. The proof of Riemann's hypothesis would have profound consequences on the theory of prime numbers and on our understanding of the properties of complex numbers.

2. **Proof of the Goldbach Conjecture** Christian Goldbach (1690–1764) conjectured that every even positive integer is equal to the sum of 2 prime numbers (numbers not divisible by any integer greater than 1 except themselves). There are many examples where this is true, such as $10 = 5 + 5$ and $100 = 47 + 53$. Is it always true? He also conjectured that every positive integer greater than 2, even or odd, is equal to the sum of 3 primes. Although the first conjecture has been verified for all even integers at least as high as 100,000,000, no definitive proof for it has been found. Only a partial proof of the second conjecture was presented in 1937 by the Soviet mathematician Ivan Matveyevich Vinogradov.

3. **Poincaré Conjecture** French mathematician Henri Poincaré (1854–1912) conjectured that a simply connected closed 3-dimensional manifold is a 3-dimensional sphere. ("Simply connected" means that any closed path can be contracted to a point. In general, a manifold may mean any collection or set of objects. It is sometimes convenient to think of a manifold as an abstract generalization of a surface. Despite many attempts, no one has proven this conjecture, and it remains a cause célèbre in mathematics. The conjecture is important in

the history of mathematics partly because it focused attention on manifolds as objects of study. As a result, Poincaré influenced much of 20th-century mathematics, which emphasizes geometric objects.

4. Langlands Philosophy In January 1967, Robert Langlands, a 30-year-old Princeton mathematics professor, wrote a letter to the famous number theorist André Weil. Langlands asked for Weil's opinion about two new conjectures: "If you are willing to read [my letter] as pure speculation, I would appreciate that. If not—I'm sure you have a waste basket." According to the February 4, 2000, issue of *Science*, Weil never wrote back, but Langland's letter turned out to be a "Rosetta stone" linking two different branches of mathematics [See Dana Mackenzie, "Fermat's Last Theorem's First Cousin," *Science* 287(5454): 792–793, 2000]. In particular Langlands posited that there was an equivalence between Galois representations (relationships among solutions to equations studied in number theory) and automorphic forms (highly symmetric functions). Part of the problem is to work out what the correct formulation of the conjecture should be. Langlands philosophy asserts that one can associate automorphic representations to Galois representations, and that for irreducible representations one obtains cuspidal representations. This is also sometimes referred to as the Langlands program. Robert Langlands's vision is to bring group representation methods into the arithmetic theory of automorphic forms. (Does this all sound like mathematical gibberish? See "Further Exploring.")

5. Various Prime Number and Perfect Number Questions For centuries, mathematicians have tried to explain the underlying pattern behind the primes. Perhaps no pattern exists. Certain prime numbers occur in pairs, just two apart; these are called twin primes. Here are some twin primes: (3,5), (5,7), (11,13,) (17,19), (29,31). A long-standing conjecture of mathematics holds that there are an infinite number of twin primes. So far, no proof or disproof has come forth. (Notice that twin primes differ by only 2, which is as close as primes can be to each other. If they differed by 1, one of the numbers would be even and therefore divisible by 2.) Will we ever develop a formula that generates all prime numbers or that counts the number of primes up to a particular large number?

As discussed in Chapter 95, a perfect number's proper divisors add up to the number itself. For example, 6 is perfect because $6 = 1 + 2 + 3$, and 1, 2, and 3 divide into 6. (A "proper divisor" is simply a divisor of a number N excluding N itself.) Are all perfect numbers even? Is there an inexhaustible supply of perfect numbers?

6. The Structure of Pi Pi (expressed by the Greek letter π) has been calculated to billions of digits. What significant patterns, if any, exist in the seemingly-random digits of pi? (In 1991, the Chudnovsky brothers, two eminent pi researchers, wrote: "The decimal expansion of pi in billion plus range passes with flying colors all classical randomness tests: frequency, chi square, poker, arctan law, . . . etc." (See "Further Reading".)

7. Does P = NP? This relates to the number of steps required by computer algorithms. Very generally speaking, problems for which proposed solutions can be verified easily are sometimes referred to as NP problems. The P stands for *polynomial time*, which is the formal definition of *fast*. The N stands for *nonde-*

terministic, referring (misleadingly) to a notion that these problems could be solved easily if we were able to build computers that can nondeterministicially guess good solutions. This area is associated with the currently hot research area of "computational complexity," where problems are encountered that are solvable only by running a computer for millions of years. For example, Martin Gardner in *Gardner's Whys & Wherefores* notes that the infamous "traveling-salesman problem" asks for the shortest route that visits *n* points on the plane. Computers can hope to solve this problem when *n* is small, but as *n* increases, running time rapidly accelerates to impractical lengths. These kinds of problems belong to a category known as NP complete. Gardner suggests that these problems are connected in such a way that if a procedure is ever found for solving one of them in a reasonable time, all others will be solved.

8. **Artin Conjecture** This conjectures that Artin *L*-series are holomorphic for irreducible Galois representations. (This is the apparently simplest case of Langlands philosophy, at present unapproachable.)

9. **Various Computer/Mathematical Problems** Can we develop a rigorous mathematical theory of computer programming? (The world grows more dependent on computer software, but there is no rigorous theory that can be applied to verify program correctness.) Can a Turing machine do everything that any digital computer can do? Can we develop a rigorous theory of artificial intelligence? (For computers to replace humans in dangerous or routine tasks, the machines must be able to deal with a wide range of possibilities.)

10. **What are the limits of human and machine intelligence?** How are the brain and computer alike? Can mathematics be used to answer this question?

Outrageously difficult runners-up suggested by colleagues in e-mail: various problems in sphere packing

⊙ Which model of set theory best describes the "real world"?

⊙ Hilbert's 18th conjecture

⊙ the abc conjecture

⊙ Can numbers be factored in polynomial time (related to problem 7)?

⊙ various problems relating to the axiom of choice . . .

In the spring of 2000, the Clay Mathematics Institute (www.claymath.org) announced a most-wanted list of seven of the most intractable math problems in the world. With a reward of $1 million for each problem solved, this was the biggest math prize ever announced. The seven problems are: P versus NP, the Hodge Conjecture, the Poincaré Conjecture, the Riemann hypothesis, Yang-Mills existence and mass gap, Navier-Stokes existence and smoothness, and the Birch and Swinnerton-Dyer Conjecture. (For more information, see Charles Seife, "Is that your final equation?" *Science*, May 26, 288(5470): 1328–1329.)

❈ If you think the brief explanations for Langlands philosophy and the Artin conjecture sound like gibberish and want more information, see "Further Exploring."

Chapter 36

A Ranking of the 10 Most Influential Mathematicians Who Ever Lived

Taking mathematics from the beginning of the world to the time of Newton, what he has done is much the better half.

—*Gottfried Leibnitz*

No great discovery was ever made without a bold guess.

—*Isaac Newton*

Of the thousands of famous and important mathematicians who have affected the course of human history, which have most influenced our lives and our thoughts? Here is a ranking of the 10 most influential mathematicians who ever lived, starting with the most influential, Isaac Newton. (Anonymous persons, such as the first cave person to scratch numerical representations on cave walls, were disqualified from the list.)

Dr. Googol was surprised at the remarkable agreement among many respondents' lists. He urges you to experiment by composing your own list, which will no doubt differ from the one presented here.

1. **Sir Isaac Newton** (1643–1727) Brilliant English mathematician, physicist, and astronomer. He and Gottfried Leibniz invented calculus independently. Isaac Newton was so influential that some extra background on his odd life may appeal to you. Newton was a posthumous child born with no father on Christmas Day 1642. In his early 20s, he invented calculus, proved that white light was a mixture of colors, explained the rainbow, built the first reflecting telescope, discovered the binomial theorem, introduced polar coordinates, and showed that the force causing apples to fall is same as the force driving planetary motions and producing tides. Many of you probably don't realize that Newton was also a biblical fundamentalist, believing in the reality of angels, demons, and Satan. He subscribed to a literal interpretation of Genesis and believed the Earth to be only a few thousand years old. In fact, Newton spent much of his life trying to prove that the Old Testament is accurate history. One wonders how many more problems in physics Newton would have solved if he had spent less time on his biblical studies. Newton said many of his physics discoveries resulted from random playing rather than directed and planned exploration. He likened himself to a little boy "playing on the seashore, and diverting myself now and then in

finding a smoother pebble or a prettier shell than ordinary whilst the great ocean of truth lay all undiscovered before me." Newton, like other great scientific geniuses (such as Nikola Tesla or Oliver Heaviside), had a rather strange personality. For example, he had not the slightest interest in sex, never married, and almost never laughed (although he sometimes smiled). Newton suffered a massive mental breakdown, and some have conjectured that throughout his life he was a manic depressive with alternating moods of melancholy and happy activity. Today we would classify this as bipolar disorder.

2. **Johann Carl Friedrich Gauss** (1777–1855) Worked in a wide variety of fields of math and physics including algebra, probability, statistics, number theory, analysis, differential geometry, geodesy, magnetism, astronomy, and optics. His work has had an immense influence in many areas. As a boy his great mathematical precocity came to the attention of the Duke of Brunswick, who paid for his education. A notebook kept in Latin by Gauss as a youth was discovered in 1989 showing that, from the age of 15, he had conjectured many remarkable results, including the prime number theorem and ideas of non-Euclidean geometry. He published papers in astronomy, the theory of errors, differential equations, optics, and magnetism. Manuscripts unpublished until long after his death show that he had made many other discoveries including the theory of elliptic functions.

3. **Euclid** (365–300 B.C.) Greek geometer, number theorist, astronomer, and physicist, famous for his treatise on geometry *The Elements*, a 13-book extravaganza and the earliest substantial Greek mathematical treatise to have survived. The enduring nature of *The Elements* makes Euclid the leading mathematics teacher of all time. Euclid's *Elements* has essentially served as the standard means of teaching geometry for some 2,500 years and has taught the world how to think systematically. When Abraham Lincoln wanted to learn the meaning of *demonstrate* in practicing law, he turned to reading Euclid by candlelight in his Kentucky log cabin. Euclid also wrote other works on geometry, astronomy, optics, and music, many of which are lost forever.

4. **Leonhard Euler** (1707–1783) Swiss mathematician, and the most prolific mathematician in history. Even when he was completely blind, he made great contributions to modern analytic geometry, trigonometry, calculus, and number theory. Euler published over 8,000 books and papers, almost all in Latin, on every aspect of pure and applied mathematics, physics, and astronomy. In analysis he studied infinite series and differential equations, introduced many new functions (e.g., the gamma function and elliptic integrals), and created the calculus of variations. His notations such as e and π are still used today. In mechanics, he studied the motion of rigid bodies in 3 dimensions, the construction and control of ships, and celestial mechanics. Leonhard Euler was so prolific that his papers were still being published for the first time 2 centuries after his death. His collected works have been printed bit by bit since 1910 and will eventually occupy more than 75 large books.

5. **David Hilbert** (1862–1943) German mathematician and philosopher, judged by many as the foremost mathematician of the 20th century, who

contributed to the theory of algebra, number fields, integral equations, functional analysis, and applied mathematics. Measured in terms of influence, Hilbert's work in geometry is second after Euclid's. Hilbert published *Grundlagen der Geometrie* in 1899, which added immensely to the field of geometry. His famous 23 Paris problems still challenge mathematicians with important mathematical questions. (At the second International Congress of Mathematicians, which met at Paris in 1900, Hilbert reviewed the basic contemporary trends of mathematical research, then formulated 23 problems, extending over all fields of mathematics, that he believed should occupy the attention of mathematicians in the following century.) Because of Hilbert's prestige, mathematicians spent a great deal of time tackling the problems, and many were solved. Some, however, have been solved only very recently, and still others continue to daunt us.

6. **Jules Henri Poincaré** (1854–1912) Great French mathematician, mathematical physicist, astronomer, and philosopher. He was the originator of algebraic topology and of the theory of analytic functions of several complex variables. In applied mathematics, he studied optics, electricity, telegraphy, capillarity, elasticity, thermodynamics, potential theory, quantum theory, and the theory of relativity and cosmology. In the field of celestial mechanics, he studied the 3-body problem and theories of light and electromagnetic waves. He is acknowledged as a codiscoverer, with Albert Einstein and Hendrik Lorentz, of the special theory of relativity. In his work on planetary orbits, Poincaré was first to consider the possibility of chaos in a deterministic system.

7. **Georg Friedrich Bernhard Riemann** (1826–1866) German mathematician who made important contributions to geometry, complex variables, number theory, topology, and mathematical physics. His ideas concerning the geometry of space had an important influence on modern relativity theory. He clarified the notion of integrals by defining what we now call the Riemann integral. His first publication, in 1851, was on the theory of complex-variable functions including the result now known as the Riemann mapping theorem. In this and a later paper (1857) on abelian functions, he introduced the concept of "Riemann surface" to deal with multivalued algebraic functions; this was to become a major idea in the development of analysis. His famous lecture "On the Hypotheses That Underlie Geometry," given in 1854 in the presence of the aged Gauss, first introduced the concept of a "manifold," an n-dimensional curved space, greatly extending the non-Euclidean geometry of Janos Bolayai and Nikolai Lobachevski. Riemann's ideas led to the modern theory of differentiable manifolds, playing an important part in current attempts to unify relativity and quantum theory. Riemann's name is associated with the Riemann hypothesis, a famous unsolved problem concerning the zeta function, which is central to the study of the distribution of prime numbers.

8. **Évariste Galois** (1811–1832) Responsible for Galois theory. Famous for his contributions to group theory, Évariste Galois produced a method of determining when a general equation could be solved by radicals. Although he obviously knew more than enough mathematics to pass the Lycée's examinations,

Galois's solutions were often so innovative that his university examiners failed to appreciate them. Also, Galois would perform so many calculations in his head that he would not bother to outline his arguments clearly on paper. These facts in addition to his temper and rashness denied him admission to the Ecole Polytechnique.

When he was taunted into a duel, he accepted, knowing he would die. The circumstances that led to Galois's death have never been fully explained. It has been variously suggested that it resulted from a quarrel over a woman, that he was challenged by royalists who detested his republican views, or that an agent provocateur of the police was involved. In any case, preparing for the end, he spent the entire night feverishly writing his mathematical ideas and discoveries in as complete a form as he could. Figure 36.1 shows a page from his last night's writing on quintic equations (equations with the term x^5).

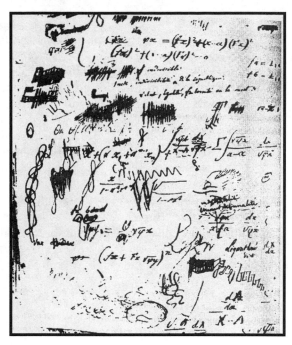

36.1 The frantic mathematical scribblings Galois made during the night before his fatal duel. On this page, on the left below the center, are the words *Une femme*, with *femme* crossed out—a reference to the woman at the center of the duel.

The next day, Galois was shot in the stomach. He lay helpless on the ground. There was no physician to help him, and the victor casually walked away, leaving Galois to writhe in agony.

Not until 1846 had group theory progressed sufficiently for his discoveries to be appreciated. Galois never received recognition for his extraordinary work and advanced ideas, but his legacy has had a great impact on 20th-century mathematics. His mathematical reputation rests on fewer than 100 pages of posthumously published work of original genius.

9. René Descartes (1596–1650) French philosopher and mathematician whose work *La géométrie* became one of the most influential geometry books in history. Descartes was a Catholic all his life, and he was careful to modify or even suppress some of his later scientific views—for example, his sympathy with Copernicus—perhaps fearing the wrath of the Inquisition. Nevertheless, he made important contributions in astronomy, including his theory of vortices, and more especially in mathematics, where he reformed algebraic notation and helped found coordinate geometry. Descartes had a lifetime habit of staying in bed meditating and reading until 11 A.M.

10. Blaise Pascal (1623–1662) French geometer, probabilist, combinatorist, physicist, and philosopher. Pascal and Pierre de Fermat founded probability theory independently. Pascal also invented the first calculating machine, studied conic sections, and produced important theorems in projective geometry. His father, a mathematician, was responsible for his education. Blaise was not allowed to begin a subject until his father thought he could easily master it. As a result, the 11-year-old boy worked out for himself in secret the first 23 propositions of Euclid. At 16, he published essays on conics that Descartes refused to believe were the handiwork of a teenager. In 1654, Blaise Pascal decided that religion was more to his liking, so he joined his sister in her convent and gave up mathematics and social life.

Runners-up: **Gerolamo Cardano**, **Kurt Gödel**, **Georg Cantor**, and **John Napier**. Napier's invention of logarithms was a major advance and freed people from a considerable amount of mathematical drudgery.

Chapter 37

What is Gödel's Mathematical Proof of the Existence of God?

> Were theologians to succeed in their attempt to strictly separate science and religion, they would kill religion. Theology simply must become a branch of physics if it is to survive. That even theologians are slowly becoming effective atheists has been documented.
> —*Frank Tipler,* The Physics of Immortality

Perhaps the most interesting example of a mathematician studying cosmic questions is Austrian mathematician Kurt Gödel, who lived from 1906 to 1978. Sometime in 1970, Gödel's mathematical proof of the existence of God began to circulate among his colleagues. The proof was less than a page long and caused quite a stir:

Gödel's Mathematical Proof of God's Existence

Axiom 1. (Dichotomy) A property is positive if and only if its negation is negative.

Axiom 2. (Closure) A property is positive if it necessarily contains a positive property.

Theorem 1 A positive property is logically consistent (i.e., possibly it has some instance.)

Definition. Something is God-like if and only if it possesses all positive properties.

Axiom 3. Being God-like is a positive property.

Axiom 4. Being a positive property is (logical, hence) necessary.

Definition. A property P is the essence of x if and only if x has P and P is necessarily minimal.

Theorem 2 If x is God-like, then being God-like is the essence of x.

Definition. $NE(x)$: x necessarily exists if it has an essential property.

Axiom 5. Being NE is God-like.

Theorem 3. Necessarily there is some x such that x is God-like.

How shall we judge such an abstract proof? How many people on Earth can really understand it? Most logicians and mathematicians that Dr. Googol consulted were not able to explain all aspects of the proof, and so it is difficult to assess its full nature. Is the proof a result of profound contemplation or the raving of a lunatic? Recall that Gödel's academic credits were impressive. For example, he was a respected mathematician and a member of the faculty of the University of Vienna starting in 1930. He emigrated to the United States in 1940 and became a member of the Institute of Advanced Study in Princeton, New Jersey. Gödel is most famous for his theorem that demonstrated there must be true formulas in mathematics and logic that are neither provable nor disprovable, thus making mathematics essentially incomplete. (This theorem was first published in 1931 in *Monatshefte für Mathematik und Physik*, volume 38.) Gödel's theorem had quite a sobering effect upon logicians and philosophers because it implies that within any rigidly logical mathematical system there are propositions or questions that cannot be proved or disproved on the basis of axioms within that system, and therefore it is possible for basic axioms of arithmetic to give rise to contradictions. The repercussions of this fact continue to be felt and debated. Moreover, Gödel's article in 1931 put an end to a centuries-long attempt to establish axioms that would provide a rigorous basis for all of mathematics.

Over the span of his life, Gödel kept voluminous notes on his mathematical ideas. Some of his work is so complex that mathematicians believe many decades will be required to decipher all of it. Author Hao Wang writes on this very subject in his *Reflections on Kurt Gödel* (Cambridge, Mass.: MIT Press, 1987):

The impact of Gödel's scientific ideas and philosophical speculations has been increasing, and the value of their potential implications may continue to increase. It may take *hundreds of years* for the appearance of more definite confirmations or refutations of some his larger conjectures.

Gödel himself spoke of the need for a physical organ in our bodies to handle abstract theories. He also suggested that philosophy will evolve into an exact theory "within the next hundred years or even sooner." He even believed that humans will eventually disprove propositions such as "there is no mind separate from matter."

Alas, Dr. Googol is not a logician and cannot appreciate Gödel's 11-step proof of God's existence. Dr. Googol welcomes comments from more erudite readers on this proof, which he obtained from: Wang's *Reflections*, page 195.

Chapter 38

A Ranking of the 10 Most Influential Mathematicians Alive Today

Music is the pleasure the human mind experiences from counting without being aware that it is counting.
—*Gottfried Leibnitz*

Here is a ranking of the 10 most influential mathematicians alive today, based on surveys and interviews with mathematicians.

1. Andrew Wiles (b. 1953) Wiles is Eugene Higgins Professor of Mathematics at Princeton. His famous paper proving Fermat's Last Theorem is titled "Modular Elliptic Curves and Fermat's Last Theorem," published in the 1995 *Annuals of Mathematics*. (Fermat's Last Theorem says that if x, y, z, and n are positive integers, then $x^n + y^n = z^n$ cannot be solved for any n greater than 2.)

During his 8-year search for a proof, Wiles had brought together most of the breakthroughs in 20th-century number theory and incorporated them in one stupendous proof. Along the way, he created completely new mathematical methods and combined them with traditional ones in novel ways. In doing this, Wiles opened up novel lines of attack on many other mathematical problems and made tremendous contributions toward the resolution of long-standing fundamental problems in number theory. The problems that he has addressed on his own and jointly with others include the Birch and Swinnerton-Dyer conjectures, the main conjecture of Iwasawa theory, and the Shimura-Taniyama-Weil conjecture. Wiles has been awarded the Schock Prize in Mathematics from the Royal Swedish Academy of Sciences and the Prix Fermat from the Université Paul Sabatier. He also received the 1995–96 Wolf Prize "for spectacular contributions to number theory and related fields, for major advances on fundamental conjectures, and for settling Fermat's Last Theorem." In 1995, mathematician John Coates announced:

> In mathematical terms, the final proof is equivalent of splitting the atom or finding the structure of DNA. A proof of Fermat is a great intellectual triumph, and one shouldn't lose sight of the fact that it has revolutionized number theory in one fell swoop.

Mathematician Ken Ribet notes:

> There's an important psychological repercussion, which is that people now are able to forge ahead on other problems that they were too timid to work on before.

2. **Harold (Donald) Coxeter** (b. 1907) Coxeter made significant advances in geometry. In particular he made contributions of major importance in the theory of polytopes (polygons in higher dimensions) and non-Euclidean geometry. His hundreds of books and articles cover diverse areas. Coxeter met M.C. Escher in 1954, and the two became close friends. Coxeter also had an influence on architect Buckminister Fuller.

3. **Roger Penrose** (b. 1931) This British mathematician and physicist predicted singularities in black holes. A professor of mathematics at the University of Oxford in England, Penrose also pursues an active interest in recreational math, which he shares with his father. While most of his work pertains to relativity theory and quantum physics, he is fascinated with a field of geometry known as tessellation, the covering of a surface with tiles of prescribed shapes. Penrose received his Ph.D. at Cambridge in algebraic geometry. While there, he became interested in a geometrical puzzle involving the covering of a flat surface with tiles so that there were no gaps and no overlaps. In particular, Penrose found shapes that could tile a surface but did not generate a repeating pattern (known as quasi-symmetry). These tilings are useful in understanding certain chemical substances that form crystals in a quasi-periodic manner. A French company has recently found a practical application for substances that form these quasi-crystals: they make excellent nonscratch coating for frying pans. Professor Penrose was knighted in 1994 and awarded the prestigious Wolf Prize for Physics in 1988, which he shared with Professor Stephen Hawking.

4. Edward Witten (b. 1951) Witten, an American, is one of the foremost leaders in reviving the symbiosis between physics and mathematics and is famous for his work with superstring theory and other areas of mathematical physics. When modern science was born in the 1600s, physics and mathematics were united in one discipline. They gradually evolved into different fields, and by the middle 1900s research in these fields had little in common. Witten's work on string theory has inspired a new generation of theoretical physicists and also led to new research in pure mathematics. He is the first and only physicist to be awarded the Fields Medal; the mathematical equivalent of the Nobel Prize, it is awarded to mathematicians under 40 years of age for outstanding, seminal research in mathematics.

5. William Thurston (b. 1946) Thurston conducted pioneering work in geometry, particularly 3-dimensional topology and foliations. He was appointed a full professor at Princeton University only 2 years after receiving his Ph.D. and is widely regarded as being among the most creative mathematicians in the world. He has solved or clarified dozens of fundamental problems in geometry and topology. In his work on foliations, Thurston transformed an existing field of mathematics. Thurston is a member of the National Academy of Sciences and a Fields medalist. At the International Congress of Mathematicians in 1983, Professor C.T.C. Wall spoke of Thurston's work: "Thurston has fantastic geometric insight and vision; his ideas have completely revolutionized the study of topology in 2 and 3 dimensions and brought about a new and fruitful interplay between analysis, topology and geometry."

6. Stephen Smale (b. 1930) A Fields Medal recipient, Smale is famous for work on the Poincaré conjecture, Morse theory, topology, and various aspects of chaos theory. In June 1996, he received the National Medal of Science, the highest honor in science and technology awarded in the United States. Although Smale has worked in many areas, including differential topology, nonlinear analysis, economic theory, computation, and mechanics, his work in chaos and dynamical systems will be best known to many readers. Smale's 1960s work on the structure stability of vector fields led to the construction of the horseshoe map and his early study of chaotic phenomena. (To make a simple version of Smale's horseshoe, you take a bar and repeatedly fold, stretch, and squeeze it, like a mechanical taffy-maker with rotating arms stretching taffy, doubling it up, stretching it again, and so on. This topological transformation provided a basis for understanding the chaotic properties of dynamical systems.) In 1967, he published a landmark survey article on hyperbolic systems, which outlined a number of outstanding problems, stimulating much of the work that followed in the next 20 years. Subsequently he applied dynamical systems ideas to various physical processes, including the n-body problem and electric circuit theory, and to economics.

7. Robert P. Langlands (b. 1950) This pioneering mathematician and 1982 Cole Prize recipient works on automorphic forms, Eisenstein series, and product formulas. Langlands, of the Institute for Advanced Study, in Princeton, and Andrew J. Wiles of Princeton University shared the 1995–1996 Wolf Prize in Mathematics; Langlands received it for "his path-blazing work and extraordinary insight in the fields of number theory, automorphic forms, and group representation." Langlands shaped the modern theory of automorphic forms with foundational work on Eisenstein series, group representations, *L*-functions and the Artin conjecture, the principle of functoriality, and the formulation of the far-reaching Langlands program. His contributions inspire present and future researchers in these fields.

8. **Michael Freedman** (b. 1951) Freedman has received many honors, including a Fields Medal in 1986 for his work on a dimension 4 analogue of the Poincaré conjecture—one of the famous problems of 20th-century mathematics, which asserts that a simply connected closed 3-dimensional manifold is a 3-dimensional sphere. He was California Scientist of the Year in 1984, and in that same year he was made a MacArthur Foundation Fellow and elected to the National Academy of Sciences. In 1985, he was elected to the American Academy of Arts and Sciences. In June 1987, Freedman was presented with the National Medal of Science at the White House by President Ronald Reagan. The following year, he received the Humboldt Award, and in 1994 he received the Guggenheim Fellowship Award.

9. **John Horton Conway** (b. 1937) John von Neumann Professor of Mathematics at Princeton University. Author of numerous publications in mathematics, and the inventor of The Game of Life. (His coauthored books include *On Numbers and Games; Winning Ways for Your Mathematical Plays; Sphere Packing, Lattices and Groups;* and *The Book of Numbers.*) To develop Life, Conway used the basic premise of von Neumann's automata and created a checkerboard world. This world is inhabited by single cells who live or die based on the state of the cell and its neighbors. Therefore, at any instant, a Life universe can be described completely by specifying which cells are on and which are off. This type of world has come to be known as a cellular automaton and is an important tool for artificial life research.

10. **Alexander Grothendieck** (b. 1928) Grothendieck provided unifying themes in geometry, number theory, and topology. He was born in Berlin, where his Russian father was murdered by the Nazis. Grothendieck moved to France in 1941 and later entered Montpellier University. After graduating from Montpellier he spent the year 1948–49 at the Ecole Normale Supérieur in Paris, then moved to the University of Nancy, where he worked on functional analysis and became one of the Bourbaki group of mathematicians.

Grothendieck's years between 1959 and 1970 are described as a golden age during which mathematics flourished under his energetic leadership. During this period, Grothendieck's work provided unifying themes in geometry, number theory, topology and complex analysis. He received the Fields Medal in 1966.

Martin Gardner (b. 1914) appeared on many lists. While he is not a mathematician, many respondents felt that his regular columns in *Scientific American* and numerous books have done more to heighten modern interest in mathematics than any other writing in history. Therefore, his influence is great and may warrant his inclusion on this list. In 1996, Martin Gardner received the Forum Award of the American Physical Society. The citation read:

. . . for his popular columns and books on recreational mathematics which introduced generations of readers to the pleasures and uses of logical thinking; and for his columns and books which exposed pseudoscientific and antiscientific bunk and explained the scientific process to the general public.

Kendrick Frazier in the April 1998 *Skeptical Inquirer* notes:

Gardner's mind is highly philosophical, at home with the most abstract concepts, yet his thinking and writing crackle with clarity—lively, crisp, vivid. He achieved worldwide fame and respect for the three decades of his highly popular mathematical games column for *Scientific American*, yet he is not a mathematician. He is by every standard an eminent intellectual, yet he has no Ph.D. or academic position. He has a deep love of science and has written memorable science books (*The Ambidextrous Universe* and *The Relativity Explosion*, for instance), and yet he has devoted probably more time and effort to—and has been more effective than any thinker of the twentieth century in—exposing pseudoscience and bogus science.

Runners-up: **Jean-Pierre Serre** (b. 1926; number theory, algebraic geometry), **Vladimir Arnold** (b. 1937; dynamical systems, geometry), **Richard Borcherds** (b. 1959; group theory), **William Timothy Gowers** (b. 1963; Banach space theory and combinatorics), **Maxim Kontsevich** (b. 1964; algebraic curves and manifolds), and **Curtis T. McMullen** (b. 1958; theory of computation, dynamical systems, 3-manifolds).

Chapter 39

A Ranking of the 10 Most Interesting Numbers

The tantalizing and compelling pursuit of mathematical problems offers
mental absorption, peace of mind amid endless challenges, repose in
activity, battle without conflict, refuge from the goading urgency of contingent happenings, and the sort of beauty changeless mountains present
to senses tried by the present-day kaleidoscope of events.
—*Morris Kline*

Here is a ranking of the 10 most interesting numbers, based on surveys and interviews with mathematicians.

1. **0** See Chapter 2 on the importance of 0 in history and positional notation. 0 is the additive identity for $a + 0 = a$.

2. **π** Normally we think of π (3.1415 . . .) as the ratio of the circumference of a circle to its diameter. So did pre-17th-century humanity. However, in the 17th century, π was freed from the circle. Many curves were invented and studied (e.g., various arches, hypocycloids, witches), and it was found that their areas could be expressed in terms of π. Finally π ruptured the confines of geometry altogether. For example, today π relates to uncountably many areas in number theory, probability, complex numbers, and series of simple fractions such as $\pi/4 = 1 - 1/3 + 1/5 - 1/7$. . . . As another example of how far π has drifted from its simple geometrical interpretation, consider the book *Budget of Paradoxes*, where Augustus De Morgan explains an equation to an insurance salesman. The formula, which gives the chances that a particular group of people would be alive after a certain number of days, involves the number π. The insurance salesman interrupts and exclaims, "My dear friend, that must be a delusion. What can a circle have to do with the number of people alive at the end of a given time?"

Even more recently, π has turned up in equations that describe subatomic particles, light, and other quantities that have no obvious connection to circles. John Polkinghorne (a physicist at Cambridge University before he became an Anglican priest in 1982) believes this points to a very deep fact about the nature of the universe, namely that our minds, which "invent" mathematics, conform to a reality of the universe. We are tuned to its truths. (See Sharon Begley's "Science Finds God" in the July 20, 1998, issue of *Newsweek*.)

3. **e** The base of the natural system of logarithms; the limit of $(1 + 1/n)^n$ as n increases without limit. Its numerical value is 2.7182 . . . (note that if we use $n = 10$ in the formula we get $(1 + 1/10)^{10} = 2.59$ . . . , and if we use $n = 20$ we get $(1 + 1/20)^{20} = 2.65$ . . .). The constant e is related to the other important numbers, 1, π, and i, by $e^{\pi i} = -1$. Additionally, e, like π, is an example of a transcendental number (see Chapter 44 for more information on transcendentals, which cannot be expressed as the root of any algebraic equation—for example, a polynomial—with rational coefficients.) Numerous growth processes in physics, chemistry, biology, and the social sciences exhibit exponential growth typified by the formula $y = e^x$. This function is exactly the same as its derivative, a fact that partially explains e's frequent occurrence in calculus. Many hanging shapes in nature (like a rope suspended at two points and sagging in the middle) follow a catenary curve defined by $(a/2)(e^{x/a} + e^{-x/a})$.

4. **i** Imaginary unit. If you were asked to find an x such that $x^2 + 1 = 0$, you would quickly realize that there was no real solution. This fact led early mathematicians to consider solutions involving the square root of negative numbers. Heron of Alexandria (c. A.D. 100) was probably the first individual who formally presented a square root of a negative number as a solution to a problem. (For trivia aficionados, it was $\sqrt{-63}$). These numbers were considered quite meaningless, and hence the term *imaginary* was used. When imaginary numbers were

first considered, many people were not sure of their validity. What real-world significance could they have? Today, amazingly, imaginary numbers are everywhere in science—from hydrodynamics to electrical theory. The space shuttle's flight software uses them for navigation. They're used by protein chemists for spatially manipulating models of protein structure. Ted Kaczynski, the Unabomber, spoke of imaginary numbers fondly throughout his highly theoretical mathematical journal articles. Carl Friedrich Gauss coined the word *complex* in 1832 to describe numbers with both real and imaginary components. Humanity's expansion into the realm of complex numbers turned many difficult problems into relatively easy ones.

5. $\sqrt{2}$ The square root of 2 has a numerical value of 1.414214. . . . When it was first proved to be irrational (that is, it could not be expressed as the ratio of two integers like 7/5), a whole new area of mathematics was developed. The Pythagoreans, a mystical brotherhood based on the philosophical teachings of Pythagoras, discovered that the diagonal of a square with sides of length 1 is not a rational number. This was considered so shocking that those who knew about it were sworn to secrecy for fear that it might disrupt the fabric of society! It is often said that when Hippasus discovered that the ratio between the side and the diagonal of a rectangle cannot be expressed in integers, this shattered the Pythagorean worldview. The problem caused an existential crisis in ancient Greek mathematics. The digits of 1.4142 . . . go on forever without any known pattern. Pythagoreans dubbed these irrational numbers *alogon,* or unutterable.

6. **1** A factor of all numbers, 1 has no factors but itself. It is the multiplicative identity for $1 \times a = a$.

7. **2** The only even prime number. In the words of Richard Guy, this makes it the oddest prime of all. Notice that $2 + 2 = 2 \times 2$, which gives it a unique arithmetic property. Importantly, it is the basis for the binary system of numbers upon which all computers are built. Powers of 2 appear more frequently in mathematics and physics than those of any other number.

8. **Euler's Gamma** *(γ)* Numerical value, 0.5772. . . . This number links the exponentials and logs to number theory, and it is defined by the limit of $(1 + 1/2 + 1/3 + \ldots + 1/n - \log n)$ as *n* approaches infinity. In addition to many infinite series, products, and definite integral representations, Euler's constant γ also plays a role in probability. Calculating γ has not attracted the same public interest as calculating π, but γ has still inspired many ardent devotees. While we presently know π to billions of decimal places, only several thousand places of γ are known. The evaluation of γ is considerably more difficult than that of π.

9. **Chaitin's constant** (Ω) An irrational number which gives the probability that a "universal Turing machine" (for any set of instructions) will halt. The digits in Ω are random and cannot be computed prior to the machine halting. (A Turing machine is a theoretical computing machine that consists of an infinitely long magnetic tape on which instructions can be written and erased, a single-bit register of memory, and a processor capable of carrying out certain simple instructions. The machine keeps processing instructions until it reaches a particular state, causing it to halt.) Chaitin's constant has implications for the development of human and natural languages and gives insight into the ultimate potential of machines.

10. $\aleph_0$ (Aleph naught) A "transfinite" number. Even though there are an infinite number of rational and irrational numbers, the infinite number of irrationals is in some sense greater than the infinite number of rationals. To denote this difference, mathematicians refer to the infinity of rationals as $\aleph_0$ and the infinite number of irrationals as C, which stands for *continuum*. There is a simple relationship between C and $\aleph_0$. It is $C = 2^{\aleph_0}$. The "continuum hypothesis" states that $C = \aleph_1$; however, the question of whether or not C truly equals $\aleph_1$ is considered undecidable. In other words, great mathematicians such as Kurt Gödel proved that the hypothesis was a consistent assumption in one branch of mathematics. However, another mathematician, Paul Cohen, proved that it was also consistent to assume the continuum hypothesis is false! Interestingly, the number of rational numbers is the same as the number of integers. The number of irrationals is the same as the number of real numbers. (Mathematicians usually use the term *cardinality* when talking about the "number" of infinite numbers. For example, true mathematicians would say that the "cardinality" of the irrationals is known as the continuum.) Thinking about the number of elements in infinite sets led to the discovery of transfinite numbers and the fact that there are different "levels" of infinity.

Chapter 40

The Unabomber's 10 Most Mathematical Technical Papers

> Let F be our finite skew field, F^* its multiplicative group. Let S be any Sylow subgroup F^*, of order, say, p^a. Choose an element g of order p in the center of S. If some $h \in S$ generates a subgroup of order p different from that generated by g, then g and h generate a commutative field containing more than p roots of the equation $x^p = 1$, an impossibility. Thus S contains only one subgroup of order p and hence is either a cyclic or general quaternion group.
> —*T. J. Kaczynski,* "Another Proof of Wedderburn's Theorem"

> The majority of people engage in a significant amount of naughty behavior.
> —Unabomber Manifesto

Ted Kaczynski, also known as the Unabomber, was a mathematically adept Harvard graduate (see Chapter 31). After teaching for 2 years and publishing mathematical papers that put him on a tenure track at one of the nation's most prestigious universities, he suddenly quit, spent nearly half his life in the woods, and used homemade bombs to kill 3 strangers and injure 22 others.

Ted Kaczyinski's research into the properties of functions of circles was by all accounts brilliant, but when he sent his papers to journals for publications, he did so quietly, without telling his professors or classmates. (This occurred *before* Kaczyinski became a hermit in the woods and started killing people.) When his articles began appearing in respected mathematics journals, professors and students were amazed. According to Joel Shapiro, a fellow student now a mathematics professor:

> While most of us were just trying to learn to arrange logical statements into coherent arguments, Ted was quietly solving open problems and creating new mathematics. *It was if he could write poetry while the rest of us were trying to learn grammar.*

Various mathematicians have said that Kaczynski's papers, such as "Boundary Functions for Functions Defined in a Disk" and "On a Boundary Property of Continuous Functions," were cutting-edge mathematics when they were published. In order to help judge Kaczynski's work, Dr. Googol acquired his papers and spread them out on the table. Alas, despite some mathematical training, Dr. Googol cannot understand Kaczynski's works. Here are some of Kaczynski's erudite titles:

1. Kaczynski, T. J. (1967) *Boundary Functions* (doctoral dissertation). Ann Arbor: University of Michigan. (This 80-page thesis won "best thesis of the year" in the math department at the University of Michigan.)

2. Kaczynski, T. J. (1964) Another proof of Wedderburn's theorem. *American Mathematical Monthly.* 71: 652–653.

3. Kaczynski, T. J. (1964.) Distributivity and $(-1)x = -x$ (proposed problem). *American Mathematical Monthly.* 71: 689.

4. Kaczynski, T. J. (1965) Boundary functions for functions defined in a disk. *Journal of Mathematics and Mechanics.* 14(4): 589–612.

5. Kaczynski, T. J. (1965) Distributivity and $(-1)x = -x$(problem and solution). *American Mathematical Monthly.* 72: 677–678.

6. Kaczynski, T. J. (1966) On a boundary property of continuous functions. *Michigan Mathematics Journal.* 13: 313–320.

7. Kaczynski, T. J. (1969) Note on a problem of Alan Sutcliffe. *Mathematics Magazine.* 41: 84–86.

8. Kaczynski, T. J. (1969) The set of curvilinear convergence of a continuous function defined in the interior of a cube. *Proceedings of the American Mathematics Society.* 23: 323–327.

9. Kaczynski, T. J. (1969) Boundary functions and sets of curvilinear convergence for continuous functions. *Transactions of the American Mathematics Society.* 141: 107–125.

10. Kaczynski, T. J. (1969) Boundary functions for bounded harmonic functions. *Transactions of the American Mathematics Society.* 137: 203–209.

Dr. Googol shouldn't feel too bad about not comprehending any of the Unabomber's papers. According to the hype, there are few who can fully appreciate Kaczynski's work without considerable mathematical education. Professor Maxwell O'Reade, who was on Kaczynski's dissertation committee, noted, "I would guess that maybe 10 or 12 people in the country understood or appreciated it."

Chapter 41

The 10 Mathematical Formulas That Changed the Face of the World

> Perhaps an angel of the Lord surveyed an endless sea of chaos, then troubled it gently with his finger. In this tiny and temporary swirl of equations, our cosmos took shape.
>
> —*Martin Gardner*

A few years ago, Nicaragua issued 10 postage stamps bearing *las 10 fórmulas matemáticas que cambiaron la faz de la tierra.* (the 10 mathematical formulas that changed the face of the world). Isn't it admirable that a country so respects mathematics that it devotes a postage-stamp series to a set of abstract equations? Have other countries produced a similar series?

Dr. Googol is not sure how the Nicaraguan government determined which particular formulas should be elevated to so high a status. Perhaps a survey was conducted among the mathematicians in the country. In addition to scientific merit, perhaps such practical issues as space limitations were considered so as to avoid long formulas on small stamps.

Dr. Googol conducted his own informal survey as to which formulas scientists considered "the 10 mathematical formulas that changed the face of the world." The survey was conducted via electronic mail networks, and a majority of the respondents were mathematicians (professors, other professionals, and graduate students). Here's the answer to this question from approximately 50 interested individuals who gave Dr. Googol their opinions as to the most important and influential equations. The equations are ordered from most influential to least influential, based on the number of different people who listed the same formulas when they sent their lists to Dr. Googol. For example, $E = mc^2$ received the most votes.

How many of the following formulas can you identify? If you can identify more than 5, you are probably more knowledgeable than 99% of the people on Earth. If you can identify all equations in the top 10 and all the equations in the runners-up list, you are worthy of cavorting with the antediluvian gods. Dr. Googol identifies these equations later in the chapter.

THE TOP 10

Here are the 10 most influential and important mathematical expressions, listed in descending order of importance:

1. $E = mc^2$

2. $a^2 + b^2 = c^2$

3. $\varepsilon_0 \oint \vec{E} \cdot d\vec{A} = \Sigma q$

4. $x = (-b \pm \sqrt{b^2 - 4ac}) / (2a)$

5. $\vec{F} = m\vec{a}$

6. $1 + e^{i\pi} = 0$

7. $c = 2\pi r,\ a = \pi r^2$

8. $\vec{F} = Gm_1 m_2 / r^2$

9. $f(x) = \Sigma c_n e^{in\pi x/L}$

10. $e^{i\theta} = \cos\theta + \sin\theta$, tied with $a^n + b^n = c^n,\ n \geq 2$

THE RUNNERS-UP

These mathematical expressions did not score high enough to be included in the top 10 but scored favorably. They are listed in no particular order but are numbered for reference.

1. $f(x) = f(a) + f'(a)(x - a) + f''(a)(x - a)^2/2! \ldots$

2. $s = vt + at^2/2$

3. $V = IR$

4. $z \rightarrow z^2 + \mu$ (for complex numbers)

5. $e = \lim_{n \rightarrow \infty}(1+1/n)^n$

6. $c^2 = a^2 + b^2 - 2ab\cos C$

7. $\int K dA = 2\pi \times x$

8. $d/dx \int^a f(t)dt = f(x)$

9. $1/(2\pi i)\oint f(z)/(z-a)dz = f(a)$

10. $dy/dx = \lim_{h \rightarrow 0}(f(x+h)-f(x))/h$

11. $\partial^2 \psi/\partial x^2 = -[8\pi^2 m/h^2(E-V)]\psi$

NICARAGUA LIST

Here is a list of Nicaragua's postage-stamp equations. Note how many of these formulas agree with the top 10 list based on Dr. Googol's own informal survey.

1. $1 + 1 = 2$

2. $\vec{F} = Gm_1 m_2 / r^2$

3. $E = mc^2$

4. $e^{\ln N} = N$

5. $a^2 + b^2 = c^2$

6. $S = k \log W$

7. $V = V_c \ln m_0/m_1$

8. $\lambda = h/mv$

9. $\nabla^2 E = (Ku/c^2)(\partial^2 E/\partial t^2)$

10. $F_1 x_1 = F_2 x_2$

Do you recognize several of these formulas?

❁ ❁ ❁

IDENTIFICATION OF EQUATIONS

Here are the solutions for the Nicaragua stamp list: (1) Basic addition formula. (2) Isaac Newton's law of universal gravitation. If the two masses m_1 and m_2 are separated by a distance, r, the force exerted by one mass on the other is F, and G is a constant of nature. (3) Einstein's formula for the conversion of matter to energy. (4) John Napier's logarithm formula. This allows us to do multiplication and division simply by adding or subtracting the logarithms of numbers. (5) Pythagorean theorem relating the lengths of sides of a right triangle.

(6) Bolzmann's equation for the behavior of gases. (7) Konstantin Tsiolkovskii's rocket equation. It gives the speed of a rocket as it burns the weight of its fuel. (8) De Broglie's wave equation, relating the mass, velocity, and wavelength of a wave-particle. h is Planck's constant. De Broglie postulated that the electron has wave properties and that material particles have an associated wavelength. (9) Equation relating electricity and magnetism, derived from Maxwell's equations, which form the basis for all computations involving electromagnetic waves including radio, radar, light, ultraviolet waves, heat radiation, and X rays. (10) Archimedes' lever formula.

Here are explanations for some of the formulas in Dr. Googol's own lists. (3) One of Maxwell's equation for electromagnetism. (4) The quadratic formula for solving equations of the form $ax^2 + bx + c = 0$. (5) Newton's second law, relating force, mass, and acceleration. (7) Gives the circumference and area of a circle. (9) represents a Fourier series. Complicated wave disturbances may be represented as the sum of a group of sinusoidal-like waves. (10) The first formula is Euler's identity relating exponential and trigonometric functions; the second formula represents Fermat's Last Theorem. (Runner-up 7) The Gauss-Bonnet formula, where χ is the Euler characteristic. (Runner-up 9) Cauchy's integral formula in complex analysis.

A few respondents suggested Fermat's Last Theorem be included among the top 10 influential mathematical expressions because a significant amount of research and mathematics has been a direct result of attempts to prove the theorem. This theorem by Pierre de Fermat (1601–1665) states that there are no whole numbers a, b, and c such that $a^n + b^n = c^n$ for $n > 2$. (In 1995, Andrew Wiles published a famous paper in the *Annuals of Mathematics* that finally proved Fermat's Last Theorem.) In 1769, Leonhard Euler stated that he thought the related formula $a^4 + b^4 + c^4 = d^4$ had no possible integral solutions. Two centuries later, Noam Elkies of Harvard University discovered the first solution: $a = 2,682,440$, $b = 15,365,639$, $c = 18,796,760$, and $d = 20,516,673$. (For more information, see: Elkies, N. (1988) On $a^4 + b^4 + c^4 = d^4$. *Mathematics of Computation*. Oct. 51(184): 825–35.)

COMMENTS FROM COLLEAGUES

Clifford Beshers of Columbia University suggested adding a fixed loan payment formula to the top 10 because populations that govern industrial economies have had a great impact on our world. The fixed loan payment formula involves variables such as the monthly interest rate, principal, and duration of the loan.

Roy Smith of the Public Health Research Institute in New York noted the following about $c^2 = a^2 + b^2$ (the Pythagorean formula for right triangles):

> This formula is vital to any vector problem, and hence vital to most of physics. Any field of study using complex numbers, such as electronics, involves the conversion between polar and rectangular forms, and this formula has application here. This formula is one of the first things the Scarecrow in *The Wizard of Oz* recited when he got his brain. If you consider the formula's logical extension, the

law of cosines for non-right triangles ($C^2 = A^2 + B^2 - 2AB\cos(\theta)$), then you have the basic formula that surveyors use to measure land. The related formulas for solving spherical triangles were used for celestial navigation, which allowed people to explore the entire world by sea.

Charles Ashbacher, editor of the *Journal of Recreational Mathematics*, wrote to Dr. Googol with "significant disagreements with the list." For the record, Charles's top 10, with a few of his explanations, follow:

1. $1 - 2 = -1$ (The positive integers are intuitively obvious. This formula establishes the existence of negative integers, the first "nonintuitive" set of numbers imagined by humans.)

2. $\sqrt{2} \neq m/n$ (This formula established the existence of irrational numbers and was the first instance where it was proven that some things, like "all" of the digits of $\sqrt{2}$, will never be known.)

3. $a0\,b = a \times \text{base} \times \text{base} + 0 \times \text{base} + b$ (This formula establishes the concept of positional notation and the use of 0 as a place-holder. This eliminated enormously cumbersome systems such as Roman numerals and greatly sped up all manner of computation. It also allowed arithmetic to be mechanized.)

4. $\vec{F} = m\vec{a}$

5. $E = mc^2$

6. $V = I\,R$

7. $\lambda = h/mv$

8. $\vec{F} = Gm_1 m_2 / r^2$

9. $c = 2\pi r$

10. $e^{\ln N} = N$

❀ For more reader comments on formulas that changed the world, see "Further Exploring."

Chapter 42

The 10 Most Difficult-to-Understand Areas of Mathematics

> In heterotic string theory . . . the right-handed bosons (carrier particles) go counterclockwise around the loop, their vibrations penetrating 22 compacted dimensions. The bosons live in a space of 26 dimensions (including time) of which 6 are the compacted "real" dimensions, 4 are the dimensions of ordinary space-time, and the other 16 are deemed "interior spaces"—mathematical artifacts to make everything work out right.
> —*Martin Gardner,* The Ambidextrous Universe

> String theory may be more appropriate to departments of mathematics or even schools of divinity. How many angels can dance on the head of a pin? How many dimensions are there in a compacted manifold thirty powers of ten smaller than a pinhead? Will all the young Ph.D.'s, after wasting years on string theory, be employable when the string snaps?
> —*Sheldon Glashow*

> String theory is twenty-first-century physics that fell accidentally into the twentieth century.
> —*Edward Witten*

We can hardly imagine a gorilla's understanding the significance of prime numbers, yet the gorilla's genetic makeup—its DNA sequence—differs from ours by only a few percentage points. These minuscule genetic differences in turn produce differences in our brains. Additional alterations of our brains would admit a variety of profound concepts to which we are now totally closed. What mathematics is lurking out there that we can never understand? What new aspects of reality could we absorb with extra cerebrum tissue? And what exotic formulas could swim within the additional folds? Philosophers of the past have admitted that the human mind is unable to find answers to some of the most important questions, but these same philosophers rarely thought that our lack of knowledge was due to an organic deficiency shielding our psyches from higher knowledge.

If the yucca moth, with only a few ganglia for its brain, can recognize the geometry of the yucca flower from birth, how much of our mathematical capacity is hardwired into our convolutions of cortex? Obviously, specific higher mathematics is not inborn, because acquired knowledge is not inherited, but our mathematical capacity is a function of our brain. There is an organic limit to our mathematical depth.

How much mathematics can we know? The body of mathematics has generally increased from ancient times, although this has not always been true. Mathematicians in Europe during the 1500s knew less than Grecian mathematicians at the time of Archimedes. However, since the 1500s humans have made tremendous excursions along the vast tapestry of mathematics. Today there are probably around 300,000 mathematical theorems proved each year.

In the early 1900s, a great mathematician was expected to comprehend the whole of known mathematics. Mathematics was a shallow pool. Today the mathematical waters have grown so deep that a great mathematician can know only about 5% of the entire corpus. What will the future of mathematics be like as specialized mathematicians know more and more about less and less until they know everything about nothing?

The following is a ranking of the 10 most difficult areas of mathematics today, from most difficult to least. Of course, the question is inevitably biased. As French mathematician Olivier Gerard notes, a theory can be "difficult" for many reasons. It can be poorly written. It can also be temporarily difficult because some pieces are lacking (such as in a jigsaw puzzle). Most mathematics does not seem to be "eternally" difficult, but many areas are difficult because one must go through a lengthy initiation, review, and training process before hoping to say anything useful or new.

The following are the 10 most difficult-to-understand areas of mathematics, as ranked by mathematicians around the world. Items 1, 2, 3, and 10 are very closely related.

1. *Motivic cohomology (cohomology theory)*

2. *Special cases of the Langlands functoriality conjecture*—examples include $U(2,1)$, as applied to Hilbert modular varieties and stabilization of the trace formula)

3. *Advanced Number Theory* —includes the mathematics used in the proof of Fermat's Last Theorem (by Andrew Wiles).

4. *Quantum groups*

5. *Infinite-dimensional Banach spaces*

6. *Local and micro-local analysis of large finite groups*

7. *Large and inaccessible cardinals*

8. *Algebraic topology*

9. *Superstring theory*

10. *Non-abelian reciprocity (Langlands philosophy), automorphic representations, and modular varieties*

Don't ask Dr. Googol to define these for you. Most seasoned mathematicians don't understand much about these detailed specialties or insanely difficult areas. Those few mathematicians that do understand these areas can't explain them to a general audience. Given all this, here is a brief explanation of superstring theory. Various modern theories of "hyperspace" suggest that dimensions exist beyond the commonly accepted dimensions of space and time. The entire universe may actually exist in a higher-dimensional space. This idea is not science fiction; in fact, hundreds of international physics conferences have been held to explore the consequences of higher dimensions. From an astrophysical perspective, some of the higher-dimensional theories go by impressive-sounding names such as Kaluza-Klein theory and supergravity. In Kaluza-Klein theory, light is explained as vibrations in a higher spatial dimension. Among the most recent formulations of these concepts is superstring theory that predicts a universe of 10 dimensions—3 dimensions of space, 1 dimension of time, and 6 more spatial dimensions. In many theories of hyperspace, the laws of nature become simpler and more elegant when expressed with these several extra spatial dimensions.

The basic idea of string theory is that some of the most basic particles, like quarks and fermions (which include electrons, protons, and neutrons), can be modeled by inconceivably tiny, 1-dimensional line segments, or strings. Initially, physicists assumed that the strings could be either open or closed into loops, like rubber bands. Now it seems that the most promising approach is to regard them as permanently closed. Although strings may seem to be mathematical abstractions, remember that atoms were once regarded as "unreal" mathematical abstractions that eventually became observables. Currently, strings are so tiny there is no way to "observe" them; perhaps we will never be able to. In some string theories, the loops of string move about in ordinary 3-space, but they also vibrate in higher spatial dimensions perpendicular to our world. As a simple metaphor, think of a vibrating guitar string whose "notes" correspond to different, "typical" particles such as quarks and electrons along with other mysterious particles that exist only in all 10 dimensions, such as the hypothetical graviton, which conveys the force of gravity. Think of the universe as the music of a hyperdimensional orchestra. And we may never know if there is a hyper-Beethoven guiding the cosmic harmonies.

In the last few years, theoretical physicists have been using strings to explain all the forces of nature—from atomic to gravitational. Although string theory describes elementary particles as vibrational modes of infinitesimal strings that exist in 10 dimensions, many of you may be wondering how such things exist in our 3-dimensional universe with an additional dimension of time. String theorists claim that 6 of the 10 dimensions are "compactified"—tightly curled up (in structures known as Calabi-Yau spaces) so that the extra dimensions are essentially invisible.

As technically advanced as superstring theory sounds, it could have been developed a long time ago. This is according to string-theory guru Edward Witten, a theoretical physicist at the Institute for Advanced Study in Princeton. For example, he indicates that it is quite likely that other civilizations in the universe discovered superstring theory, then later derived Einstein-like formulations (which in our world predate string theory by more than half a century). Unfortunately for experimentalists, superstrings are so small that they are not likely ever to be detectable by humans. Consider the ratio of the size of a proton to the size of the solar system; this is the same ratio that describes the relative size of a superstring to a proton.

John Horgan, an editor at *Scientific American*, recently published an article describing what other researchers have said of Witten and superstrings in 10 dimensions. One researcher interviewed exclaimed that in sheer mathematical mind power, Edward Witten exceeds Einstein and has no rival since Newton. So complex is string theory that when a Nobel Prize–winning physicist was asked to comment on the importance of Witten's work, he said that he could not understand Witten's recent papers; therefore, he could not ascertain how brilliant Witten is.

Recently, humanity's attempt to formulate a "theory of everything" includes not only string theory but membrane theory, also known as M-theory. In the words of Edward Witten (who *Life* magazine dubbed the sixth most influential American baby boomer), "*M* stands for Magic, Mystery, or Membrane, according to taste." In this new theory, life, the universe, and everything may arise from the interplay of membranes, strings, and bubbles in higher dimensions of spacetime. The membranes may take the form of bubbles, or they may be stretched out in 2 directions like a sheet of rubber, or they may be wrapped so tightly that they resemble a string. The main point to remember about these advanced theories is that modern physicists continue to produce models of matter and the universe requiring extra spatial dimensions.

Chapter 43

The 10 Strangest Mathematical Titles Ever Published

I once asked Gregory Chudnovsky if a certain impression I had of mathematicians was true, that they spend immoderate amounts of time declaring each other's works trivial. "It is true," he admitted.
—*Richard Preston,* The New Yorker, *1992*

Strange. Weird. Intriguing. The following is a list of 10 serious mathematical papers with strange, indecipherable, and/or amusing titles. Candidates for this list were nominated by students, educators, and researchers around the world.

THE TOP 10

The award for all-time strangest title goes to Dr. A. Granville:

Granville, A. (1992) Zaphod Beeblebrox's brain and the fifty-ninth row of Pascal's Triangle. *American Mathematical Monthly.* April, 99(4): 318–331.

Second place goes to Dr. Forest W. Simmons of Portland Community College for:

Simmons, F. (1980) When homogeneous continua are Hausdorff circles (or yes, we Hausdorff bananas). In *Continua Decompositions Manifolds (Proceedings of Texas Topology Symposium 1980)*. University of Texas Press. (Not too surprisingly, the illustrations are reminiscent of bananas!)

Third place goes to the romantic S. Strogatz for:

Strogatz, S. (1988) Love affairs and differential equations. *Mathematics Magazine.* 61(1): 35. (This is an analysis of the time-evolution of the love affair between Romeo and Juliet.)

Fourth place goes to A. Berezin for:

Berezin, A. (1987) Super super large numbers. *Journal of Recreational Mathematics.* 19(2): 142–143. (This paper discusses the mathematical and philosophical implications of the "superfactorial" function defined by the symbol $, where $N\$ = N!^{N!^{N!\cdots}}$ The term $N!$ is repeated $N!$ times.)

Fifth place goes to Alan Mackay for:

Mackay, A. (1990) A time quasi-crystal. *Modern Physics Letters B.* 4(15): 989–991.

Sixth place goes to J. Tennenbaum for:

Tennenbaum, J. (1990) The metaphysics of complex numbers. *21st Century Science.* Spring 3(2): 60.

Seventh place goes to Tom Morley for:

Morley, T. (1985) A simple proof that the world is 3-dimensional. *SIAM Review.* 27: 69–71. (The article starts, "The title is, of course, a fraud. We prove nothing of the sort. Instead we show that radially symmetric wave propagation is possible only in dimensions one and three.")

Eighth place goes to the encyclopedic Professor Akhlesh Lakhtakia, from the Department of Engineering Science and Mechanics at Pennsylvania State University, for:

Lakhtakia, A. (1990) Fractals and The Cat in the Hat. *Journal of Recreational Mathematics.* 23(3): 161–164. (Reprints available from: Prof. A. Lakhtakia, Dept. of Engineering Science, Pennsylvania State University, University Park, PA 16802.)

Ninth place goes to R. C. Lyness for:

Lyness, R. C. (1941) Al Capone and the Death Ray. *Mathematical Gazette.* 25: 283–287.

Tenth place is shared by several individuals:

Englebretsen, G. (1975) Sommers' proof that something exists. *Notre Dame Journal of Formal Logic* 16: 298–300. (The review [MR 51 #7803] by K. Inoue, says "The author points out that F. Sommers's proof that something exists is invalid.")

Hale, R. (1978) Logic for morons. *Mind.* 87: 111–115.

Braden, B. (1985) Design of an oscillating sprinkler. *Mathematics Magazine.* 58: 29–33.

Taylor, C. (1990) Condoms and cosmology: The "fractal" person and sexual risk in Rwanda. *Social Science and Medicine.* 31(9): 1023–1028. (This entry might have been higher up on the list had it been published in a mathematics journal.)

Hoenselaers, C, and Skea, J. (1989) Generating solutions of Einstein's field equations by typing mistakes. *General Relativity Gravity.* 21: 17–20. (The authors made some typing mistakes entering the problem into a computer and came out with new solutions to the equations.)

Marchetti, C. (1989) On the beauty of sex and the truth of mathematics. *Rivista di Biologia — Biology Forum.* 82(2), 209–216.

Khrennikov, A. Yu. (1999) Description of the operation of the human subconscious by means of p-adic dynamical systems. [Russian] *Doklady Akademii Nauk.* 365(4): 458–460.

Chapter 44

The 15 Most Famous Transcendental Numbers

"Math is a perfection in expression, like ballet or a shaolin class martial art.
—*V. Guruprasad*

In this book's introduction, Dr. Googol explained his love of integers and hinted he would seldom discuss complicated numbers like π with an infinite number of digits. Please forgive him for this brief yet fascinating digression into transcendental numbers.

The mathematical constant pi, denoted by the Greek letter π, represents the ratio of the circumference of a circle to its diameter. It is the most famous ratio in mathematics both on Earth and probably for any advanced civilization in the universe. The number π, like other fundamental constants of mathematics such as $e = 2.718 \ldots$, is a transcendental number. The digits of π and e never end, nor has anyone detected an orderly pattern in their arrangement. Humans know the value of π to over a billion digits.

Transcendental numbers cannot be expressed as the root of any algebraic equation with rational coefficients. This means that π could not exactly satisfy equations of the type $\pi^2 = 10$ or $9\pi^4 - 240\pi^2 + 1492 = 0$. These are equations involving simple integers with powers of π. The numbers π and e can be expressed as an endless continued fraction or as the limit of an infinite series. The remarkable fraction 355/113 expresses π accurately to 6 decimal palaces.

Many of you have probably heard of π and e. But are there other famous transcendental numbers? After conducting a brief survey of readers, Dr. Googol made a list of the 15 best-known transcendental numbers. Can you list these in order of relative fame and/or usage?

1. $\pi = 3.1415 \ldots$

2. $e = 2.718 \ldots$

3. *Euler's constant,* $\gamma = 0.577215 \ldots = \lim_{n \to \infty}(1 + 1/2 + 1/3 + 1/4 + \ldots 1/n - \ln(n))$ (Not proven to be transcendental, but generally believed to be by mathematicians.)

4. *Catalan's constant,* $G = \Sigma(-1)^k/(2k + 1)^2 = 1 - 1/9 + 1/25 - 1/49 + \ldots$ (Not proven to be transcendental, but generally believed to be by mathematicians.)

5. *Liouville's number,* .110001 . . . (This is an example of a transcendental number "discovered" much later than pi or e. It was first discussed in 1851 and named after its "inventor," French mathematician J. Liouville. You can compute this fascinating number with $\Sigma_{k=1}^{\infty} a_k r^{-k!}$ where $a \le a_k \le r$. The numbers a_k are integers. The resulting number is a Liouville number of base r. If the values for a_k are all 1, and $r = 10$, we get: $1/10 + 1/10^{1 \times 2} + 1/10^{1 \times 2 \times 3} + \ldots$ Believe it or not, the decimal value can easily be written down: 0.110001000000000000000001000 . . . which has a 1 in the 1st, 2nd, 6th, 24th, etc., places and 0s elsewhere.)

6. *Chaitin's constant,* the probability that a random algorithm halts. (Noam Elkies of Harvard notes that this number is not only transcendental but also incomputable.)

7. *Champernowne's number,*
0.12345678910111213141516171819202122232425 . . . (This is constructed by concatenating the digits of the positive integers. Can you see the pattern?)

8. *Special values of the zeta function,* such as $\zeta(3)$ (Transcendental functions can usually be expected to give transcendental results at rational points. Technically speaking, $\zeta(3)$ is known to be irrational but not yet proven to be transcendental, although it is generally believed to be my mathematicians.)

9. *ln(2)*, 0.6931 . . .

10. *Hilbert's number,* $2^{\sqrt{2}}$ (This is called Hilbert's number because the proof of whether or not it is transcendental was one of Hilbert's famous 100 problems. In fact, according to the Gelfond-Schneider theorem, any number of the form a^b is transcendental where a and b are algebraic, a is neither 0 nor 1, and b is not rational. Many trigonometric or hyperbolic functions of nonzero algebraic numbers are transcendental.)

11. e^{π}

12. π^e (Not proven to be transcendental, but generally believed to be by mathematicians.)

13. *Morse-Thue's number,* 0.01101001 . . . (See Chapter 17 for more information.)

14. i^i (Here i is the imaginary number $\sqrt{-1}$. If a is algebraic and b is algebraic but irrational, then a^b is transcendental. Since i is algebraic but irrational, the theorem applies. Note also: i^i is equal to $e^{-\pi/2}$ and several other values. Consider $i^i = e^{i \log i} = e^{i \times i\pi/2}$. Since log is multivalued, there are other possible values for i^i.)

15. *Feigenbaum numbers,* e.g. 4.669 . . . (These are related to properties of dynamical systems with "period-doubling" oscillations. The ratio of successive differences between period-doubling bifurcation parameters approaches the number 4.669. . . . Period doubling has been discovered in many physical systems before they enter the chaotic regime. Feigenbaum numbers have not been proven to be transcendental but are generally believed to be.)
Keith Briggs from the Mathematics Department of the University of Melbourne in Australia computed what he believes to be the world record for the number of digits for the Feigenbaum number:

4.66920160910299067185320382046620161725818557747576863274565134300413433021131473713868974402394801381716598485518981513440862714202793252231244298889089085994493546323671341153248171421994745564436582379320200956105833057545861765222207038541064674949428498145339172620056875566595233987560382563722

Briggs carried out the computation using special-purpose software designed by David Bailey of NASA Ames running on an IBM RISC System/6000. The computation required a few hours of computation time. For more information, see: Briggs, K. (1991) A precise calculation of the Feigenbaum constants. *Mathematics of Computation*. 57: 435.

Some final questions: Is -1^{-i} a transcendental number? Is there a compact formula relating e, pi, i, and phi, the golden ratio described in Chapter 96? Drum roll please . . . one answer is $e^{i\pi} + 2\phi = \sqrt{5}$.

Chapter 45

What is Numerical Obsessive-Compulsive Disorder?

> The rationality of our universe is best suggested by the fact that we can discover more about it from any starting point, as if it were a fabric that will unravel from any thread.
> —*George Zebrowski*, "Is Science Rational," OMNI, June 1994

> When we learn more about the function gone wrong in obsessive-compulsive disorder, we will also learn more about the most mysterious secrets of the nature of man.
> —*Judith Rapoport, M.D.*, The Boy Who Couldn't Stop Washing

Individuals afflicted with obsessive-compulsive disorder are often compelled to commit repetitive acts that are apparently meaningless such as persistent hand washing, counting objects, checking to see if doors are locked, and avoiding oddly stressful situations such as stepping on the cracks of the sidewalk. Obsessive-compulsive disorder involving numbers is particularly sad and fascinating. The great inventor Nikola Tesla had "arithromania," or "numerical obsessive-compulsive disorder." He demanded precisely 18 clean towels each day. If asked why, Tesla provided no explanation. Table accoutrements and towels were not the only items he demanded come in multiples of 3. For example, he often felt compelled to walk around the block 3 times, and he always counted his steps while walking. He chose room number 207 in the Alta Vista Hotel, because 207 is divisible by 3. When dining, he always stacked 18 napkins in a neat little pile because he favored numbers divisible by 3.

Does numerical obsessive-compulsive disorder often involve *particular* numbers? Are obsessions with odd numbers more likely than even? Do obsessions ever involve numbers larger than 10? To better understand numerical obsessive-compulsive disorder, Dr. Googol pored through many case histories and created a list sorted by the number with which the person was obsessed:

⊙ 1. No cases found.

⊙ 2. No cases found.

⊙ 3. A 13-year-old girl (see "9") is compelled to knock 3 times on the edge of the window and 3 times on a nearby door before unlocking the door.

⊙ 4. Case 1: An 11-year-old boy's life is ruined because the number 4 dominates his existence. Case 2: A teenage boy must have everything in 4s and avoids 6s. (He also has the compulsion to see the bottoms of his feet whenever he looks at the clock in his room.)

⊙ 5. No cases found

⊙ 6. Case 1: A college boy avoids repeating any actions 6, 13, 60, 66, or 130 times. Multidigit numbers (such as 42 or 33) adding up to 6, 13, or 130 must be avoided. Case 2: A teenage boy begins his day normally, then suddenly the only thing on his mind is the repeating numbers "6, 6, 6, 6" or "8, 8, 8, 8." He reports, "I had no control over these numbers; they had a mind of their own—*my* mind."

⊙ 7. The 11-year-old boy listed under "4" suddenly switches to a heptaphiliac when, after a brain operation, he has the very time-consuming compulsion to touch everything 7 times and ask for everything in 7s. He swallows 7 times. (His heptaphilia is cured by Anafranil, a drug that helps many afflicted with obsessive-compulsive disorder.)

⊙ 8. Case 1: A 12-year-old boy is compelled to turn around exactly 8 times in a coat room in order to calm himself. Case 2: A boy in the shower strokes the right side of his head 8 times, applies shampoo, then strokes another 8 times, rinses 8 times, and strokes 8 times again. He repeats the process for the left side.

⊙ 9. A 13-year-old girl must lift her feet and tap 9 times on the edge of her bed before climbing into it.

⊙ 22. An 18-year-old boy is compelled to count to 22 over and over again. He taps on the wall 22 times or in multiples of 22. He walks through doorways 22 times and gets in and out of his chair 22 times. The boy becomes addicted to drugs, which have interesting effects on his 22-ness. For example, while on amphetamines and cocaine, his 22-tapping increases to the point where all his time is spent tapping out 22 all over his walls. LSD makes the ritual completely disappear.

⊙ 50. A 7-year-old girl must count to 50 in between reading or writing each word. This time-consuming ritual makes her an extremely slow reader in the second grade.

⊙ 13, 60, 66, 130. See the college boy listed under "6."

⊙ 100. The 13-year-old girl listed under "9" must also count to 100 after brushing her teeth.

One wonders if the incidence of "numerical obsessive-compulsive disorder" is lower in societies with less emphasis on numbers or in pre-literate societies. Most of the time, people with obsessive-compulsive disorder know that their behavior is illogical or self-destructive, but like someone with a super-strong addiction, they find it impossible to stop. For example, a person may spend 5 hours a day washing himself and still feel dirty. Another sufferer must check the door hundreds of times a day to make sure it is locked. Some pluck out every hair on their heads. Children may play endlessly with strings or pick up objects with their elbows to avoid getting their hands dirty. These children usually continue to suffer the same symptoms as adults if untreated.

One of the cures for obsessive-compulsive disorder is the drug Anafranil (clomipramine), which affects the metabolism of serotonin in the brain. Other drugs such as fluoxetine (Prozac) and fluvoxamine (Luvox) also are useful. The success with these drugs leads many researchers to believe that obsessive-compulsive disorder has a physical basis—just like manic depression (biploar disorder) or epilepsy. LSD is another drug that increases serotonin levels and appears to "cure" obsessive-compulsive disorder. (The use of LSD outside of the laboratory may be dangerous. Mood shifts, time and space distortions, and impulsive or aggressive behavior are complications possibly hazardous to an individual who takes the drug.) Amphetamines make the disease worse, probably because they affect the dopamine system, which acts against the serotonin system.

Medical imaging studies suggest that obsessive-compulsive disorder is caused by an abnormality in a part of the brain known as the basal ganglia, buried deep within the brain and in the frontal lobes. In particular, one portion of the basal ganglia called the caudate nucleus appears to behave differently in people with obsessive-compulsive disorder. People with Tourette's syndrome and Parkinson's disease also have abnormalities in these areas. Evidence continues to mount for obsessive-compulsive disorder's biological basis. For example, obsessive-compulsive disorder appears to have a genetic component and often runs in families. Some obsessive-compulsive disorder starts after a first epileptic seizure. The fact that many obsessive-compulsive disorder sufferers have movement disorders, such as facial tics, at one point during the course of their disease further suggests a biological cause.

Chapter 46

Who Is the Number King?

Mathematics is the only infinite human activity. It is conceivable that humanity could eventually learn everything in physics or biology. But humanity certainly won't ever be able to find out everything in mathematics, because the subject is infinite. Numbers themselves are infinite.

—*Paul Erdös*

Erdös' driving force was his desire to understand and to know. You could think of it as Erdös' magnificent obsession. It determined everything in his life.

—*Ronald Graham, AT&T Research*

Dr. Paul Erdös was a legendary mathematician who was so devoted to math that he lived as a wanderer with no home and no job. Like Dr. Googol, his best friends were the integers.

Paul Erdös kept working on mathematics until his death from a heart attack in 1996 at the age of 83 while attending a mathematics meeting in Warsaw. Erdös (pronounced *air*-dosh) was one of the greatest, most eccentric, and most original mathematicians of all time. His passion was to both pose and solve difficult problems in number theory (the study of properties of integers) and in other areas like discrete mathematics, which is the foundation of computer science. He was also one of the most prolific mathematicians in history, with more than 1,500 papers to his name.

Erdös is often remembered as being stooped and slight, and wearing socks and sandals. In order to pursue his mathematical obsession, he stripped himself of all the usual burdens of daily life—finding a place to live, driving a car, paying income taxes, buying groceries, and writing checks—and relied on his friends to look after him. "Property is nuisance," he said.

Because he focused so intently on numbers, his name is not well known outside the mathematical community. He wrote no bestselling books and showed little interest in worldly success and personal comfort. In fact, he lived out of a suitcase (which he never learned to pack) for much of his adult life. He gave away all the money he made from lecturing and mathematical prizes in order to help mathematics students or to offer cash prizes for solving problems he had posed. Erdös left behind only $25,000 when he died.

Sexual pleasure revolted him; even an accidental touch by anyone made him feel uncomfortable. Like many others in this book, Erdös never married or had a family. He often referred to husbands as "slaves" and wives as "bosses." Yet he was not a hermit like Ted Kaczynski (the Unabomber and brilliant mathematician discussed in Chapter 31) or Henry Cavendish (the eccentric but genius physicist and chemist). In fact, Erdös hated to be alone, and almost never was; he loved to attend conferences and enjoyed the attention of mathematicians. His main aim in life was "to do mathematics: to prove and conjecture."

Concentrating fully on mathematics, Erdös traveled from meeting to meeting, carrying a half-empty suitcase and staying with mathematicians wherever he went. Ronald Graham of AT&T Research often handled Erdös's money for him, setting aside an "Erdös room" in his house for helping to manage Erdös's life. In fact, many of Erdös's colleagues took care of him, lending him money, feeding him, buying him clothes, and even doing his taxes. In return, he inundated them with brilliant ideas and challenges—with problems to be solved and masterful ways of attacking them.

Dr. Joel H. Spencer, a mathematician at New York University's Courant Institute of Mathematical Sciences, once noted: "Erdös had an ability to inspire. He would take people who already had talent, that already had some success, and just take them to an entirely new level. His world of mathematics became the world we all entered."

Erdös was born into a Hungarian-Jewish family in Budapest in 1913, the only surviving child of parents who were mathematics teachers. He was educated mostly at home until 1930, when he entered the Peter Pazmany University in Budapest, where he was soon at the center of a small group of outstanding young Jewish mathematicians. Erdös practically completed his doctorate as a second-year undergraduate. He received his doctorate in mathematics from the University of Budapest and in 1934 came to Manchester on a postdoctoral fellowship.

By the time he finished at Manchester in the late 1930s, it became clear that it would be an act of suicide for a Jew to return to Hungary. (Most members of his family who remained in Hungary were killed during the war.) Therefore, in 1938 Erdös sailed for the United States, where he was to stay for the next decade. During his first year, at the Institute for Advanced Study in Princeton, he wrote groundbreaking papers that founded probabilistic number theory. He also solved important problems in approximation theory and dimension theory. When his fellowship at the institute was not renewed, he started his wanderings, with longer stays at the University of Pennsylvania, Notre Dame, Purdue, and Stanford.

The great mathematical event of 1949 was an elementary proof of the Prime Number Theorem, given by Atle Selberg and Erdös. The result, which predicts the distribution of primes with some accuracy, was first proved in 1896 by sophisticated methods, and it had been thought that no elementary proof could be given. Erdös was only 20 when he discovered this elegant proof for the famous theorem in number theory. The theorem says that for each number greater than 1, there is always at least 1 prime number between it and its double.

Erdös never saw the need to limit himself to a single university. He needed no laboratory for his work, no library or equipment. Instead he traveled across America and Europe from one university and research center to the next, inspired by making new contacts. When he arrived in a new town he would present himself on the doorstep of the most prominent local mathematician and announce: "My brain is open." While a guest, Erdös would often work furiously for a few days and then leave when he had exhausted the ideas or patience of his host. (He was quite capable of falling asleep at the dinner table if the conversation was not about mathematics.) He would end sessions with "We'll continue tomorrow, if I live."

Although his research spanned a variety of areas of mathematics, Erdös continued his interest in number theory for the rest of his life, posing and solving problems that were often simple to state but notoriously difficult to solve and that involved relationships between numbers. Erdös believed that if one can state a problem in mathematics that is unsolved and over 100 years old, it is probably a problem in number theory.

Erdös, like many other mathematicians, believed that mathematical truths are discovered, not invented. He mused about a Great Book in the sky, maintained by God, that contained the most elegant proofs of every mathematical problem. He used to joke about what he might find if he could just have a glimpse of that book.

It is commonly agreed that Erdös is the second most prolific mathematician of all times, being superseded only by Leonhard Euler, the great 18th-century mathematician whose name is spoken with awe in mathematical circles. In addition to Erdös's roughly 1,500 published papers, another 50 or more are still to be published after his death. (Erdös was still publishing a paper a week in his 70s.) Erdös undoubtedly had more coauthors (around 500) than any other mathematician in history. He collaborated with so many mathematicians that the phenomenon of the "Erdös number" evolved. To have an Erdös number 1, a mathematician must have published a paper with Erdös. To have a number of 2, he or she must have published with someone who had published with Erdös, and so on. Four and a half thousand mathematicians have an Erdös number of 2.

At the end of 1999, researchers discovered that winners of the Fields medal— math's Nobel equivalent—all have an Erdös number of 5 or less. In addition, more than 60 Nobel laureates, many in fields far removed from mathematics, can brag of single-digit Erdös numbers. Watson and Crick, for example, have numbers of 7 and 8. (See Constance Holden, "Analyzing the Erdös star cluster" [*Science,* February 4, 287(5454): 799, 2000].)

Erdös would always make use of whatever time was available to do mathematics. It was common for him to listen to music and to do mathematics at the same time; he would even bring notebooks to concerts and start solving problems. According to legend, on a long train journey he wrote a joint paper with the conductor.

During his lifetime Erdös had many close friends and faithful and cherished disciples, but he had the deepest emotional contact with his mother, who started to accompany him on his incessant travels when she was in her mid-'80s and continued to do so till her death at the age of 91 in 1971. The death of his mother was an incredible blow from which he never fully recovered. After she died he found solace in doing even more mathematics than before. Erdös launched himself into his work with greater vigor, regularly working a 19-hour day. He fueled his efforts almost entirely by coffee, caffeine tablets, and benzedrine. He looked more frail, gaunt, and unkempt than ever, and often wore his pajama top as a shirt.

In spite, or perhaps because, of his eccentricities, mathematicians revered him and found him inspiring to work with. He was regarded as the wit of the mathematical world, the one man able to produce short, clever solutions to problems on which others had suffered through pages of equations.

Chapter 47

What 1 Question Would You Add?

The mathematical spirit is a primordial human property that reveals itself wherever human beings live or material vestiges of former life exist.
—*Willi Hartner*

After reading through all of these chapters with lists and outrageous questions, what question would *you* ask Dr. Googol?

Dr. Googol asked mathematicians from around the world, "What 1 question would you add to my list of questions?" Here is a sampling of replies:

1. What effect would doubling the salary of every mathematics teacher have on education and the world at large?

2. How important are new mathematical findings to the advance of science (including astronomy and physics) in the 21st century?

3. What are the 5 most beautiful ideas in mathematics? (Suggestions included Riemann surfaces, Fermat's Last Theorem, Euler's equation, the Fundamental Theorem of Algebra, and the Fundamental Theorem of Calculus.)

4. What are the 5 mathematical theories you find overrated or so publicized that serious work in the fields are hindered by the hype?

5. What are the 5 mathematical theories you find most underrated, unknown, or underused?

6. Who are the 5 mathematicians and nonmathematicans who best communicate their mathematical ideas to nonspecialists?

7. The discovery of a number or property of a number sometimes opens up entirely new areas of mathematics. Which discovery had the greatest impact on the development of modern society?

8. What is your favorite proof of Pythagoras's theorem? (Note that this theorem is the most-proved theorem. A book published in 1940 entitled *The Pythagorean Proposition* contained 370 different proofs of it.)

9. Should the National Science Foundation fund a project whose goals were to determine 1 trillion digits of π?

10. How profound would it be if mathematicians discovered strange and unusual patterns in the first trillion digits of π?

11. Would you rather marry the best mathematician in the world or the best chess player? Why?

12. How important is the mathematical concept of fractals? Should Benoit Mandelbrot receive the Nobel Prize? Will strange and totally new patterns be discovered in the Mandelbrot set, or have we already seen all the basic structural themes and patterns?

13. Answer this question with *yes* or *no*. Will your next word be *no*?

Chapter 48

Cube Maze

> We should take care not to make the intellect our god; it has, of course, powerful muscles, but no personality.
> —*Albert Einstein,* Out of My Later Life, *1950*

> ⊛ **Number Maze 2,
> a visual intermission before
> the next book part. . . .**

Dr. Googol created a cubical array of metallic spheres with numbers painted on each sphere. Figure 48.1 is a schematic illustration of the original 3-dimensional model. Although it's more fun to hold the model in your hand and turn it

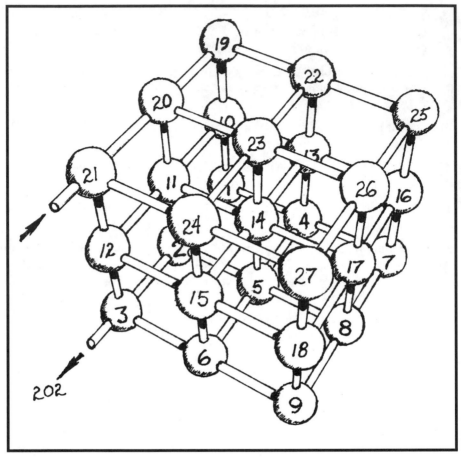

48.1 Cube Maze. Can you reach the arrow at the end with a sum of 202? (Drawing by Brian Mansfield.)

around as needed, you can still have a terrific challenge by trying to trace a path from the arrow at the top to the arrow at the bottom such that the sum of all the spheres you pass through is 202.

Dr. Googol originally intended to give the model to all potential girlfriends to solve before he would even consider marrying them. However, after Monica threw the puzzle at his head, Dr. Googol reconsidered and has instead published this illustration here for the first time. He has also sent models to the leaders of several nations as gifts, and, alas, none could solve the puzzle. In fact, no one on the planet has yet solved the puzzle. Can you?

❀ For a solution, see "Further Exploring."

Part iii

Fiendishly
Difficult
Digital Delights

I don't need to know where I'm going
to enjoy the road I'm on.
—*Deepak Chopra,* Ageless Body, Timeless Mind

The mathematician's eye is a mystic mirror,
not only reflecting reality but absorbing it.
—*Dr. Francis O. Googol*

Chapter 49

Hailstone Numbers

What could be more beautiful than a deep, satisfying relation between whole numbers. How high they rank, in the realms of pure thought and aesthetics, above their lesser brethren: the real and complex numbers.
—*Manfred Schroeder*, Number Theory in Science and Communication, *1984*

The external world exists; the structure of the world is ordered; we know little about the nature of the order, nothing at all about why it should exist.
—*Martin Gardner*, "Order and Surprise," *paraphrasing Bertrand Russell, 1985*

On a recent trip to the Himalayas, Dr. Googol found himself walking in a blinding hailstorm! The hailstones drifted up and down in the wisps and eddies of wind. Sometimes the stones shot up for as far as his eye could see and then came plummeting back to Earth, smashing into the ground like little meteorites. Dr. Googol smiled because he realized hailstones provide a wonderful metaphor for one of the most famous and unusual problems in number theory. He whipped out a piece of paper from his pocket and began scribbling a strange sequence of numbers: 7, 22, 11, 34, 17.

"Hailstone number" problems have fascinated mathematicians for several decades and are studied because they are so simple to calculate yet apparently intractably hard to solve. To compute a sequence of hailstone numbers, start by choosing any positive integer you like.

> **If your number is even, divide it by 2.**
> **If it is odd, multiply by 3 and add 1.**

Next, take your answer and repeat the rule. Again, if your answer is even, divide it by 2. If it is odd, multiply by 3 and add 1. Repeat this process for as long as you like. For example, the hailstone sequence for 3 is 3, 10, 5, 16, 8, 4, 2, 1, 4, ... (The " . . . " indicates that the sequence continues forever as 4, 2, 1, 4, 2, 1 4, etc.) Dr. Googol sometimes like to draw little melting hailstones indicating the numbers in the sequence:

Like hailstones falling from the sky through storm clouds, this sequence drifts down and up, sometimes in seemingly haphazard patterns. Also like hailstones, hailstone numbers always seem eventually to fall back down to the ground (the integer 1, represented as a single ◆). In fact, most mathematicians believe that *every* hailstone sequence ends in the cycle 4, 2, 1, 4, . . . , no matter what number the sequence starts with:

(repeating over and over again)

This hailstone conjecture (about settling back to 1) has been numerically checked for a large range of starting points, and the current record has been set by N. Yoneda, who has checked all integers less than 1,000,000,000,000.

Do you think that all hailstone numbers fall back down to 1? Various large cash awards have been offered to anyone who can prove or disprove this. The hailstone sequence, also known as the $3n + 1$ sequence, gives rise to a mixture of regularity and disorder: it is definitely not random, but the pattern resists interpretation. (This problem in number theory can be placed in a much larger context of chaos theory, which involves the study of a range of mathematical and physical phenomena exhibiting a sensitive and often irregular dependence on initial conditions.) Computer graphics can be used to reveal patterns in this hailstone sequence so that mathematical structures are made more obvious to the mathematician. Unfortunately, computer graphics has been little exploited in $3n + 1$ research.

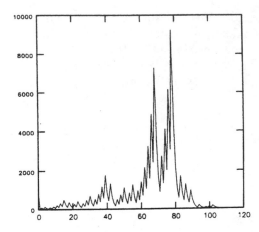

49.1 Hailstone numbers produced by just 1 starting number, 54.

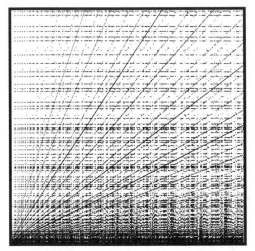

49.2 Hailstone numbers for all starting numbers between 1 and 1,000 on the x axis. The paths of the hailstone numbers are along the y axis.

Figure 49.1 shows a hailstone sequence for just one starting value, 54. Its path length (before settling back down to 1) is 112, and the maximum value reached is 9,232. The plot suggests a seemingly chaotic trajectory that eventually settles back down to 1.

Figure 49.2 is a plot for all the starting numbers between 1 and 1,000 along the x axis. As the hailstone numbers drift up and down, Dr. Googol plots a dot along the y axis. (So that he can represent the information in a small plot, he excludes all y values of the hailstone numbers greater than 1,000.) Notice the plot reveals a pattern of diagonal lines of varying density that pass through the origin, a pattern of horizontal lines, and a diffuse "background" of chaotically positioned dots. Can you figure out why there are these patterns? The diffuse horizontal lines represent certain values that are much more likely than others. An outstanding example is state 9,232. Of the first 1,000 integers, more than 350 have their maximum at 9,232. Why are there other patterns? The hailstone numbers clearly display preferred values, but exactly why these values and clusters of values exist is unclear. Every possible integer state and trajectory length (path before returning to 1) can be produced—but again some numbers appear more often than others. As Paul Erdös commented on the complexity of $3n + 1$ numbers, "Mathematics is not yet ready for such problems."

❁ For more information on hailstone numbers, see "Further Exploring."

▣ See [www.oup-usa.org/sc/0195133420] for a computer program to generate hailstone numbers.

Chapter 50

The Spring of Khosrow Carpet

The essences are each a separate glass, through which the sun
of being's light is passed—each tinted fragment sparkles in the sun:
a thousand colors, but the light is one.
—Jami (15th century)

A mathematician, like a painter or a poet, is a maker of patterns.
If his patterns are more permanent than theirs, it is because
they are made with ideas.
—G. H. Hardy, A Mathematician's Apology

I believe the geometric proportion served the creator as an idea
when He introduced the continuous generation of similar objects
from similar objects.
—Johannes Kepler

Dr. Googol was touring Teheran, Iran, when a carpet dealer showed him a fragment of the famous Spring of Khosrow Carpet, an ancient Persian rug.

"Gorgeous," Dr. Googol said. "What is it?"

"It's the most costly and magnificent carpet of all time—made for the Ctesiphon palace of the Sasanian King Khosrow I. He was king between A.D. 531 and 579."

The merchant told Dr. Googol that the carpet was called the Spring Carpet because it represented, in silk, gold, silver, and jewels, the blossoming splendor of spring. It was also called the Winter Carpet because it was used in bad weather, when real gardens were unavailable. In this way, it symbolized the king's power to command the return of the seasons.

Dr. Googol stooped down to take a closer look and saw a paradise with streams, paths, rectangular plots of flowers, and flowering trees. Water was represented by crystals, soil by gold, and fruits and flowers by precious stones. The merchant told Dr. Googol that when the Arabs captured Ctesiphon in 637, the carpet, which measured about 84 square feet (7.8 square meters), was cut into fragments and distributed to the troops as booty.

As Dr. Googol contemplated the sad tale, he recalled his favorite algorithm for generating Persian carpet designs. Figure 50.1 results from the simplest of algorithms and shows self-similar patterns, that is, repeated patterns at different size scales. One part of the recursion recipe requires us to start with a large rectangle, subdivide it into 4 equal rectangles, and continue the process until we cannot go any further. The carpets look beautiful in color, but even in this

50.1 Self-similar carpet synthesized using a simple algorithm.

black-and-white representation, we can begin to appreciate the infinite reservoir of structures. In a computer program, the algorithm colors each cell in a matrix by assigning a cell a number from 0 to m-1 where m is the number of colors available.

❀ For more details on the algorithm, see "Further Exploring."
🖪 See [www.oup-usa.org/sc/0195133420] for a BASIC code listing.

Chapter 51

The Omega Prism

> It may well be doubted whether human ingenuity can construct
> an enigma of the kind which human ingenuity may not, by proper
> application, resolve.
> —*Edgar Allan Poe,* The Gold-Bug

On a cool night in November, Dr. Googol sees a streaking across the sky. After a few minutes, there is a glowing in a nearby cornfield. Upon closer inspection, he finds an object resembling a Rubik's cube protruding from the ground.

When he picks up the crystalline object, he finds that it is nearly cubical, and its 6 faces are tiled in a colorful substance that luminesces. On the ground by the object is a note, which reads:

You hold in your hands an Omega Prism, a 230 mm × 231 mm × 232 mm brick whose faces are tiled with 1 mm × 1 mm squares. If you were to draw a straight line on the rectangular faces from one corner to another, on which face does the diagonal line cross the most tiles? Can you determine the number of tiles crossed for any face? To solve this puzzle, you are not permitted to trace a diagonal on a prism face and count the number of tiles crossed. We are watching. If you fail to solve the puzzle within a week, we will colonize the Earth and use humans as food for further thought.

Dr. Googol stares at the Omega Prism for several minutes, clenches his fists, and throws the prism to the ground. Even if he were allowed to trace the diagonal with a marker, the colors are blinking so rapidly that it would be nearly impossible for him to count the crossed tiles. A wind begins to blow through the field—a cold wind that sounds like the chanting of monks.

Simultaneously, Omega Prisms land in New York City, London, Tokyo, Moscow, and Calcutta. Unfortunately, none of the people who find the prisms can solve the problem. Can you help save the Earth? Given just the side lengths

**51.1 Small version of the Omega Prism. Humans find it difficult to count tiles inter-
sected by a diagonal line without actually using a straightedge and drawing a line.
When the colors blink, it is impossible for humans to count "intersected" tiles by
eye alone.**

of Omega Prisms, can you determine the number of square tiles through which
a diagonal crosses? How do solutions change as the faces grow?

Figure 51.1 shows a computer graphics rendition of a smaller Omega Prism.
Renditions of the actual 230-by-231-by-232 prism contain facets so small that
they are impossible to distinguish when printed on a page. The purpose of
Figure 51.1 is to emphasize the difficulty individuals have when they attempt to
count tiles intersected by a diagonal line without actually using a straightedge
and drawing the line. When the colors blink, it is impossible for humans to
count "intersected" tiles by eye alone.

✸ For a solution and additional speculation, see "Further Exploring."

▣ See [www.oup-usa.org/sc/0195133420] for a BASIC code listing that is
explained in "Further Exploring."

Chapter 52

The Incredible Hunt for Double Smoothly Undulating Integers

> The essence of mathematics resides in its freedom.
> —*Georg Cantor*

Dr. Googol was exploring the African jungles when he came upon a large snake whose body undulated up and down, up and down, like waves on the water. He had to watch out before the snake encircled him in its muscular twists and turns! Slowly, Dr. Googol began to ponder mathematical undulation.

The term *undulation* in mathematics has a similar meaning to the up-and-down bends in the snake's body. For example, if an integer's digits are alternately greater or less than the digits adjacent to them (consider 4,253,612), then the number is called an undulating integer. The term *smoothly undulating integer* refers to numbers whose adjacent digits oscillate, as in 79,797,979.

A *double* smoothly undulating integer is one that undulates in both its decimal and binary representations. (Binary numbers are defined in the "Further Exploring" section of Chapter 21.) For example, 1010 is an undulating binary number. There are some trivially small smoothly undulating integers, such as 21 (with binary representation 10101). Dr. Googol calls this trivial because a 2-digit oscillation can hardly be called an oscillation. However, he asks you if there are any multidigit double smoothly undulating integers. He has searched for such an integer and never found one, and he has long doubted that such numbers exists. Of course, his brute-force computer searches provide no real answer to the question, and it would be interesting to prove the conjecture that there is no double smoothly undulating integer. It is also interesting to speculate whether there is anything special about the arrangement of digits within a decimal number corresponding to a binary undulating number. Casual inspection suggests that the arrangement is random.

Note that if an n-digit decimal number is selected at random, the chance that it will be smoothly undulating is $81/9 \times 10^{n-1}$, which is approximately equal to

$1/10^n$ for large n. This means that if the decimal equivalent of a smoothly undulating binary integer could be considered as a random arrangement of digits, the probability of it being smoothly undulating becomes exceedingly small. Note also the interesting fact that there is a constant number, 81, of possible undulating integers for any given n-digit decimal number. This speeds the search for double smoothly undulating integers using a computer. You may wish to use computer graphics to find patterns in the undulation of even/odd numbers in the decimal equivalents.

❀ For more information on undulating numbers, see "Further Exploring."

Chapter 53

Alien Snow: A Tour of Checkerboard Worlds

> He became aware of a kind of sparkle in the air ahead. Fairy lights
> blinking on and off. Cal saw three-dimensional patterns within the cloud,
> geometric ratios building and rebuilding in dazzling arrays.
> —*Piers Anthony,* Ox

So begins a science-fiction saga that describes humanity's first encounter with ephemeral entities resembling points of light on a 3-dimensional checkerboard—lights that move and change shapes according to mathematical laws. Some readers will recognize the cloud as a 3-dimensional cellular automaton. The theme saturates *Ox*, even to the point where each chapter begins with a small cellular grid decorated by dots. The presence or absence of a dot in a grid cell indicates which of 2 states a cell is in (that is, the cell is either on or off).

In general, a cellular automaton is an array of cells and a finite collection of possible states. At any given moment, each cell of the array must be in one of the allowed states. The rules that determine how the states of its cells change with time are what determine the cellular automaton's behavior. There is an infinite number of possible cellular automata, each like a checkerboard world in its own right. The world can be a 1-dimensional strip of cells, a 2-dimensional grid, or, as in *Ox*, a 3-dimensional array. Of course, even higher dimensions are possible, but they are difficult to represent as clearly as the lower dimensions.

In this chapter, Dr. Googol explores some personal favorites among the 2-dimensional cellular automata, and he passes along some algorithms that you can feed your personal computer.

Cellular automata comprise a class of simple mathematical systems that are fast becoming important as models for a variety of physical processes. Some cellular automata act in bizarre and random-looking ways, while others exhibit highly ordered behavior. It all depends on the rules of the game. Cellular automata have been used to model the spread of plants, animals, and even forest fires. They have mimicked fluid flow and chemical reactions. They have even been investigated as computers in their own right! Cellular automata are also referred to variously as "homogenous structures," "cellular structures," and "iterative arrays."

The concept of the cellular automaton was introduced in the 1950s by John von Neumann and Stanislaw Ulam. They saw in cellular automata an idealized system capable of modeling fundamental qualities of life itself. Self-reproduction seemed possible. By the 1960s, as computers became widespread in academic institutions, the Cambridge mathematician John Horton Conway grew interested in cellular automata. Conway discovered a particular cellular automaton he called Life, not only because its two states resembled life and death but because computer experiments with certain configurations of cells produced behavior that could only be called lifelike. The game was first publicized by Martin Gardner in his "Mathematical Games" column in the October 1970 issue of *Scientific American*. Since that time, cellular automata have become a very popular area of research for physicists, computer scientists, and mathematicians. They have particular appeal because any differential equation can be converted into a corresponding cellular automaton. This one simple fact opened the door to a brand-new exploration of many differential equations, most of them being models for various physical processes of great interest to scientists.

The Game of Life makes an ideal introduction to the subject of cellular automata. It is "played" on a 2-dimensional grid of cells, each cell being in 1 of 2 states (alive or dead) at any one time. During each new generation at a particular time t, each cell "decides" whether it will be alive or dead. All cells use exactly the same rules. In particular, each cell considers its own state and the state of its 8 neighbors, 4 along edges and another 4 at the corners. The rules themselves are simple:

1. If a cell is alive at time t, it will remain alive at time $t + 1$ if it has no more than 3 neighbors (otherwise it is too crowded) and no fewer than 2 living neighbors at time t (which would make it too isolated).

2. If a cell is dead at time t, it will remain dead unless it has exactly 3 living neighbors. These act as parents.

Using these rules, Life can exhibit fantastically complicated and hard-to-predict behavior. The cellular game has spawned a software-publishing industry and hundreds of papers, books, and computer experiments. After exploring different sets of Lifelike rules, some scientists have suggested that, given a large enough array of cells in random states, and given a long enough time, very complicated, self-replicating entities would merge. They might even evolve to produce intelligent societies that develop and compete. It would be hard not to call such entities "alive."

If you believe that only flesh and blood can support consciousness, then you are probably wondering how Dr. Googol could consider cellular automata alive—even the supercomplex cellular entities evolving on huge checkerboard worlds. To his way of thinking, there's no reason to exclude the possibility of nonorganic sentient beings. If our thoughts and consciousnesses do not depend on the actual substances in our brains but rather on the structures, patterns, and relationships among parts, than the automata "beings" could think. If you could make a copy of your brain with the same structure but using different materials, the copy would think it was you.

CELLS THAT LIVE FOREVER

Now let's consider a cellular automaton developed by cellular-automata pioneer Stanislaw Ulam. Although the automaton grows according to certain rules, it differs from the Game of Life because the Ulam automaton has no rules for death. The rules dictate that any configuration will grow without limit as time progresses. Once a cell is on, it lives forever.

If Dr. Googol represents the two states of growth by 0 and 1, the fate of a single 1, isolated amid 0s, is interesting to watch. In fact, you can simulate what happens on a sheet of paper. A 5-by-5 grid suffices to demonstrate the first 2 generations of growth. A black circle represents a 1. An unfilled square represents a 0. Here is how it all starts:

The Ulam automaton is easy to set up, yet the behavior is intriguing. Given the nth generation, the $n+1$ generation arises from just 1 rule: a new cell is "born" (changes its state from 0 to 1) if it is orthogonally adjacent to 1 and only 1 living (1) cell of the nth generation. (*Orthogonal* implies the up, down, right, and left directions.) Thus, if the previous pattern is counted as generation 1, then generations 1 through 4 are easy to work out:

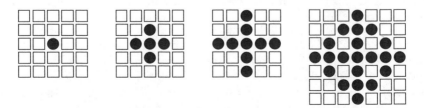

53.1 Ulam's automaton at generation 200.

What would the pattern look like at the 200th generation? The answer lies in the illustration in Figure 53.1.

It is particularly fascinating to watch this pattern grow. It shows fractal ambitions, each corner elaborating a square of its own. A close examination of its structure reveals a highly orderly tree structure in which each tiny black dot represents a cell in state 1. It is possible to travel from the center of the configuration to any black cell along a "branch" of black cells.

Dr. Googol's favorite cellular automaton is called Alien Snow. He invented this automaton, which has a time-dependent rule. Give a cell in state in the nth generation, the cell will enter state 1 if,

1. when n is even, the cell is orthogonally adjacent to exactly 1 cell in state 1;

2. when n is odd, the cell *touches* exactly 1 cell in state 1.

By *touches* Dr. Googol means that the cell is adjacent to the cell in state 1 along *either* an edge or a corner. The rule could be framed in the form of an algorithm:

Alien Snow Algorithm
for each pair (i,j)
if A(i,j) = O then
    if n even then add up 4 neighbors A(k,l)
           else add up 8 neighbors A(k,l)
    if sum = 1 then B(i,j) ← 1 else B(i,j) ← 0
for each pair (i,j)
A(i,j) ← B(i,j)

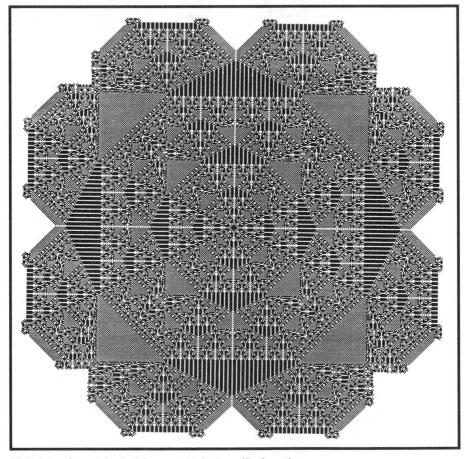

53.2 *Alien Snow started from a single "seed" at center.*

53.3 A hexacylcic version of the Alien Snow cellular automaton.

(A is the current array used for plotting, and B is a temporary array used to hold new cell states. The ← symbol indicates an assignment. Even cycles use orthogonal neighbors. Odd cycles use all 8 neighbors.)

The Alien Snow algorithm uses the statement "for each pair (i,j)" to indicate a double loop in a computer program. The size of the loop will depend on the dimensions of your display as measured in pixels or some other graphic element. The statement "add up neighbors $A(k,l)$" refers to a looped or direction enumeration of the cells in each of the two kinds of neighborhood. In the case of even cycles, the neighborhood of nearby cells will have values (k, l) equal to ($i, j-1$), ($i, j+1$), ($i-1, j$), ($i+1, j$). In odd cycles, k and l will vary from $i-1$ to $i+1$ and from $j-1$ to $j+1$, respectively. The same basic program structures can be used to produce other cellular automata with similar neighborhoods.

The illustration in Figure 53.2 shows what happens when the Alien Snow rules are applied to a single 1 in the center of the screen. Note the elaborate festoons and barred patterns that predominate.

A second variation on Alien Snow can be found in Figure 53.3. Here, Dr. Googol has used the same time-dependency rule but with a delayed cycle. On the *n*th generation, the cellular automata use the orthogonal neighborhood provided that *n* is a multiple of 6 (i.e., *n* = 0 mod 6). If not, the cellular automata use the full neighborhood of 8 neighbors for each cell. One would think that if this second variation on Alien Snow spends the great majority of its time behaving like the original, unmodified Alien Snow, then it would generate patterns that resembled it much more closely.

❀ For other experiments with Alien Snow, see "Further Exploring."
▣ See [www.oup-usa.org/sc/0195133420] for program hints.

Chapter 54

Beauty, Symmetry, and Pascal's Triangle

A mathematician is someone who can take a cup of coffee
and turn it into a theory.

—*Paul Erdös*

Dr. Googol was climbing Cheops's pyramid in Egypt when he became mesmerized by the triangular faces created by row upon row of large, rectangular bricks. He began to imagine a number painted on each face of the pyramid as a grin of pure delight lit up his face. He was dreaming of Pascal's triangle—one of the best-known integer patterns in the history of mathematics. The famous mathematician Blaise Pascal was the first to write a treatise about this progression of numbers, in 1653—although the pattern had been known by Omar Khayyam as far back as A.D. 1100. The first 7 rows of Pascal's triangle can be represented as

```
1                                                    1
1  1                                              1   1
1  2   1                                        1   2   1
1  3   3   1            or                   1   3   3   1
1  4   6   4   1                           1   4   6   4   1
1  5  10  10   5   1                     1   5  10  10   5   1
1  6  15  20  15   6   1              1   6  15  20  15   6   1
```

Take a look at the triangle at right. You can see that each entry, other than the 1s, is the sum of the 2 numbers immediately above. For example, to get the 2 in the third row, we add the two 1s above in the second row. This pattern continues indefinitely. Do you think there are any rows that have all odd entries?

There are infinitely many fascinating patterns in the triangle. For example, start at any 1 at the left side and go along the diagonal, and you'll find that the sum is a Fibonacci number (see Chapter 71 for information on the Fibonacci sequence: 1, 1, 2, 3, 5, 8, 13, . . . where each number is the sum of the previous two). If it's hard to see how to create a diagonal, pick a 1 at left, move 1 number to the right, and then move up. For example, in this figure add the underlined numbers 1 + 5 + 6 + 1 = 13. Try some others: 1 + 4 + 3 = 8, 1 + 3 + 1 = 5, 1 + 2 = 3. (The role that Pascal's triangle plays in probability theory, in the expansion of equations of the form $(x + y)^n$, and in various number theory applications has been discussed extensively by Martin Gardner—see "Further Reading.")

Mathematician Donald Knuth indicated that there are so many relations in Pascal's triangle that when someone finds a new identity, there aren't many people who get excited about it anymore, except the discoverer. Many researchers have found fascinating geometric patterns in the diagonals, discovered the existence of perfect square patterns with various hexagonal properties, and extended the triangle and its patterns to negative integers and higher dimensions.

Computer graphics is a good method by which patterns in Pascal's triangle can be made obvious. The figures in this chapter represent Pascal's triangle computed with modular arithmetic. For example, Figures 54.1 and 54.2 are Pascal's triangles, mod 2; that is, points are plotted for all even numbers occurring in the triangle. (Figure 54.2 is a photographic negative of Figure 54.1.) Figure 54.3 is the triangle, mod 3. Using the BASIC programming language, you can create the even/odd triangle by scanning all entries in the triangle using the conditional "IF I MOD K = 0", where I is the numerical entry in Pascal's triangle and $K = 2$. The "Smorgasboard" section at [www.oup-usa.org/sc/0195133420] includes computational hints. Patterns computed in this way reveal a visually striking and intricate class of patterns that make up a family of regular fractal networks. The patterns are self-similar fractals; that is, if we look at any one of

54.1 Pascal's triangle mod 2. The arrows indicate a size change in the central triangles every k^m rows ($m = 0, 1, 2, 3 \ldots$) where k is the mod index (in this figure, $k = 2$). The size-change relation holds for all triangles mod p where p is a prime number. The arrows shown indicate 2^3, 2^4, 2^5, 2^6, and 2^7.

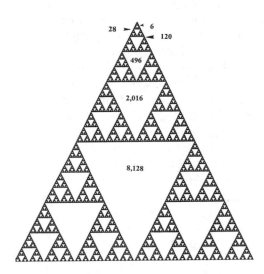

54.2 Photographic negative of Figure 54.1. The numbers on the figure indicate the number of dots that make up each triangle in the central stack. All perfect numbers appear in this central pattern.

the triangular motifs within Pascal's triangle we notice that the same pattern is found at another place in another size. (These patterns are also called Sierpinski gaskets, as discussed in the "Further Reading" references.)

In Figure 54.1 we observe that the central triangles undergo a size change (starting at the top triangle with 1 dot) every 2^m rows where m is an integer. If you plot these triangles for other modulus numbers, you'll find that the higher the modulus index k the more intricate and harder-to-define are the symmetries. Figure 54.4 shows Pascal's triangle for $k = 666$.

By visually familiarizing oneself with Pascal's triangle for various modulus indices, it is possible at a glance to determine the prime factors of k for many Pascal's triangles. (For training methods, see the various Pickover references.) Also notice that if you were to count the number of dots in the central triangles starting from the top of Figure 54.2, you would find that each is made up entirely of an even number of dots. At the top is 6, then 28, 120, 496, . . . dots. 6, 28, and 496 are perfect numbers because each is the sum of all its divisors excluding itself (6 = 1 + 2 + 3). The formula for the number of dots in the nth central triangle, moving along the central axis, is $2^{n-1}(2^n - 1)$. Because every even perfect number is of the form $2^{n-1}(2^n - 1)$, where $2^n - 1$ is prime, all even perfect numbers appear in the central stacked triangular pattern in Figure 54.2. Look closely. Can you find other patterns in the triangle?

Not only are the patterns pretty to look at, but these kinds of self-similar patterns have been discovered and applied in condensed matter physics, diffusion, polymer growth, and percolation clusters. One example given by Professor Leo Kadanoff is petroleum-bearing rock layers. These typically contain fluid-filled pores of many sizes, which, as Kadanoff points out, might be effectively understood as Sierpinski gaskets. These figures also have a practical importance in that they can provide models for materials scientists to produce new structures with novel properties. For example, several scientists have created wire gaskets on the micron-size scale almost identical to the mod 2 structure in Figure 54.2. The area of their smallest triangle was 1.38 ±0.01 μm^2, and researchers have investigated many unusual properties of their superconducting Sierpinski gasket network in a magnetic field.

❀ For other wonderful examples of practical fractals—such as fractal antennas, reaction chambers, Internet traffic, and optical devices —see "Further Exploring."

🖬 See [www.oup-usa.org/sc/0195133420] for Pascal triangle program hints.

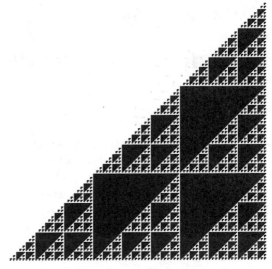

54.3 Pascal's triangle mod 3.

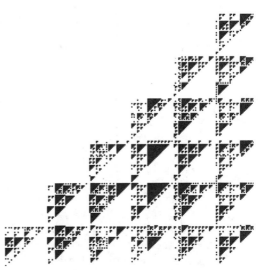

54.4 "Pascal's Beast"—Pascal's triangle mod 666.

Chapter 55

Audioactive Decay

> If we wish to understand the nature of the Universe we have an inner hidden advantage: we are ourselves little portions of the universe and so carry the answer within us.
> —*Jacques Boivin,* The Heart Single Field Theory

Dr. Googol was walking along a picturesque street in the German hinterland when he suddenly came upon a small boy. He handed the boy a slip of slightly soiled paper.

"*Was ist das?*" the boy asked.

On the paper was the following sequence of numbers:

Row	Sequence
1	1
2	1 1
3	2 1
4	1 2 1 1
5	1 1 1 2 2 1
6	?

Dr. Googol smiled at the boy and asked if he could guess the numerical entries in the next row. The boy looked at the paper for a few seconds and said, "This is the Gleichniszahlen-Reihe Monster. But I'm sorry. I don't know what comes next. And anyway, I'm not supposed to talk to strangers."

The boy went running down the street.

If you do not read German, the boy's comment may conjure up visions of a strange animal from a science fiction movie. However, "the Gleichniszahlen-Reihe Monster" refers to a number sequence with some rather strange and compelling properties, and the German name will be explained shortly. Because the sequence never seems to contain a number greater than 3, you don't need sophisticated computers to begin exploring.

You probably can't guess the numerical entries for the next row. However, the answer is actually simple, when viewed in hindsight. To appreciate the answer, it helps to speak the entries in each row out loud. Note that row two has two "ones," thereby giving the sequence 2 1 for the third row. Row three has one "two" and one "one." Row four has one "one," one "two," and two "ones." From this, an entire sequence can be generated. This interesting sequence was described in a German article, where Mario Hilgemeier called it *die Gleichniszahlen-Reihe*, which translates into English as "the likeness sequence." To the best of my knowledge, the sequence first appeared in the early 1980s, at an international student competition held in Belgrade, Yogoslavia. The sequence was also extensively studied by mathematician John H. Conway, who called the growth process "audioactive decay." The sequence grows rather rapidly. For example, row 16 is:

```
13211321322113311213211331121113122112132113121113222112
311311222211311123113321112132113222113121113211
```

Row 27 contains 2,012 entries (see Table 55.1).

```
311311222211311123113321112132113111231121113311211131221121321131211113
222112311311221112132112311311222112111331121113112221121113122112132113
211213211321322123211211131211121332211231131112231121113311211131221121
121113222112311311221113111231133221121113311211131122112111321122112321113
322112111312211312111322212321121111312111213322112132112132132211211231131322
112111312212321121113122122211211123222112311311221113111231133211121113113
211311231121113311211113131221121321131211132221232112111312211213211331211231
122111312121113221231311221121111312211312111322211213211321322123113121132132211321
322112111312212221211211232221121311312211312111231133321112132113213221123311121
2111321121111312212131121132211231131112221133211213211321322112132113232211231311
221112132112311311222112111312211213211322221322112131211231311231121113221121231
332111311231222112131121113113322113111231222112111312211312111231133221121113122112
311311222113111231133221112132113213221211131211231132211221113122121322113111121311
221112131211231221212211121321132132221112132113232221211212321132131321321113211312112132113312
221121321131211132221123311211232211331112112131332211121311221311321322113213112111311
221112131132111312212131121132211231131112221133211213211321322113121211231131322112
1321321132132131211123222112132113213221231113111231221121113112221121321131211132221232112132
111312211312111322211331121321322123113131122211311231133211121113211
31221123311312111322211231113122211311312211312111312211213211321322112132
1132132211331122211332111213322112132113213221123112311221112131122113221123113221113122121321
111312211312111322211232112111312111213322112132112132112311131211231131322112132113213213221
131213211121133322113311231222112131121113113322112131321221123311312111331112112131332211213211
131211112311211231132211221113122121322113113221112312112311311222112132113312
311311222112311231133221112111312211311321322113211231121113221121321131211132221232112131213211
13213211121132211212311331121311121131211231131222113131221211311321122113221122311311121131221
121321132113213221231131121211121322211331121321322123113112122113122211331121321132211232211311321322112132113213221
```

Table 55.1 Likeness sequence for row 27.

If you were to study the sequence carefully you would find a predominance of 1s, with 2s and 3s less common. For rows between 6 and 27, there are about 50% 1s, 30% 2s, and 20% 3s. As Hilgemeier proved, the largest number the sequence contains is a 3. Is it possible to prove that 3-3-3 can never occur? Dr. Googol has looked for three 3s in a row up to row $r = 33$, which has over 10,000 entries. You can see from the following representation of row 11 (in which 3s are represented by circles in squares) that 3's occurrence seems erratic, like lost ships on an infinite sea:

Wouldn't it be fun to sail on such a sea, sipping from a good bottle of wine, searching for adventure amidst the chaos?

In this chapter, Dr. Googol is particularly interested in the *distribution* of 1s, 2s, and 3s. While you can simply compute the percentage of occurrence of each digit for a given row, this does not tell us anything about any interesting clusters or peculiar areas of concentration of one digit over another. To overcome this drawback, Dr. Googol transforms the digit strings into 2-dimensional patterns that characterize the sequence. A single digit is inspected and assigned a direction of movement on a plane. To visualize this (and other) ternary sequences, use a 3-way vectorgram where the occurrence of a 1 directs the trace one unit at 0°, a 2 causes a walk at 120°, and a 3 a walk at 240°. Each of these angles is with respect to the x axis. Figures 55.1 to 55.3 show patterns for row 15, containing 102 digits, row 25, containing 1,182 digits, and row 33, containing over 10,000 digits. Dr. Googol used different scales to fit the graphs on a page. Notice that if the string contained only 1s, the walk would be only to the right.

As you can see from the figures, the sequence is far from random. The upward diagonal trend in Figure 55.2 and Figure 55.3 indicates a mixture of predominantly 1s, some 2s, and relatively few 3s. The fact that the trends are fairly linear

55.1 Vectorgram for row 15.

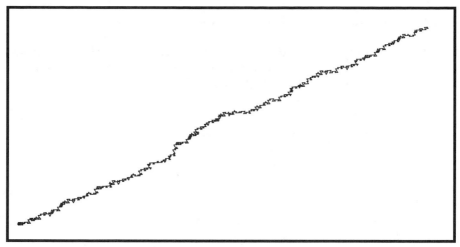

55.2 Vectorgram for row 25.

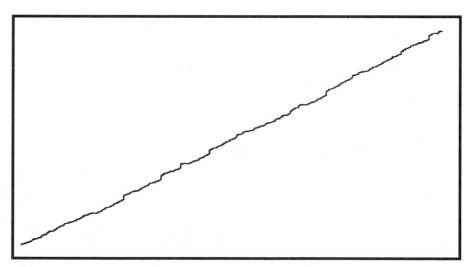

55.3 Vectorgram for row 33.

suggests that the ratios are relatively constant throughout the row. Figure 55.2 and 55.3 show the occurrence of sudden upward bumps, which eventually return to the diagonal baseline. These bumps indicate a temporary change in the trend to more 2s.

You can understand the resulting patterns by considering the directions traveled by various combinations of entries in the sequence. For example, the sequence 1-1-1 is totally *x*-directed. 1-2-3, 1-3-2, and various cyclic permutations return to the original point (as if traveling along the edges of a △).

For future experiments, you may wish to compute the slope of the vectorgram's mean (trend) line as a function of the row number, or make a plot of the slope of the mean line versus the number of entries in a row. It appears, from just a few sample points, that the slope of the mean line increases as a function of row number.

Dr. Googol hopes that you will uncover or solve additional mysteries with this unusual sequence. If you are interested in the use of 8-way vectorgrams in the characterization of genetic sequences, see "Further Reading."

❀ Want a quick way to determine how many digits the nth term in this sequence has? Want to know the largest likeness sequence ever computed? Want to learn about other related sequences? See "Further Exploring."

Chapter 56

Dr. Googol's Prime Plaid

The Universe is a grand book which cannot be read until one first learns
to comprehend the language and become familiar with the characters in
which it is composed. It is written in the language of mathematics.
—*Galileo,* Opere Il Saggiatore

Dr. Googol was walking down a back road in the beautiful Scottish countryside, wearing a dapper plaid kilt, when he started contemplating various patterns produced by prime numbers. He sat down on a bench at Ardoe House, his hotel, which was a few miles outside Aberdeen. As he gazed longingly at the Scottish baronial hall with turrets, heraldic inscriptions, and ornate ceilings, he began to draw dots on a piece of paper.

A prime is a positive integer that cannot be written as the product of 2 smaller integers. The number 6 is equal to 2 times 3; therefore it is not prime. On the other hand, 7 cannot be written as a product of factors; therefore, 7 is a prime number or prime. Here are the first few prime numbers: 2, 3, 5, 7, 11, 13, 17, 19, 23, 29, 31, 37, 41, 43, 47, 53, 59. Notice that the gaps between successive prime numbers varies; for example, in these first few primes, the gaps are 1, 2, 2, 3, 2, 4, 2, 4, 6, 2, 6, 4, 2, 4, 6, 6. . . . . The Greek mathematician Euclid proved that there are an infinite number of prime numbers. But these numbers do not occur in a regular sequence, and there is no formula for generating them. Therefore, the discovery of large new primes requires generating and testing millions of numbers. (See Chapter 76's "Further Exploring" for some of the largest known prime numbers and how they were calculated.)

Consider the prime numbers p_i, where i = 0,1,2,3, . . . and where p_0 = 2, p_i = 3, etc. Dr. Googol made a plot of p_i vs. p_{i+1} for p_i < 2,000 (not shown) that yielded a "dusty" (approximately) diagonal line with a slope of about 1. It is dusty because there are gaps in the prime number sequence, and roughly diagonal since p_i is roughly equal to p_{i+1} at the size scale of the plot. Try making this plot yourself.

Dr. Googol generated a visually interesting "plaid" structure (Figure 56.1) by using different shift values a and superimposing plots of p_i vs. p_{i+a} where a = 1, 2, 3, . . . , 200 for p_i < 2,000. The bottom diagonal edge of the plaid corresponds to p_i vs. p_{i+1}. The gaps indicate gaps in the prime number sequence.

As we go to larger and larger integers, the primes become increasingly rare, so the plaid also becomes more diffuse. When Dr. Googol attempts to compute a fairly good approximation for the number of primes smaller than or equal to x, usually designated $\pi(x)$, he prefers to use $\pi(x) \sim x/(\ln x - 1.08366)$. (This formula, given by Legendre in 1778, is much simpler to implement on a computer than other methods, like the Gauss and Riemann methods, although this Legendre formula should be used only for prime numbers less than 5 million. Above 5 million, Legendre's formula becomes less accurate.)

Dr. Googol computed the prime numbers needed to form the plaid pattern in this chapter in just a second or 2, using the Sieve of Eratosthenes method.

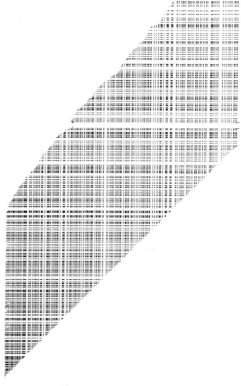

56.1 A prime plaid. The x axis is P_i. The y axis is P_{i+a}.

⚛ For other odd facts about prime numbers, see "Further Exploring."

▣ See [www.oup-usa.org/sc/0195133420] for a computer program that generates prime numbers.

Chapter 57

Saippuakauppias

> I think that modern physics has definitely decided in favor of Plato. In fact the smallest units of matter are not physical objects in the ordinary sense; they are forms, ideas which can be expressed unambiguously only in mathematical language.
>
> —*Werner Heisenberg*

Dr. Googol was in Helsinki, Finland, when he glanced at a local newspaper and saw a curious-looking word:

> **saippuakauppias**

Dr. Googol turned toward the tall blond woman beside him. "Madam, can you tell me what this wonder word means?"

Her eyebrows raised. "Certainly. It is the Finnish word for 'soap dealer.'"

"Wonderful!"

"Sir, why are you so delighted by this word?"

"Because it is the largest palindrome I have ever seen in any language!"

Dr. Googol merrily walked away.

A *palindrome* is usually defined as a word, sentence, or set of sentences that spells the same backward and forward. Dr. Googol doesn't think there are any common English words of more than 7 letters that are palindromic. Examples of 7-letter palindromes are *rotator* and *reviver*. An interesting example of a palindromic sentence in which words, not letters, are the units is:

> **"You can cage a swallow, can't you,
> but you can't swallow a cage, can you?"**

In this chapter Dr. Googol is more interested in palindromic *numbers* than palindromic words or sentences. Palindromic numbers are positive integers that "read" the same backward or forward. For example, 12,321, 11, 261162, and 454 are all palindromic numbers.

Figure 57.1 is an interesting plot showing the distribution of the first 200 palindromes when multiplied by a constant. To create the plot, start with an integer x between 1 and 200, multiply it by a constant a, and determine if the result is a palindrome. The "multiplier" a on the y axis of the plot goes from 1 to 200. A dot on the graph indicates a palindromic number. The various patterns produced are quite interesting, and Dr. Googol is fond of making a few casual observations. Note that there is clearly a dense structure below some "hyperbol-

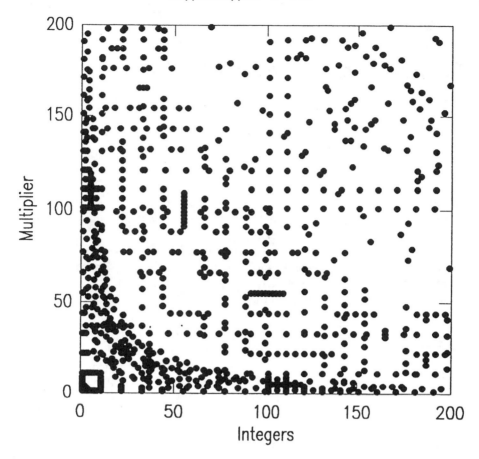

57.1 Distribution of palindromes. The horizontal axis gives the integers x, and the vertical axis indicates the integral multiplier a. A dot on the graph indicates that a × x is palindromic.

ic" boundary. There is a conspicuous vertical line of closely spaced dots at $x = 55$ corresponding to 10 consecutive odd a values that produce palindromes. The products are 55×91, 55×93, 55×95, 55×97, 55×99, 55×101, 55×103, 55×105, 55×107, and 55×109. Also, when the x-axis value is an even multiple of 5, there are no y data. When the x-axis value is a nonpalindromic odd multiple of 5, the y data are scarce. When x is palindromic, there are many y-data points. Notice the plot has symmetry: if $x \times y$ is palindromic, $y \times x$ is also palindromic.

Can you find other patterns in this plot? Can you extend this to a 1,000-by-1,000 plot?

❀ See "Further Exploring" for more fun palindromic sentences and for some wild challenges.

Chapter 58

Emordnilap Numbers

There is no excellent beauty that hath not some strangeness in the proportion.
—*Francis Bacon*

About a year ago, Dr. Googol was lecturing a class at Harvard University. "I want someone in the audience to pick any integer, reverse its digits, add the 2 numbers together, and continue to reverse and add."

A boy with punk hair and a pierced nose raised his hand. "Sir, I'll start with 19. I reverse its digits to make 91 and then add. 19 + 91=110. I reverse the digits of 110 to make 011, and then I add. 110 + 011 = 121."

Dr Googol stomped his foot on the floor. "Yes!"

"Sir, excuse me?"

"You just ended up with a palindromic number—that is, the number reads the same in both directions. With some numbers, this happens in a single step. For example, 18 + 81 = 99, which is a palindrome. Other numbers may require more steps. This process of reversing, adding, and looking for palindromes (also called an Emordnilap process) is quite wonderful. Of all the numbers under 10,000, only 249 fail to form palindromes in 100 steps or less. In 1984, Fred Gruenberg noted that the smallest number that *seems* never to become palindromic by this process is 196. (It has been tested through hundreds of thousands of steps.)"

"Sir, have you done tests yourself?"

"Certainly. Moreover, I have tested the starting number 879 for 19,000 steps, producing a 7,841-digit number—with no palindrome resulting. Isn't that impressive? The 7,841-digit number starts with the digits 58084187 . . . and ends with . . . 139075! My statistical tests indicate an approximately equal percent occurrence of digits 0 through 9 for this large number. Similarly, I have tested 1,997 for 8,000 steps, with no palindrome occurring."

The class loudly applauded Dr. Googol's mathematical accomplishments.

❀ ❀ ❀

Are there any patterns underlying this reverse-and-add process? Can we make any predictions? The number of steps needed to make a palindrome (called the "path length" and represented by p) is often under 5 steps. Figure 58.1 shows all path lengths for starting integers n between 1 and 1,000. To produce a convenient graphical representation, Figure 58.1 is truncated in the y-axis direction; in particular, the search for palindromes is stopped after 25 steps. Notice the interesting periodicity in the path lengths made apparent in the graph. Also notice that while patterns exist, they are not perfect or entirely regular. A *power spec-*

trum can be computed from a mathematical method called the Fourier transform in order to quantify periodic patterns.

The graph poses dozens of questions that are more difficult to answer. For example, why are the periodic large path lengths absent in the 400–500 integer range (Figure 58.1)? Also, if we were to list the palindrome values for the moderate-size path lengths, we would find a high percentage of occurrence of the digit 8. Why 8? Table 58.1 shows the palindromic end points for some of the moderate-size path lengths for the first 300 starting integers.

Finally, you may wish to look for patterns for larger starting integers. For example, the path-length graph corresponding to

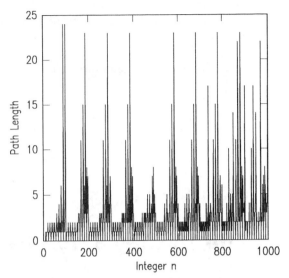

58.1 Path lengths for the first 1,000 starting integers. To produce a convenient graphic representation, the figure is truncated in the *y*-axis direction by stopping the search for palindromes after 25 steps).

Figure 58.1 for (1000 ≤ *n* ≤ 10000), while displaying similar interesting periodic patterns, looks quite different. There are many fewer 0-length paths because there are fewer starting palindromes. There are various gaps and peaks. The resultant graph is left as a curious exercise for you. For those of you who wish to learn more about this palindrome problem, see Martin Gardner and Charles

n	Palindrome	Path Length
9	8813200023188	24
98	8813200023188	24
167	88555588	11
177	8836886388	5
187	8813200023188	23
266	88555588	11
276	8836886388	15
286	8813200023188	23

Table 58.1 Palindromic end points for some of the moderately-sized path lengths.

Trigg in "Further Reading." Gardner also discusses the problem for other number systems (e.g., binary numbers).

❀ For just a smidgen more mathematical analysis, see "Further Exploring."

Chapter 59

The Dudley Triangle

> One cannot escape the feeling that these mathematical formulae have an
> independent existence and an intelligence of their own, that they are
> wiser than we are, wiser even than their discoverers, that we get more out
> of them than we originally put in to them.
>
> —*Heinrich Hertz*

After studying Pascal's triangle in Chapter 54, Dr. Googol became interested in other infinite triangular arrays. He spent many hours contemplating the beauty and intricacy of the less-known and less-understood Dudley triangular array, proposed in 1987 and represented as follows:

```
                              2
                           2     2
                        2     1     2
                     2     0     0     2
                  2     6     5     6     2
               2     6     4     4     6     2
            2     6     3     2     3     6     2
         2     6     2     0     0     2     6     2
      2     6     1     9     8     9     1     6     2
   2     6     0     8     6     6     8     0     6     2
2     6    12     7     4     3     4     7    12     6     2
2  6   12    6     2     0     0     2     6    12     6     2
2  6   12    5     0    12    11    12     0     5    12     6     2
```

Can any of you figure how this triangle was generated? Study it before reading further. Is there any human on Earth who could write down the next row of the triangle without reading the next paragraph?

In 1987, Dr. Underwood Dudley conducted extensive research on this triangle. We can denote the location of each array element by its diagonal coordinates (m,n), where m signifies the mth diagonal descending left to right and n signifies the nth diagonal descending right to left. Every value in the array a is in the range from 0 to the sum of its coordinates, $m+n$. One way the array can be reproduced is by the following formula:

$$a_{m,n} = (m^2 + mn + n^2 - 1) \bmod n + m + 1$$

Try experimenting with different values for m and n. The mod function, or modulo function, yields the remainder after division. A number $x \bmod n$ gives the remainder when x is divided by n. This number is anywhere from 0 to $n-1$. For example, 200 mod 47 = 12 because 200/47 has 12 as a remainder.

Like Pascal's triangle represented graphically in Figure 54.1, the Dudley triangle is bilaterally symmetric. (That means a mirror plane could be drawn down the center of the triangle.) Notice that the triangle's values grow more slowly than those of Pascal's triangle and that the Dudley triangle has fewer odd-valued entries. Figure 59.1 shows the positions of even entries.

Can you find any other marvelous patterns in the Dudley triangle? Experiment! Search for structure and rapid ways to generate the triangle. Can you extend the triangle to a 3-dimensional pyramid? See the references in "Further Reading" for more information on the properties of this triangle.

59.1 Dudley's triangle mod 2.

Chapter 60

Mozart Numbers

> Of course, we would like to study Mozart's music the way scientists
> analyze the spectrum of a distant star.
> —*Marvin Minsky,* Computer Music Journal

Dr. Googol was listening to his favorite Mozart piece, Symphony no. 40 in G
minor, while contemplating mathematics. As the mellifluous music filled the air
like a fragrant scent, he soon realized that in order to estimate any Mozart symphony
number S from its Köchel number K you can use

$$S = 0.027465 + 0.157692 \times K + .000159446 \times K^2$$

(The Köchel catalogue is a chronological list of all of Mozart's works, and any
work of Mozart's may be referred to uniquely by its Köchel number. For example,
the Symphony no. 40 in G minor is K.550.) The formula will give an answer
not more than 2 off, 85% of the time.

Mozart once wrote a waltz in which he specified 11 different possibilities for
14 of the 16 musical bars of the waltz, and 2 possibilities for another bar. How
exciting that Mozart gave us such freedom! This gives 2×11^{14} variations of the
waltz. What percentage of the number of these waltzes have humans heard?
What percentage of the waltzes *could* a human hear in a lifetime?

For more information on the formula for Mozart symphony numbers, see
"Further Reading."

Chapter 61

Hyperspace Prisons

Wise Mystic: What is the best possible question,
and what's the best answer to it?
Dr. Googol: You've just asked the best possible question,
and I'm giving the best possible answer.

He showed me a little thing, the quantity of a hazelnut, in the palm of
my hand, and it was round as a ball. I looked thereupon with the eye of
my understanding and thought: What may this be? And it was answered
generally thus: It is all that is made.
—*Julian of Norwich, 14th century*

Dr. Googol enjoys simple-looking geometrical puzzles that require you to estimate the number of overlapping triangles within a diagram such as the one in Figure 61.1a. Can you guess how many triangles are in this figure? Stop. Take a guess before reading further. This figure contains a walloping 87 triangles.

Sometimes it is possible to come up with rules that specify the number of triangles in an ever-growing sequence of diagrams, such as the sequence in 61.1b. Impress your friends with your ability to compute the number of triangles in the nth triangular figure: $[n(n + 2)(2n + 1)]/8$, for even n, and $[n(n + 2)(2n + 1) -1]/8$ for odd n.

Can you count the number of triangles in Figure 61.1c, a more difficult diagram? Actually, this figure will consume too much of your time; let Dr. Googol give you the answer—653 triangles—so that you will be free to ponder the more interesting enigmas that follow. Why not give these 3 triangle puzzles to a friend to ponder?

One August, while catching fireflies in a jar, Dr. Googol began to develop puzzles of a similar geometrical variety, and he calls them "flea cages" or "insect prisons" for reasons you will soon understand. He enjoys these flea cages because

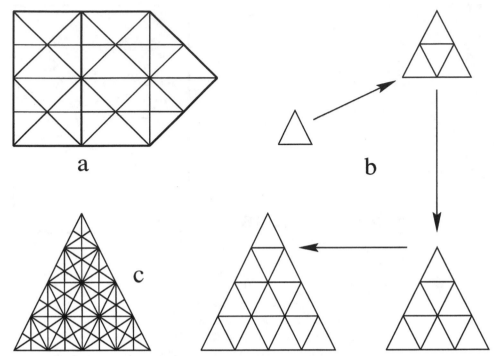

61.1 Triangle madness. (a) How many overlapping triangles are in this figure? **(b)** Can you determine a rule that gives the number of triangles in the nth figure in this sequence? **(c)** How many more triangles does this figure have than the one figure in (a)?

they are simpler to analyze than the triangle figures. Also, since the figures consist of a network of perpendicular lines, they are much easier for you (or your computer program) to draw. Consider a lattice of 4 squares that form 1 large square (Figure 61.2).

How many rectangles and squares are in this picture? Think about this for a minute. There are the 4 small squares marked "1," "2," "3," and "4," plus 2 horizontal rectangles containing "1 and 2" and "3 and 4", plus 2 vertical rectangles, plus the 1 large surrounding border square. Altogether, therefore, there are 9 4-sided overlapping areas. The *lattice number* for a 2-by-2 lattice is therefore 9, or $L(2) = 9$. What is $L(3)$, $L(4)$, $L(5)$, and $L(n)$? It turns out that these lattice numbers grow very quickly, but you might be surprised to realize just how quickly. The formula describing this growth is fairly simple for an *n*-by-*n* lattice: $L(n) = n^2(n + 1)^2/4$. The sequence is 1, 9, 36, 100, 225, 441, ... For a long time, Dr. Googol liked to think of the squares and rectangles (quadrilaterals) as little containers or cages in order to make interesting analogies about

61.2 How many overlapping quadrilaterals does this figure contain?

how the sequence grows. (Obviously they wouldn't really make very desirable cages, because they overlap, but even Dr. Googol can dream.) For example, if each quadrilateral were considered a cage that contained a tiny flea, how big a lattice would be needed to cage 1 representative for each different variety of flea (Siphonaptera) on earth? To solve this, consider that siphonapterologists recognize 1,830 varieties of fleas. Using the equation Dr. Googol has just given you, you can determine that a mere 9-by-9 lattice could contain 2,025 different varieties, easily large enough to contain all varieties of fleas. (For *Siphonaptera* lovers, the largest known flea was found in the nest of a mountain beaver in Washington State in 1913. Its scientific name is *Hystirchopsylla schefferi*, and it measured up to 0.31 inches in length, about the diameter of a pencil.)

It is possible to compute the number of cages for 3-D cage assemblies as well. The formula is $L(n) = ((n^3) (n + 1)^3)/8$. The first few cage numbers for this sequence are 1, 27, 216, 1000, 3375.

Can you determine the number of cages for *4-dimensional* assemblies?

How many cubes in a 3-D cage assembly would you need to contain 1 of each species of insect on Earth today? To contain all the people on Earth?

❀ See "Further Exploring" for further analyses and information on amazing *4-dimensional* cages.

Chapter 62

Triangular Numbers

> Au fond de l'Inconnu pour trouver du nouveau. (Into the depths
> of the Unknown in quest of something new.)
> —*Charles Baudelaire*, Le Voyage

Dr. Googol was lecturing the Spice Girls, a famous all-girl British rock band popular in the late 1990s. The sun shone brightly as they sat together on a bench beside Abbey Road.

"Let's talk about triangular numbers," Dr. Googol says to Baby Spice, the blond-haired woman in the band. (Dr. Googol speculates she received her nickname because of her innocent, youthful appearance.)

She casually flicks her hair to the side. "A triangular number?"

"Yes." Dr. Googol drops his voice half an octave and assumes a professorial demeanor. "Triangular numbers form a series, 1, 3, 6, 10, . . . , corresponding to the number of points in ever-growing triangles." He takes a piece of chalk and sketches an array of triangular dots on Abbey Road:

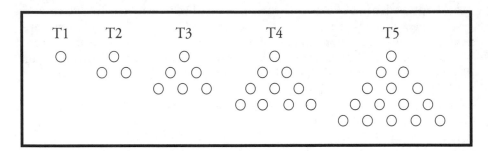

"The early Greek mathematicians noticed that if groups of dots were used to represent numbers, they could be arranged so as to form geometric figures such as these."

Baby Spice nods. "Incredible, sir. The possibilities are endless. The fourth triangular number is 10. I wonder what the 100th triangular number is?" She begins to count using her fingers.

"Baby Spice, there's an easier way. The nth triangular number is given by a simple formula: $n(n + 1)/2$. The variable n is called the *index* of the formula. If you want the 100th triangular number, just use $n = 100$ for the index. You'll find that the answer is 5,050."

Perhaps Dr. Googol detects admiration in the Spice Girls' eyes, no doubt elicited by his mathematical prowess.

"Sir, can we use a computer to determine the 36th triangular number?"

Next to Dr. Googol is a marble statue of Paul McCartney. He reaches into the statue's stomach, where he has secretly stashed a notebook computer. A hinged door swings out, and he removes the computer and tosses it to Baby Spice.

Unfortunately, his aim is inaccurate, forcing the Spice Girls to make a leaping dive for the computer. They catch it but, in doing so, crash into a marble frieze running along the curb, with representations of Mick Jagger of the Rolling Stones and Celine Dion. Celine crashes down upon Baby Spice.

Baby Spice struggles to free herself of the horizontal Celine and brushes herself off. "Never mind, sir. My youthful appearance can't be hurt by marble." She begins to type furiously on the computer's keyboard with her well-manicured fingers. She hands Dr. Googol a computer printout:

Triangular Numbers:

1, 3, 6, 10, 15, 21, 28, 36, 45, 55, 66, 78, 91, 105, 120, 136, 153, 171, 190, 210, . . .

"Sir, I can't believe it! The 36th triangular number is 666—the Number of the Beast in the Book of Revelation." Baby Spice begins to quote from the Bible, "Here is wisdom. Let him that hath understanding count the number of the beast; for it is the number of a man, and his number is six hundred, three score, and six."

"Just coincidence, Baby Spice."

"And the 666th triangular number is 222,111. What a strange arrangement of digits!"

"Calm down, Baby Spice. It's just coincidence."

"Sir, did you know that each square number is the sum of 2 successive triangular numbers?"

"What are you getting at?" Dr. Googol's voice is low.

"Square numbers are numbers like $5 \times 5 = 25$ or $4 \times 4 = 16$. Every time you add 2 successive triangular numbers, you get a square one. For example, $6 + 10 = 16$."

Dr. Googol is intimidated by Baby Spice's mental agility, but then he quickly snaps back with a mathematical gem of his own: "Each odd square is 8 times a triangular number, plus 1." He begins to draw a grid of squares on Abbey Road. "Look at this." He points to the diagram (Figure 62.1).

Dr. Googol looks back at Baby Spice. "The Greek mathematician Diophantus, who lived 200 years after Pythagoras, found a simple connection between triangular numbers T and square numbers K. My diagram shows this graphically. It has 169 square cells in an array. This represents the square number $K = 169$ (13×13). One dark square occupies the array's center, and the other 168 squares are grouped in 8 triangular numbers T in the shape of 8 right triangles. I've darkened 1 of the 8 right triangles."

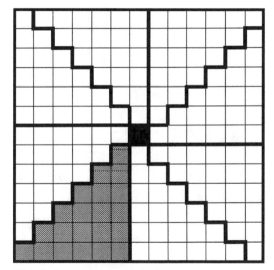

62.1 A deep connection between square numbers K and triangular numbers T. A visual proof that $8T + 1 = K$.

Baby Spice gasps, and the Spice Girls stare at one another. Dr. Googol feels as if Abbey Road is trembling with a minor earthquake.

"Sir," Baby Spice whispers with a trace of hesitation, "no wonder the Pythagoreans worshiped triangular numbers. You can find an infinite number of triangular numbers that when multiplied together form a square number. For example, for every triangular number T_n, there are an infinite number of other triangular numbers, T_m such that $T_n T_m$ is a square. For example, $T_2 \times T_{24} = 30^2$."

Dr. Googol slams his fist down, feeling a slight pain as it makes contact with the hot asphalt. He needs to outdo Baby Spice. He shouts back, "666 and 3,003 are palindromic triangular numbers. They read the same forward and backward."

Baby Spice starts singing the lyrics of her hit song "When Two Become One" as she types on the notebook computer. "It cannot be," she screams. "The 2,662nd triangular number is 3,544,453, so both the number and its index, 2,662, are palindromic."

Dr. Googol feels a strange shiver go up his spine as he looks into the rock star's glistening eyes. He feels a chill, an ambiguity, a creeping despair. The Spice Girls are still. No one moves. Their eyes are bright, their smiles relentless and practiced. Time seems to stop. For a moment, Abbey Road seems to fill with a cascade of mathematical symbols. But when he shakes his head, the formulas are gone. Just a fragment from a dream. But the infuriating Baby Spice remains.

"Baby Spice, I grow weary of our little competition."

"Sir, triangular numbers are fascinating. Are there other numbers like this? Pentagonal numbers? Hexagonal numbers? What properties might these have?"

"Baby Spice, that's the subject for another day."

❀ For other odd facts about triangular numbers, see "Further Exploring."

▦ See [www.oup-usa.org/sc/0195133420] for a computer program that generates triangular numbers.

Chapter 63

Hexagonal Cats

> Computers are useless. They can only give you answers.
> —*Pablo Picasso*

Many years ago, Dr. Googol was visiting a Middle Eastern museum. Outside, the villagers were gathered around dozens of primitive ocelot statues. One of the bearded men in the gathering began to meticulously arrange the new archeological findings on the hot sand amidst the parched and withered cacti. He arranged the cats in the shape of concentric hexagons, as shown below. After resting for a few minutes, the wizened man groaned, knelt down, and began to count the cats, starting from the center. He noted that there was 1 cat, surrounded by 6 cats, surrounded by 12, and so on:

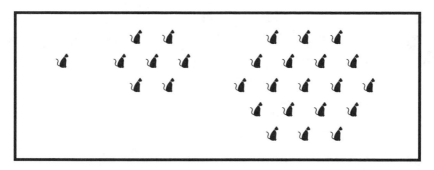

Dr. Googol stepped closer. "I can tell you how many cats there will be in each surrounding hexagonal layer."

The old man looked up. "If you do, we will be forever grateful."

Dr. Googol began his lecture and to sketch formulas in the sand.

Can you tell how many cats will be in each layer?

Before giving you the formula, here is some background to polygonal numbers, that is, numbers associated with geometric arrangements of objects. As you read in Chapter 62, the early Greek mathematicians noticed that if groups of dots were used to represent numbers, they could be arranged so as to form geometric figures, such as triangles, squares, and hexagons. For example, since 1, 3, 6, 10, and 15 dots can be arranged in the form of a triangle, these numbers are called triangular. (Polygonal numbers appeared in 15th-century arithmetic books and were probably known to the ancient Chinese, but they were of special interest to the Pythagoreans due to their mystical interest in the properties of such numbers.)

The sequence that Dr. Googol derived for the Middle Eastern men was $H_c = 3n(n-1) + 1$, $n = 1,2,3, \ldots$, which defines the *centered hexagonal numbers*. Let's go a step further and introduce a new term sure to impress your friends, and hopefully your next Friday-night date. A centered hexagonal number is called *centered hexamorphic* if its digits terminate its associated centered hexagonal integer. For example, $n = 7$ is centered hexamorphic because $H_c(7) = 127$. The number 17 is also centered hexamorphic because $H_c(17) = 817$. The centered hexamorphic sequence is fascinating to study! Table 63.1 contains a list of the first 23 centered hexamorphic integers. Note the interesting fact that all centered hexamorphic numbers end in the digits 1 and 7.

A convenient notation $a_5 = aaaaa$ can be used, where the subscript indicates the number of times the digit or group appears consecutively. Dr. Googol has found the following interesting infinite sequence: $H_c(50_k1) = 750_{k-1}150_k1$, $k = 0, 1, 2, \ldots$. Here the k subscripts indicate how many times the 0 is repeated. For example, $k = 2$ produces $H_c(5,001) = 75,015,001$ (see Table 63.1).

Centered hexagonal numbers have a different generating formula from standard hexagonal numbers: $H(n) = n(2n-1)$; (see Figure 63.1). On the other hand,

n	H(n)_{centered}	n	H(n)_{centered}
1	1	1251	4691251
7	127	1667	8331667
17	817	5001	75015001
51	7651	5417	88015417
67	13267	6251	117206251
167	83167	6667	133326667
251	188251	10417	325510417
417	520417	16667	833316667
501	751501	50001	7500150001
667	1332667	56251	9492356251
751	1689751	60417	10950460417
917	2519917		

Table 63.1 Centered hexamorphic numbers.

n	H(n)	n	H(n)
1	1	376	282376
5	45	500	499500
6	66	501	501501
25	1225	625	780625
26	1326	876	1533876
50	4950	4376	38294376
51	5151	5000	49995000
75	11175	5001	50015001
76	11476	5625	63275625
125	31125		

Table 63.2 Hexamorphic numbers.

the infinite sequences for hexamorphic and centered hexamorphic numbers are similar. For hexamorphic numbers, we have $H(50_k1) = 50_k150_k1$, $k = 0, 1, 2, \ldots$ Table 63.2 contains a list of hexamorphic numbers. Dr. Googol invites your comments on the similarities between the formulas for centered hexamorphic and hexamorphic numbers. Why are there similarities?

Additional infinite sequences in centered hexamorphic numbers are $H_c(16_k7)$

$= 83_k16_k7$, $k = 0, 1, 2, \ldots$ and
$H_c(6_k7) = 13_k26_k7$, $k = 0,1,2,\ldots$.
Hexamorphic numbers do not contain any numbers ending with 7, but they do contain numbers ending with 1, and these also exist in the centered hexamorphic sequence. Those of you who wish to learn about hexamorphic numbers in various bases will enjoy Charles Trigg's research (see "Further Reading").

In closing, Leo A. Senneville and Dr. Googol have noted that there are some interesting relations between centered hexagonal and hexagonal numbers. For example, the second differences between successive terms for centered hexagonal numbers are always 6.

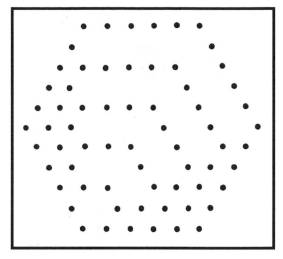

63.1 Hexagonal numbers. Derived from hexagonal points arranged as shown here, they can be generated using $X(n) = n(2n - 1)$.

The second differences between successive terms for hexagonal numbers are always 4. These statements condense to $H_c(n + 1) - 2H_c(n) + H_c(n - 1) = 6$, $H(n + 1) - 2H(n) + H(n - 1) = 4$. They also have noted the following infinite series: $H_c(n)/H(n) = 3(1/2 - 1/(4n) - 1/(8n^2) - 1/(16n^3) - \ldots)$. The sum of this series approaches 3/2 as a limit, which is also the ratio of the second differences. Finally, if you plot curves with natural numbers on the horizontal axis and the corresponding value of the hexagonal functions on the vertical axis, the difference in height between the two curves is always $(n - 1)^2$.

Can you find any additional patterns in these wondrous numbers?

❀ For other odd facts about triangular and hexagonal numbers, see "Further Exploring."

▪ See [www.oup-usa.org/sc/0195133420] for a computer program that generates polygonal numbers.

Chapter 64

The *X-Files* Number

Mulder: Hey, Scully. Do you believe in the afterlife?
Scully: I'd settle for a life in this one.

— *"Shadows,"* The X-Files

Dr. Googol was on the set of *The X-Files*, the highly acclaimed TV series involving FBI investigations of paranormal phenomena. He turned to David Duchovny, one of the lead actors in the series.

"David, people have used numbers to predict the end of the world. But predictions usually don't appear in mathematical journals." Dr. Googol raised his eyebrows. "This one appeared in the January 1947 issue of the *American Mathematical Monthly.*"

"Dr. Googol, let me see that," David said in a low voice. He grabbed the tattered article from Dr. Googol's hand and began to read:

> The famous astrologer and numerologist Professor Umbugio predicts the end of the world in the year 2141. His prediction is based on profound mathematical and historical investigations. Professor Umbugio computed the value from the formula
>
> $$\omega = 1492^n - 1770^n - 1863^n + 2141^n$$
>
> for $n = 0, 1, 2, 3$, and so on up to 1945, and found that all numbers which he so obtained in many months of laborious computation are divisible by 1946. Now, the numbers 1492, 1770, and 1863 represent memorable dates: the Discovery of the New World, the Boston Massacre, and the Gettysburg Address. What important date may 2141 be? That of the end of the world, obviously.

David lowered the slightly soiled slip of paper. "Sir, this is incredible. This is a perfect case for an *X-Files* investigation. Could all the numbers produced by the formula be divisible by 1946? Could it be that 2141 has anything to do with the end of the world?"

Dr. Googol reached into Gillian Anderson's pocketbook and tossed a programmable calculator to David. "Write a program, and see what numbers you get."

David began to type, and he soon handed Dr. Googol the results on a small printout. The E symbols are the computer's way of representing scientific notation. For instance, 1.00E + 02 would be another way of denoting 1.00×10^2, or 100.

N	W	N	W
1	0	6	3.478795E + 19
2	206276	7	9.035302E + 22
3	1.124106E + 09	8	2.246103E + 26
4	4.106015E + 12	9	5.410357E + 29
5	1.256519E + 16	10	1.272996E + 33

"Dr. Googol, the numbers grow awfully quickly! If the units were in years, the fifth value is larger than the number of years required for all the stars to have died out." David began to pace. "How could scientists in the year 1946 determine that the results were all divisible by 1946? What is the W value for $n = 100$? Are the W numbers always divisible by 1946, or do they cease to have that property after $n = 1945$?"

Dr. Googol nodded. "David, these are all very interesting unanswered questions. But they'll have to wait." Dr. Googol pointed down the street to an enigmatic man in black, smoking a cigarette. "David, you're about to have a close encounter of the third kind."

❀ For more information on *X-Files* numbers, see "Further Exploring."

▪ See [www.oup-usa.org/sc/0195133420] for a computer program that generates these numbers.

Chapter 65

A Low-Calorie Treat

The mathematician's patterns, like the painter's or the poet's, must be beautiful; the ideas, like the colours or the words, must fit together in a harmonious way. Beauty is the first test: there is no permanent place in the world for ugly mathematics.

—*G. H. Hardy,* A Mathematician's Apology

Dr. Googol was enjoying a piece of chocolate cake in Mel's Diner at 1840 Grand Concourse in the Bronx when he invented "cake integers"—a delicious low-calorie snack for health-conscious readers. Here's the big question. Given a circular cake, using just 4 straight vertical knife cuts, what's the maximum number of pieces you can create? Try this puzzle on a few friends. With just 1 cut, the answer is obvious: 2 pieces. With 2 cuts, you can create, at most, 4 pieces. How many pieces can you create with 4 cuts? It turns out that the answer is 11 (see Figure 65.1). Most of your friends will not get 11 pieces in their first attempt!

Let us define cake integers as having the form $Cake(n) = (n^2 + n + 2)/2$. Cake integers indicate the maximum number of pieces in which a cake can be cut with n slices. (The cake is represented as a flat disc.) The sequence goes as 2, 4, 7, 11, 16, 22, 29, 37, . . .

An integer n is *cakemorphic* if the last digits of $Cake(n) = n$. For example, if $n = 25$ and $Cake(n)$ were to equal 1,325, n would be cakemorphic because the starting number, 25, occurs as the last 2 digits. Dr. Googol has not

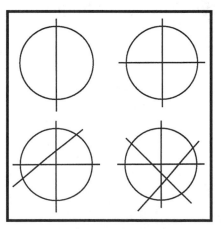

65.1 Sample dissections of several delicious cakes. You can see that for $n = 4$ (the rightmost cake), $C(n) = 11$. Can any of your friends create 11 pieces on their first attempt?

been able to find a cakemorphic integer even though he searched for all values of *n* less than 10,000,000. He therefore has conjectured that no cakemorphic integer exists.

On the other hand, you can show that *hexamorphic* and even *square pyramorphic numbers* are quite common (Figure 65.2 and Tables 65.1 and 65.2).

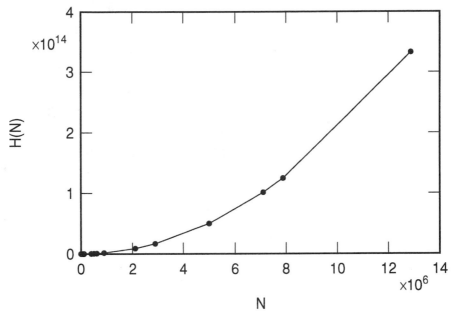

65.2 Distribution of hexamorphic numbers.

n	H(*n*)	*n*	H(*n*)
5625	63275625	609376	742677609376
9376	175809376	890625	1586424890625
40625	3300740625	2109376	8898932109376
50000	4999950000	2890625	16711422890625
50001	5000150001	5000000	49999995000000
59376	7050959376	5000001	50000015000001
90625	16425690625	7109376	101086447109376
109376	23926109376	7890625	124523917890625
390625	305175390625	12890625	332336412890625
500000	499999500000		
500001	500001500001		

Table 65.1 Large hexamorphic numbers. The table here continues the table in the previous chapter which lists the hexamorphic numbers less than 63,275,625. Note: this table may contain the most comprehensive list of hexamorphic numbers to date. In 1987, the late Charles Trigg searched only as far as *n* < 10,000.

n	S(n)	n	S(n)
1	1	960	295372960
5	55	1185	555371185
25	5525	2560	5595682560
40	22140	2625	6032742625
65	93665	4000	21341334000
80	173880	5185	46478345185
160	1378160	6560	94121656560
225	3822225	6625	96947076625
385	19096385	8000	170698668000
400	21413400	9185	258337319185
560	58695560	9376	274790059376
625	81575625	10625	399877410625
785	161553785		
800	170986800		

Table 65.2 Square pyramorphic numbers.

Hexagonal numbers have the form $H(n) = n(2n-1)$ (see Chapter 62). A number is hexamorphic if $H(n)$ terminates with n. The number 125 is hexamorphic because $H(125) = 31,125$. Square pyramidal numbers are related to 3-D objects rather than 2-D polygons. If cannonballs are piled so that each layer is a square, then the total number of balls in successive piles will be $S(n) = 1, 5, 14, 30, \ldots$ $n(n + 1)(2n + 1)/6$. Just like hexamorphic numbers, a number is square pyramorphic if $S(n)$ terminates with n.

A crazy challenge: are there any cakemorphic numbers? Another challenge: Dr. Googol hands you a doughnut. What's the greatest number of pieces you can create with n cuts?

❋ See "Further Exploring" for additional findings and for challenges requiring doughnut and pretzel cutting.

Chapter 66

The Hunt for Elusive Squarions

All the pictures which science now draws of nature and which alone seem
capable of according with observational fact are mathematical
pictures. . . . From the intrinsic evidence of his creation, the Great
Architect of the Universe now begins to appear as a pure mathematician.
—*James H. Jeans,* Mysterious Universe

Dr. Googol has always been fascinated by square numbers like 4, 9, and 25.
(They're called square numbers because $2^2 = 4$, $3^2 = 9$, and $5^2 = 25$.) What
follows are 4 fiendishly difficult questions regarding "squarions," a general-
purpose term signifying very elusive arrangements of square numbers in a variety
of settings.

THE HUNT FOR SQUARION ARRAYS

One question that Dr. Googol has pondered is whether or not it is possible to fill
an infinite square array with distinct integers such that the sum of the squares of
any 2 adjacent numbers is also a square. To illustrate, the following is a 4-by-4
array with the desired property:

1836	105	252	735
1248	100	240	700
936	75	180	525
273	560	1344	3920

For example, $75^2 + 180^2 = 195^2$. Is it possible to create bigger arrays of this
kind? Can you?

THE HUNT FOR MAGIC SQUARIONS

While on the subject of square numbers, it's not known if there exists a 3-by-3
magic square of square numbers, that is, a 3-by-3 arrangement of 9 distinct integer
squares such that the sum of each row, column, and main diagonal is the same.
However, it is possible to build arrangements that satisfy the 6 orthogonal sums so
that the row and column sums are equal. The following is from Kevin Brown:

4^2	23^2	52^2
32^2	44^2	17^2
47^2	28^2	16^2

Remarkably, each row and column of this arrangement sums to a square number: $3,249 = 57^2$. Here's a wondrous magic square of this kind constructed using prime number squares:

11^2	23^2	71^2
61^2	41^2	17^2
43^2	59^2	19^2

THE HUNT FOR STRONG SQUARIONS

What is the smallest square with leading digit 1 that remains a square when the leading 1 is replaced by a 2? In other words, if $x^2 = 1 \ldots$, is there a $y^2 = 2 \ldots$? For example, consider the square number 16. If 26 were also a square, then we would have found a solution.

We can ask a similar question. What is the smallest square with leading digit 1 that remains a square when the leading 1 is replaced by a 2 and also remains a square when the leading digit is replaced by a 3?

What is the smallest square with leading digit 1 that remains a square when the leading 1 is replaced by a 2, and also remains a square when the leading digit is replaced by a 3, and also remains a square when the leading digit is replaced by a 4?

THE HUNT FOR PAIR SQUARIONS

Certain pairs of numbers when added or subtracted give a square number. For example, 10 and 26 are *pair squarions* or *double squarions* since $10 + 26 = 36$ (a square number) and $26 - 10 = 16$ (a square number). Stated mathematically, n and p are pair squarions if $n - p = a^2$ and $n + p = b^2$ where a and b are integers. This section indicates interesting patterns in the pair squarions and also provides you a simple computer program with which to generate these numbers.

How are pair squarions distributed? Are they easy to find? What can we know about their properties? Table 66.1 lists several pair squarions, denoted by n and p. These were generated using an algorithm like the following (and like the code at [www.oup-usa.org/sc/0195133420]), which hunts for all pair square numbers less than 1,000.

```
1   do p = 0 to 1000
2   do n = p+1 to 1000
3   a = sqrt(n+p)
4   b = sqrt(n-p)
5   if (a = trunc(a)) & (b = trunc(b)) then
6   say p n
7   end
8   end
```

n	p	n	p
4	5	22	122
6	10	24	25
8	17	24	40
10	26	24	145
12	13	26	170
12	37	28	53
14	50	28	197
16	20	30	34
16	65		
18	82		
20	29		
20	101		

Table 66.1 Pair squarions.

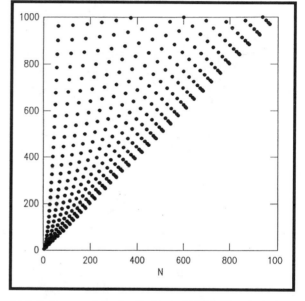

66.1 Pair squarions for $0 \leq n,p \leq 1000$. The distribution is symmetric about the line $n = p$, and the lower part is not plotted.

Line 5 is used to ensure that both *a* and *b* are integers. Figure 66.1 plots the positions of all pair squarions less than 1,000 (that is, $0 \leq n,p \leq 1,000$). The distribution is symmetric about the line $n = p$, and the lower part is not plotted. The straight line of points at $n = p$ corresponds to $n = b^2/2$. Other curves seen in the plot correspond to equations such as $n^2 - p^2 = a^2b^2$. Try connecting the dots to make a beautiful net-like structure. Can you think of any ways to speed up the hunt for pair squarions?

❁ For a partial solution to the strong squarion problem, and for more analyses regarding pair squarion numbers, see "Further Exploring."

▐ For BASIC code used to search for pair squarions, see [www.oup-usa.org/sc/0195133420].

Chapter 67

Katydid Sequences

> No live organism can continue for long to exist sanely under conditions of absolute reality. Even larks and katydids are supposed, by some, to dream.
> —*Shirley Jackson,* The Haunting of Hill House

One day while dining at an elegant restaurant in Westchester, New York, Dr. Googol found a dead katydid in his spinach soufflé. He examined the grasshopper-like insect, using his fork.

"Disgusting," his friend Monica said to him.

Dr. Googol removed the insect from the spinach. "Monica, this reminds me of *katydid sequences.*"

Monica took a deep breath and rolled her eyes. "Do I want to hear about this?"

"Sure, it's a remarkable kind of number sequence."

"Okay, tell me more." There was a hesitation in her voice as she looked up toward the ceiling.

"I call them katydid sequences because they remind me of the rapid (exponentially growing) breeding that katydids and grasshoppers undergo during their mating seasons." He paused. "Katydid sequences are defined by the following 2 functions, which can be visualized as a growing tree."

Dr. Googol scribbled on a napkin:

$$x \rightarrow 2x + 2$$
$$x \rightarrow 6x + 6$$

"Here, x is an integer. Start with $x = 1$. These mappings generate two branches of a 'binary' tree. In other words, x has two children, $2x + 2$ and $6x + 6$." He scribbled again:

"Each generation requires a month to breed. For example, after 1 generation (1 month) we have 4 and 12 as 'children' of the 'parent' 1. When x is 4, the children are 10 and 30. The next month produces 10, 30, 26, 78. All the numbers that have appeared so far, when arranged in numerical order, are 1, 4, 10, 12, 26, 30, 78, ... No number seems to appear twice in a row; for example there is no 1, 4, 10, 10, ... "

Monica stared at the napkin for nearly half a minute. "So what?"

Dr. Googol looked up at Monica. "Does a number *ever* appear twice? Maybe we don't see a repetition yet, but would we see one after a year? Hundreds of years?" He paused. "If this problem is too difficult for you, consider these similar katydid sequences. Does a number ever appear twice in the following?"

$$x \rightarrow 2x + 2, \qquad x \rightarrow x + 1$$

or

$$x \rightarrow 2x + 2, \qquad x \rightarrow 5x + 5$$

Monica stared at Dr. Googol. "I'll have to think about this for a while. Now it's time for dessert."

Monica never solved the problems. Can you? Dr. Googol looks forward to hearing from anyone who can.

❀ See "Further Exploring" for further analyses and surprises.

Chapter 68

Pentagonal Pie

> The most important sequences, such as square numbers and the factorials, turn up everywhere. The Catalan sequence is in the Top Forty in popularity, even if it does not reach the Top Ten. It occurs especially often in combinatoric problems.
> —*David Wells,* Curious and Interesting Numbers

Dr. Googol was cutting a pentagonal pie with a knife. "Happy birthday, my dear," he said to Anika.

Anika pulled her blond hair back. "A pentagonal pie. I've never heard of such a thing."

"How many ways can you divide the pie into triangles, starting your straight, downward cuts at one corner and ending at another? Your cuts can't intersect one another."

After 5 minutes of thought, Anika cut the pie. "Here is one way," she said. "Let me draw all the different ways."

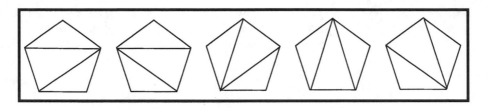

"Superb!" Dr. Googol said.

"But Dr. Googol, can't we eat it now? I don't wish to talk further about math on my birthday."

"Wait!" Dr. Googol screamed, just as Anika was about to eat a piece. "Let me ask this in a different way. How many ways can a regular n-gon—like a square, pentagon, hexagon, etc.—be divided into $n - 2$ triangles if different orientations are counted separately?"

"Different orientations?" Anika said.

"Yes. For example, in the pentagonal pie you cut, the pattern of cuts would look the same if you roated the pie, but we'll still consider them 5 separate cutting patterns."

Dr. Googol withdrew a pen from his pocket and started drawing the possibilities for a hexagon (Figure 68.1). Just as he started drawing the different cuts for a 7-sided polygon, Anika decided she'd had enough and walked out the door. Dr. Googol, deep in concentration, never noticed. He was trying to derive a formula to compute the number of ways the polygonal cakes could be cut into triangles for *any* regular polygon. Can such a formula be derived? Are there more ways to slice a 16-sided polygon then there are people on the planet?

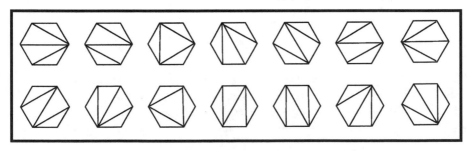

68.1 14 ways to divide a hexagon into triangles.

❀ For a solution and more graphic examples, see "Further Exploring." Hint: A sequence called the Catalan sequence can be used to solve this puzzle.

▤ For BASIC code used to study this problem, see [www.oup-usa.org/sc/0195133420].

Chapter 69

An A?

> He remembered exploring those other-worldly curves from one degree to
> the next, lemniscate to folium, progressing eventually to an ungraphable
> class of curve, no precise slope at any point, a tangent-defying mind marvel.
> —*Don DeLillo,* Ratner's Star

Dr. Googol was in London lecturing a Mensa group. Mensa has a single qualification for membership: you must score in the top 2% of the population on a standardized intelligence test. An IQ between 130 and 140 is usually acceptable.

Dr. Googol went over to a blackboard and drew a single letter:

Dr. Googol looked at his audience. "Can anyone tell me what this is?"
A distinguished gentleman with a large mustache raised his hand. "It is an *a*."
Dr Googol grinned. "Correct!" He wrote down:

"Now what is on the board?" Dr. Googol said.

A distinguished woman with a small mustache raised her hand. "It is an *a*, an *n*, and an *a*."

Dr. Googol wrote down:

anaannana

The entire audience screamed with glee and picked Dr. Googol up on their shoulders. A band started to play as confetti fell from the ceiling. The Mensa meeting was brought to a close as the members' roars of jubilant exaltation rose to fever pitch.

❀ ❀ ❀

The rule for generating Ana sequences is to begin with a letter of the alphabet and to then generate the next row by using the indefinite article *a* or *an* as appropriate. (This will probably be best understood by English-speaking readers, who should say the sequences out loud to best understand them.) The most obvious letter to start with is *a*:

Generation	Sequence
1	ana
2	ana
3	ana ann ana
4	ana ann ana ana ann ann ana ann ana

The first row contains an *a*, giving us *ana* for the second row. How many different words can you generate with this method? It turns out that only the words *ann* and *ana* occur, but there is an interesting self-similarity cascade here. (For sequences like this, *self-similarity* refers to the fact that there are repeated patterns within patterns for different sequence lengths.) One way of visually representing the sequence to find patterns is to represent *a* by a dark icon, such as an alien head, and *n* by a less dark icon, such as the figure of a man:

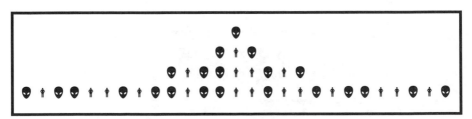

Here it's easy to see that what-ever pattern there is, after the second generation (or row) it is certainly not symmetrical about the midpoint of each sequence. A much better way to see the structure is to look at Figure 69.1, created by Dr. Googol's colleague Mike Smithson from James Cook University. Here *a* is represented as a dark rect-angle, and *n* is represented by a white space with no rectangle. In the sophisticated parlance of fractal geometry, this structure is known as an asymmetric Cantor dust.

69.1 Anabiotic Ana fractal. The letter *a* is represented by a dark bar. The letter *n* is represented by a gap. (Rendering by Mike Smithson.)

As background, a symmetrical Cantor set can be constructed by taking an interval of length 1 and removing its middle third (but leaving the end points of this middle third). The top two rows of Figure 69.1 show this removal. This leaves two smaller intervals, each one-third as long. In the symmetrical case, the middle thirds of these smaller segments are removed and the process is repeated over and over to create a symmetrical pattern:

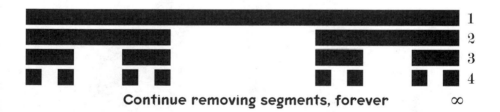

Continue removing segments, forever ∞

This symmetrical Cantor set has a "measure zero," which means that a ran-domly thrown dart would be very unlikely to hit a member of the very sparse set in higher row numbers. At the same time, it has so many members that it is in fact uncountable, just like the set of all of the real numbers between 0 and 1. Many mathematicians, and even George Cantor himself, for a while doubted that a crazy set with these properties could exist. As you have just been shown, however, such a set is possible to formulate. The dimension D of the symmetrical Cantor dust for an infinite number of iterations is less than 1 since $D = \log2/\log3 = 0.63$. You can read more about the concept of fractional dimensions, and how 0.63 was derived, in Manfred Schroeder's *Fractals, Chaos, Power Laws*. Cantor dusts with other fractal dimensions can easily be created by removing different sizes (or numbers) of intervals from the starting interval of length 1. Cantor sets are high-

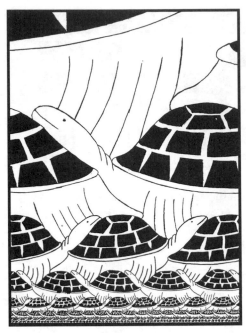

69.2 *Turtles Forever*, by Peter Raedschelders.

ly useful mathematical models for many physical phenomena, from the distribution of galaxies in the universe to the fractal Cantor-like structure of the rings of Saturn.

For those of you who are fractal experts, can you compute the dimension of the Ana fractal? Does it even have a single dimension? What happens if you start the Ana fractal sequence with a letter other than *a*? Is this new sequence fractal? Are there other verbal fractals waiting to be discovered using different rules?

After converting the *a*'s and *n*'s to tones, Mal Lichtenstein of San Diego, California, was able to listen to an 81-element Ana sequence and Morse-Thue sequence described in Chapter 17. They sounded very similar to him. He wonders if the ratios *a/n* and 0/1 approach 1

69.3 *Fractal Butterflies*, by Peter Raedschelders.

69.4 *Seal Recursion*, by Peter
Raedschelders.

69.5 *Fractal Dinosaurs*, by Peter
Raedschelders.

in both sequences. He believes that there are at most 2 of the same elements in
succession for both sequences.

Figures 69.2 through 69.5 are the intricately recursive artworks of Belgian
artist Peter Raedschelders. Like the Ana fractal and Cantor sets, these works represent a continual repetition of objects at diminishing size scales. If these had
been constructed using mathematical algorithms and computer graphics, in
principle the smaller structures could be continually magnified to reveal yet
smaller structures, like an infinite nesting of Russian dolls within dolls.

❀ For more on Ana fractals, see "Further Exploring."

Chapter 70

Humble Bits

One sign of an interesting program is that you cannot readily predict its output.
—*Brian Hayes,* "On the Bathtub Algorithm for Dot-Matrix Holograms,"
Computer Language, vol. 3, *1986*

Dr. Googol was lecturing members of YLEM, the California-based organization of artists who use science and technology. "The humble bits that lie at the very foundation of computing have a special beauty all their own. It takes just a little logical coddling to bring the beauty out. Who would guess, for example, that intricate fractal patterns lurk within the OR operation applied to the bits of ordinary numbers?"

A huge man with an orange punk hairdo and Mortal Kombat® tattoos got up out of his seat. "*Binary numbers?* Those are the ones that are made up of just the digits 1 and 0."

Dr. Googol nodded. "Some say they were invented by Leibniz while waiting to see the pope in the Vatican with a proposal to reunify the Christian churches. Here are the first 7 numbers represented in binary notation:

$$0, 1, 10, 11, 100, 101, 110, 111, \ldots$$

The sums of the digits for each number form the sequence (in decimal notation):

$$0, 1, 1, 2, 1, 2, 2, 3, 1, 2, 2, 3, 2, 3, 3, 4, \ldots$$

"Notice, just like the Morse-Thue sequence, which I lectured you about earlier (see Chapter 17), this sequence is *self-similar*: if you retain every other term you still have the same infinite sequence!"

"Amazing!" the big man yelled.

Dr. Googol put up his hand to silence the man. "For the next 10 minutes, I want to demonstrate that wonderful graphic patterns can emerge when working with binary numbers. In fact, very complex patterns with scaling symmetry can arise from the simplest of arithmetic operations that use logical operators such as **AND** and **OR**."

And for the next 10 minutes, Dr. Googol flashed image after image upon the screen, captivating his audience with his wit, beautiful visuals, and charm.

❀ ❀ ❀

Figure 70.1 was created using an **OR** operation, which Dr. Googol will now explain. For this demonstration we compute the values for a square image consisting of an array of values c_{ij}, in particular $c_{ij} = i$ **OR** j, for $(1 < i < 800)$ and $(1 < j < 800)$. For example, if $i = 6$ and $j = 1$, $c = 7$ because $111 = (110$ **OR** $001)$. Just apply a logical operation one bit at a time. (For instance: 1 **OR** 0 is 1; 0 **OR** 0 is 0; and 1 **OR** 1 is 1.) The variables i and j correspond to the x and y axes of the figures in this chapter. The value of c is represented by shades of gray. Figure 70.1 illustrates c modulo 255. The brightest picture element is therefore 254, and this corresponds to bright white. 0 is represented by black. The black, triangular, gasket-like structure represents those ($c = 255$) pixels that are made black by the modular arithmetic. The fractal nature of the entire pattern is evident. The black pattern is called a Sierpinski gasket, commonly seen in cellular automata applications. In fact, the same pattern is seen when the even entries of Pascal's triangle are colored black (see Chapter 54). Let us call this pattern a "logical" Sierpinski gasket.

Obviously, Dr. Googol has barely scratched the surface of the subject. There are endless combinations of logical (and arithmetic) operators to be tried on the humble binary numbers. In the process, some of you will discover worlds neither Dr. Googol nor anyone else has seen.

❋ For more analysis, see "Further Exploring."

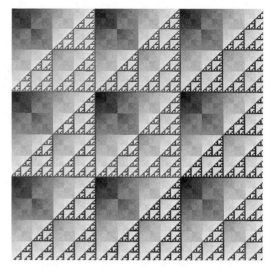

70.1 Pattern of bits. The pattern was produced by $c_{ij} = i$ **OR** j for $(1 < i < 800)$ $(1 < j < 800)$. The values of c modulo 255 are represented by shades of gray.

Chapter 71

Mr. Fibonacci's Neighborhood

> For those, like me, who are not mathematicians, the computer can be a powerful friend to the imagination. Like mathematics, it doesn't only stretch the imagination. It also disciplines and controls it.
> —*Richard Dawkins,* The Blind Watchmaker

Dr. Googol drove to Mr. Fibonacci's neighborhood pet store and bought a pair of rabbits to breed. The pair produced a pair of young after a year, and a second pair after the second year. Then they stopped breeding. Each new pair also produces 2 more pairs in the same way, and then stops breeding. How many new pairs of rabbits would you have each year? To answer this question, write down the number of pairs in each generation. First write the number 1 for the single pair you bought from the pet shop. Next write the number 1 for the pair they produced after a year. The next year both pairs have young, so the next number is 2. Continuing this process, we have the sequence of numbers: 1, 1, 2, 3, 5, 8, 13, 21, 34, 55, 89, 144, 233, 377, ... This sequence of numbers, called the *Fibonacci sequence* after the wealthy Italian merchant Leonardo Fibonacci (1170–1240) of Pisa, plays important roles in mathematics and nature. These numbers are such that, after the first 2, every number in the sequence equals the sum of the 2 previous numbers $F_n = F_{n-1} + F_{n-2}$. The code at [www.oup-usa.org/sc/0195133420] shows how to program this sequence on the computer.

THE AMAZING 1/89

Although not widely known, several mathematicians have discovered that the decimal expansion of 1/89 (0.01123 . . .) relates to the Fibonacci series when certain digits are added together in a specific way. Examine the following sequence of decimal fractions, arranged so the rightmost digit of the nth Fibonacci number is in the $n + 1$th decimal place:

n	
1	.01
2	.001
3	.0002
4	.00003
5	.000005
6	.0000008
7	.00000013
	.0112359 . . .

Unbelievably, 1/89 = .01123595505. . . . Fantastic! Why should this be so? Why on Earth is 89 so special?

REPLICATING FIBONACCI DIGITS

With these digressions, let us switch gears and discuss numerical world records with numbers related to Fibonacci numbers. (Maybe you can be the next world-record holder in the search for these numbers.) In 1989, Dr. Googol discovered 129,572,008 and 251,133,297—new *replicating Fibonacci digits* (defined in next paragraph) in the range 100 million to 1 billion. At the time, they were the largest replicating Fibonacci digits discovered, although today several people have taken up the challenge and discovered several larger numbers of this kind.

A replicating Fibonacci digit, or repfigit, has the remarkable property that it repeats itself in a sequence generated by starting with the n digits of a number and then continuing the sequence with a number that is the sum of the previous n

terms. An example should clarify this. 47 is a repfigit since the sequence (4, 7, 11, 18, 29, 47) passes through 47. Likewise 1,537 is a repfigit since the sequence (1, 5, 3, 7, 16, 31, 57, 111, 215, 414, 797, 1,537) passes through 1,537.

In 1987, Michael Keith introduced the concept of replicating Fibonacci digits. At that time the largest known repfigit was the 7-digit number 7,913,837. In November 1989, 3 larger repfigits were discovered, and the world's largest repfigit was 44,121,607.

Repfigits are interesting for several reasons. For one, the question of whether or not the number of repfigits is infinite is unsolved. It would be interesting to find that no repfigit exists for higher numbers of digits, or to discover patterns by searching for larger numbers. Moreover, progress on certain famous problems has historically been used as a yardstick for measuring the growth in computer power. How many hours would your computer require to find Dr. Googol's previous world record of 251,133,297?

Table 71.1 indicates all known repfigit numbers under 1 billion.

❋ For more information on repfigits and other Fibonacci delights, see "Further Exploring."

▣ See [www.oup-usa.org/sc/0195133420] for a computer program that generates Fibonacci numbers. Starting from this, can you create a program that computes repfigits?

2	14	19	28	47	61	75			
3	197	742							
4	1104	1537	2208	2508	3684	4788	7385	7647	7909
5	31331	34285	34348	55604	62662	86935	93993		
6	120284	129106	147640	156146	174680	183186	298320	355419	694280 925993
7	1084051		7913837						
8	11436171		33445755		44121607				
9	129572008		251133297						

Table 71.1 Replicating Fibonacci digits less than one billion. The first column indicates the number of digits.

Chapter 72

Apocalyptic Numbers

> Never dismiss the intuition of the ancients, who believed that number is
> the essence of all things. Number is the secret source of entire cultures,
> and men have been killed for their heresies and seductive credos. The
> whole history of mathematics is subterranean, taking place beneath histo-
> ry itself, a shadow-world scarcely perceived even by the learned.
> —*Don DeLillo,* Ratner's Star

Dr. Googol was contemplating the nature of reality while sitting in St. John the
Divine, the world's largest cathedral, in New York City. He started to read a
book called the Revelation (or Apocalypse) of John. Dr. Googol knew this was
the last book of the New Testament, written using fantastic imagery—blending
Jewish apocalypticism, Babylonian mythology, and astrological speculation.
Various mystics have devoted much energy to deciphering the number 666, said
by John the Apostle to designate the Number of the Beast, the Antichrist. More
recently, mystical individuals of the extreme fundamentalist right have noted
that each word in the name Ronald Wilson Reagan has 6 letters.

Dr. Googol turned to Monica, who sat quietly beside him. "Monica, isn't it
odd that 666 has especially interesting mathematical properties? For example,
the number 666 is a simple sum and difference of the first 3 natural numbers
raised to the sixth power." With very careful penmanship, Dr. Googol wrote on
the back page of a Bible:

$$666 = 1^6 - 2^6 + 3^6$$

"It is also equal to the sum of its digits plus the cubes of its digits."

$$666 = 6 + 6 + 6 + 6^3 + 6^3 + 6^3$$

Dr. Googol looked into Monica's dark eyes and whispered, "I believe that
there are only 5 other positive integers with this property. Can you find them?"

Monica looked down. "Dr. Googol, perhaps you had better not write anything more on the Bible. It's not right."

Dr. Googol nodded. "The sum of the squares of the first 7 primes is 666:

$$666 = 2^2 + 3^2 + 5^2 + 7^2 + 11^2 + 13^2 + 17^2$$

And here's a real gem: A standard function in number theory is $\phi(n)$, which is
the number of integers smaller than n and relatively prime to n. Amazingly, we
find that:

$$\phi(666) = 6 \cdot 6 \cdot 6$$

(Number theorists call 2 numbers *A* and *B* that have no common factors *relatively prime* or *coprime*.)" The first 144 decimal digits of pi add up to 666, and 144 is special because 144 = (6 + 6) × (6 + 6). Finally, the three decimals of pi that begin with the 666th are 343 = 7 × 7 × 7."

Dr. Googol said the last few words so loudly that all the people in St. John the Divine turned their heads to stare at him. Without saying another word, Dr. Googol and Monica quietly left.

About a year ago, Dr. Googol began a computer search for "apocalyptic numbers." These are Fibonacci numbers with precisely 666 digits. As described in Chapter 71, the sequence of numbers (1, 1, 2, 3, 5, 8, . . .), is called the Fibonacci sequence, and it plays important roles in mathematics and nature. These numbers are such that, after the first 2, every number in the sequence equals the sum of the 2 previous numbers $F_n = F_{n-1} + F_{n-2}$. It turns out that the 3,184th Fibonacci number is apocalyptic, having 666 digits. For numerologist readers, the apocalyptic number is:

<u>1167</u>24374081495541233435764579214184068974717443439437236331282736262082452385312960682327210312278880768244979876073455971975198631224699392309001139062569109651074019651076081705393206023798479391897000377475124471344025467950768706990550322971334370940093654442411815206857904041043400568568081<u>1194</u>37950300196766935663379234721865689613658399032791816735272116358165035957768655229310270882722424710947638<u>2115</u>4275682688200402585049861134087733332208736164591167264971986989157913558834313855569580021219281470520871752067489363661712533804220588026552914033581456195146042794653576446729028117115407601267725615728671557460702606785922979179042488538923588617<u>1163</u>

Is the number shown here the only apocalyptic Fibonacci number? Is there any significance to the fact that the first 4 digits and last 4 digits (1167 and 1163) of the apocalypse number both represent dates during the reign of Frederick I of Germany, who intervened extensively in papal politics? In fact, Frederick had set up a series of antipopes in opposition to the reigning pope, Alexander III. In 1167 Frederick attacked the Leonine City in Rome and was able to install one of the antipopes, Paschal III, on the papal throne. Notice that in the middle of the apocalyptic number we find the date 1154. On precisely this date, Frederick proceeded to Italy, where he received the Lombard crown at Pavia. Toward the end of his life, Frederick went on a crusade and drowned—sometime around the year 1194, another date that appears in the enigmatic apocalyptic number.

❀ See "Further Exploring" for more oddities involving 666.

▪ For hints on computing Fibonacci numbers, see [www.oup-usa.org/sc/0195133420] for Chapter 71.

Chapter 73

The Wonderful Emirp, 1,597

> There can be no dull numbers, because if there were, the first of them
> would be interesting on account of its dullness.
> —*Martin Gardner, 1992*

"I love 1,597!" Dr. Googol said to his friend Monica as they rode horses along the vast Montana outback.

"I thought you only had eyes for me," Monica said as she brushed back her hair, which the faint wind had teased out of place.

"I do, but don't you realize that 1,597 is both a prime number and a Fibonacci number?" Dr. Googol handed Monica a note of explanation. It read:

A prime is a positive integer that cannot be written as the product of 2 smaller integers. The number 6 is equal to 2 times 3, but 7 cannot be written as a product of factors; therefore, 7 is called a prime number or prime. Here are the first few prime numbers: 2, 3, 5, 7, 11, 13, 17, 19, 23, 29, 31, 37, 41, 43, 47, 53, 59. See Chapter 71 for background on Fibonacci numbers 1, 1, 2, 3, 5, 8, . . .) The number 1597 is also the year in which the Edict of Nantes was drafted, which gave French Protestants (Huguenots) a degree of freedom, opening public offices to them and permitting them to hold public worship in certain cities.

Monica folded the note and placed it under her saddle. "Thank you for the lovely note, but why does it say 'See Chapter 71'?"

"It's for a book I'm writing. Never mind that. More interestingly, 1,597 is fascinating because it is an 'emirp,' a prime number that turns into a different prime number when its digits are reversed (7,951)."

As the sun began to set, the meadows and hills were awash in a tangle of golden reflections. Dr. Googol began to dream.

❀ ❀ ❀

1,597 is also the basis for a number problem Dr. Googol dreamed just a year ago for which a solution seemed unlikely. Consider the formula $x = \sqrt{1{,}597y^2 + 1}$. Is x ever an integer for any integer y greater than 0? You may wish to first compute a few values of x in order to get the feel for the formula:

y	x
1	39.97
2	79.93
3	119.89

You can see that for $y = 1$, 2, or 3, x is not an integer. Is it ever an integer? The first method you might use to answer this question is to write a short computer program that would simply try thousands of values of y, starting at $y = 1$. The program would continually increment y while testing x—for as long as your patience and machine time allowed. The program could check each x value to see if it were an integer. Unfortunately, your program would run for weeks, and probably months, and you would finally toss up your hands and exclaim that there is no solution. However, it turns out there is an *infinite* number of solutions, and the first individual to solve the 1,597 problem was Noam D. Elkies of the Mathematics Department of Harvard University.

The reason it would take your computer so long to find these infinite number of solutions is the fact that the smallest integer value for x is

$x =$ 519711527755463096224266385375638449943026746249

for a y value of

$y =$ 13004986088790772250309504643908671520836229100

(Note the startling occurrence of 5,197 in x. Is this scrambling of 1,597 just a coincidence? No one knows for sure.)

Dr. Elkies, however, did not solve this through the super-CPU-intensive search methods. In fact, it has been known at least since the time of French mathematician Pierre de Fermat (1601–1665) that for any positive integer D that is not a square, there are infinitely many integers x, y such that $x^2 = Dy^2 + 1$.

Since Dr. Googol gave you the number 1,597, which is a prime number and hence cannot be a square, you know immediately that there is a solution. Furthermore, there is a known algorithm that can be used to solve problems such as these. These methods involve the use of a continued fraction representation for $\sqrt{D}$ in order to find the smallest solution. These algorithms are now implemented on several commercially available symbolic computation software packages, which is what Elkies used to solve the 1,597 problem.

❀ See "Further Exploring" for additional incredible and bizarre 1,597 challenges.

🔒 For hints on finding prime numbers that are also Fibonacci numbers, see [www.oup-usa.org/sc/0195133420].

Chapter 74

The Big Brain of Brahmagupta

As in our Middle Ages, the scientists of India, for better and for worse, were her priests.
—*Will Durant,* Our Oriental Heritage, *1954*

A person who can within a year solve $x^2 - 92y^2 = 1$ is a mathematician.
—*Brahmagupta*

Oh, the wonderful Brahmaputra River! Beautiful beyond compare! Last year, Dr. Googol was exploring the Brahmaputra, the mighty river that flows 1,800 miles from its source in the Himalayas to its confluence with the Ganges River, after which the mingled waters of the two rivers empty into the Bay of Bengal. Its upper course was long an unsolved mystery because exploration was barred by hostile mountain tribes.

The local tribes never scared Dr. Googol. He boldly went up to a young villager and said, "Have you ever heard of Brahmagupta?"

The villager backed up a hasty half-step. "Sir, do you mean Brahmaputra, the river?"

"No, *Brahmagupta.* Not to be confused with *Brahmacharia,* the vow of chastity taken by the ascetic student—a vow of absolute abstention from all sexual desire."

The young man raised his eyebrows. "No, I would not confuse those two words."

Dr. Googol nodded. "Brahmagupta was a great Indian mathematician of the 7th century and desperately interested in huge numbers. He didn't consider someone a real mathematician unless he could find an integer solution to $x^2 - 92y^2 = 1$."

The villager nodded. "Brahmagupta's brain must have been big."

Dr. Googol continued. "This kind of problem has always made me wonder about the history of large number problems. How long ago were the first huge number problems posed, solved, or even considered solvable by humans?"

Dr. Googol's eyes glazed over as the villager walked away, and when Dr. Googol returned home he began to work with colleagues, such as Chris Long from Rutgers University, on *Brahmagupta numbers*, named after the famous Hindu mathematician and astronomer who was intrigued by huge number solutions to simple-looking problems.

❀ ❀ ❀

Please don't expect to solve the following problem with pencil and paper! The solutions involve ratios of numbers so large that if you were to place a dot on a paper every second until you had a number of dots equal to the Brahmagupta numbers, our Milky Way galaxy would have rotated many times. (Did you know that the Milky Way galaxy's period of rotation is 6×10^{15} seconds?)

The problem deals with rational numbers. A *rational number* is a number that can be expressed as a ratio of two integers. Here are some fine examples: 1/2, 4/3, 7/1, 8. All common fractions and all expansions with terminating (or repeating) digits are rational. Trigonometric functions of certain angles are even rational, for example $\cos 60° = \frac{1}{2}$. (This is in contrast to irrational numbers like e and π—called transcendental numbers—and all surds such as $\sqrt{27}$. A surd is a number that is obtainable from rational numbers by a finite number of additions, multiplications, divisions, and root extractions.)

Our problem can be stated as follows. Find the smallest rational number x (smallest in the sense of smallest numerator and denominator) such that there exist rational numbers y and z and

$$x^2 - 157 = y^2, \quad x^2 + 157 = z^2$$

Jim Buddenhagen of Southwestern Bell Advanced Technology Laboratory gave a behemoth solution:

x = 50240182995338036981137754312294030993135017466889667584728816492946182669894640083390462472702407772686242505697440870727011829516260394275244183508553341864729654604103996100686780343137614 ÷ 55207127859076258183875569461342697367786240398108265147202579226331920116659466022175218717871386078381699548684974799036529476971927068616591606845144977158476992422410434693821197457720

y = 4976168309082615289459776489008494215611077198547772938
6907419538978932445636040315578821358685390299974609232
1401151168987604624257763663691302986005230429261330302 2
94516547050831196873663 9 ÷ 5520712785907625818387556946134
2697367786240398108265147202579226331920116659466022175 21
8717871386078381699548684974799036529476971927068616591 60
6845144977158476992422410434693821197457720

z = 5071416834355358956136783482025900435467719016638071 72
7211251468866841620407939138948427548409998628396610688 4
9993734660544850726462041214489884598731864517189496456
47657682653184382680416 1 ÷ 5520712785907625818387556946134
2697367786240398108265147202579226331920116659466022175 2
1871787138607838169954868497479903652947697192706861659 16
0684514497715847699242241043469382119745772 0

(Notice the division symbols buried in these large digit strings.) Buddenhagan solved this using theory provided by Don Zagier in a book titled *Introduction to Elliptic Curves and Modular Forms* (page 5) by Neal Koblitz—and by using a large-integer computer software program called Maple from the University of Waterloo.

If you substitute these huge numbers into the previous equations, you will find that $x^2 - y^2 = 157$ and also that $x^2 - z^2 = -157$, which are valid solutions to the problem. But are these the *smallest* solutions?

❋ See "Further Exploring" for an answer and for further challenges.

Chapter 75

1,001 Scheherazades

> I love to count. Counting has given me special pleasure down through the years. I can think of innumerable occasions when I stopped what I was doing and did a little counting for the sheer intellectual pleasure of it.
> —*Don DeLillo,* Ratner's Star

Since the age of 13, every 1,001 days Dr. Googol reads the *Thousand and One Arabian Nights*. (This means Dr. Googol reads the work every 2.74 years.) With the exception of the Koran, no other work of Arabic literature has been better known and more influential in the West than the *Thousand and One Arabian Nights*. This collection of stories is grouped around a central story involving a sultan and his lovers. Upon discovering that his wife has been unfaithful to him, the sultan vows to take a bride every day and have her executed at dawn.

When Scheherazade was chosen to be his new wife, each evening she told a story to the sultan but did not finish it, promising to do so the following night if she survived. This continued for a thousand and one nights, until the sultan grew deeply in love with Scheherazade and gave up his cruel execution plans.

One night after reading the *Thousand and One Arabian Nights*, Dr. Googol began to wonder about a special number called the *Arabian Nights factorial*. This number is defined as the number x such that $x!$ has 1,001 digits. (The exclamation point is the factorial sign: $n! = 1 \times 2 \times 3 \times 4 \times \ldots \times n$.) Factorials grow rather quickly: $5! = 120$, $10! = 3,628,800$, and $15! = 1,307,674,368,000$. What is the Arabian Nights factorial?

Table 79.1 shows 1,001 Scheherazade clones and a single sultan at bottom right. The sultan is getting old but wishes to kiss each woman once and return to his original position to rest. *Part I:* What path should he take to make the fewest possible turns along his amorous journey? What path should he take if he wishes to find the shortest path? *Part II:* Answer these 2 questions if the sultan does not wish to kiss a woman more than twice along his journey and also wishes to take a prime number of steps. (A "step" takes place each time the sultan goes from one woman to the next.)

❁ See "Further Exploring" for a solution and comment regarding the Arabian Nights factorial.

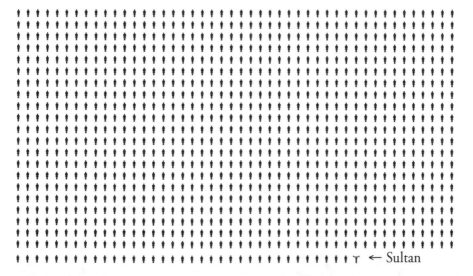

Table 79.1 Find the Sultan's path through the 1001 Scheherazades. (The Sultan is lifting weights at the bottom in preparation for his arduous journey.)

Chapter 76

73,939,133

> It's like asking why Beethoven's Ninth Symphony is beautiful. If you
> don't see why, someone can't tell you. I know numbers are beautiful.
> If they aren't beautiful, nothing is.
>
> —*Paul Erdös*

Dr. Googol was invited to the White House for a special reception honoring the country's 30 brightest minds. Reporters and journalists were everywhere.

The president and first lady began to shake hands with a line of distinguished luminaries in the world of science. CNN was airing the reception on live TV.

When it was Dr. Googol's turn, he smiled at the first lady, then turned to the President. "What is special about the number 73,939,133?"

The president's jaw dropped.

Secret service agents immediately stepped between Dr. Googol and the president. Other agents were speaking into their concealed collar-microphones, frantically trying to get the answer to Dr. Googol's question so that the president could wow the press with his mental prowess.

Can you help the president?

What is special about this number? (Hint: This number is a prime number—a positive integer that cannot be written as the product of 2 smaller integers. But something is very special about this particular prime number.)

❀ For an answer, see "Further Exploring."

Chapter 77

⊎-Numbers from Los Alamos

> As a teenager I thought that if it's at all possible, or practical, to become a
> mathematician, I would want to be one. Of course, from a practical point
> of view, it was very difficult to decide on studying mathematics at the uni-
> versity because to make a living in mathematics was very, very difficult.
> —*Stanislaw Ulam,* Mathematical People

Many years ago, Dr. Googol was working at Los Alamos, New Mexico, where he
met the great mathematician Stanislaw Ulam. Today Ulam is best known for his
theoretical calculations used for building the hydrogen bomb. However, Ulam
also worked on a range of fascinating topics in his lifetime including iteration,
strange attractors, Monte Carlo methods, the human brain, random number
generators, number theory, and genetics.

"Dr. Googol," Ulam said, "let me show you something interesting."

"Stanislaw, you make my heart race."

Ulam nodded. "Start with any 2 positive integers—for example, 1 and 2.
Next consider positive integers in increasing order that can be expressed in just 1
way as the sum of 2 distinct earlier members of the sequence."

"Stanislaw, speak in simple English!"

"Let me give you an example." Ulam began to write on a blackboard. "Here
are the first few numbers starting with 1 and 2."

Dr. Googol carefully took notes, copying down the numbers onto a card:

$⊎_{1,2}$: 1 2 3 4 6 8 11 13 16 18 26 28 36 38 47 48 53 57 62 69 72 77 82 87 97 99
102 106 114 126 131 138 145 148 155 175 177 180 182 189 197 206 209 219

Dr. Googol spoke. "I'm going to call these '⊎-Numbers' in honor of you, Dr.
Ulam. The ⊎ symbol is a U (for Ulam) with a + symbol. I pronounce the symbol
just like the letter U. I think I understand how to generate them. For example, 5
is not a ⊎-number because there is more than one way to form 5 from summing
previous sequence members: $5 = 3 + 2$ and $5 = 4 + 1$. On the other hand, 6 *is* a
⊎-number because it can only be formed by $4 + 2$."

Dr. Ulam continued. "If we draw little vehicles every time we find a ⊎-num-
ber (and leave a dash where there is no ⊎-number), it appears that the ⊎-num-
bers are getting ever sparser as we search for them among increasingly larger
numbers." Dr. Ulam drew the following:

Dr. Googol walked away from the great Ulam and began to make some inter-esting-looking plots using the following computer recipe:

```
DO For all Ulam Numbers, ⩊
    MovePenTo(⩊,0); DrawTo(⩊,⩊);
END
```

This looks like a series of unequally spaced vertical lines that gradually rise (Figure 77.1). The spacing is what Dr. Googol likes the best. It's very erratic, displaying miscellaneous gaps where no Ulam numbers exist. Many times there are visually interesting clumps and pairs. In your own computer program, you can DrawTo(⩊,C), where C is the vertical-most (y) coordinate of your graphics screen. This will give the plot a bar-code appear-ance. Looking at these kinds of graphs, can you determine if there are arbitrarily large gaps in the sequence of ⩊-numbers?

77.1 Visualization of ⩊-numbers.

Notice that on the ⩊-number graph there are pairs of consecutive ⩊-numbers corresponding to (1,2), (2,3), (3,4), and (47,48). Are there infinitely many consecu-tive pairs? In 1966, P. Muller (in his mas-ter's thesis at the University of Buffalo) calculated 20,000 terms and found no fur-ther examples! On the other hand, more than 60% of the ⩊-number terms differ from another by exactly 2.

Dr. Googol's ⩊-numbers started with 1 and 2. What are ⩊-numbers like for other starting integers?

⩊ See "Further Exploring" for other ideas and an introduction to ⊗-numbers.

Chapter 78

Creator Numbers ♌

> No definition of science is complete without a reference to terror.
> —*Don DeLillo,* Ratner's Star

On a cool April day in Athens, Greece, Dr. Googol approached a woman on the street who was selling gyro sandwiches. While waiting for the juicy meat to turn crispy brown, Dr. Googol handed her his card with the following formula:

The young woman took the card and turned it over in her well-manicured hands. "And what is *this* supposed to mean?"

$$\frac{\sqrt{\sqrt{2^{2^{2^2}}} \times \iiiint_i \frac{22\pi}{xy} \aleph_0 \, di}}{\sqrt{\frac{2\Delta\beta^2}{2\theta\gamma} \cdot \sqrt{\sum_i 2 \frac{22 \cdot \sqrt{2}}{\pi} \gamma 2}}} \; 2 \; \sqrt{\sqrt{2^{2^{2^2}}} \times \int_i 22 \frac{\pi \aleph_0}{\Psi \in 2\beta} \, di} \; \cdot \; \frac{\sqrt{\sqrt{2^{2^{2^2}}} \times \iiiint_i \frac{22\pi}{xy} \aleph_0 \, di}}{\sqrt{\frac{2\Delta\beta^2}{2\theta\gamma} \cdot \sqrt{\sum_i 2 \frac{22 \cdot \sqrt{2}}{\pi} \gamma 2}}} \; \sqrt{\sqrt{2^{2^{2^2}}} \times \iiiint_i \frac{22\pi}{xy} \aleph_0 \, di}$$

$$\frac{\sqrt{\sqrt{2^{2^{2^2}}} \times \iiiint_i \frac{22\pi}{xy} \aleph_0 \, di}}{\sqrt{\frac{2\Delta\beta^2}{2\theta\gamma} \cdot \sqrt{\sum_i 2 \frac{22 \cdot \sqrt{2}}{\pi} \gamma 2}}} \, 2 \, \sin \, \sqrt{\Delta\beta^2 \cdot \sqrt{\frac{\sum_i 2^2 2 \cdot \sqrt{2}}{2 \cdot \beta} \gamma 2}} \; \frac{\sqrt{\sqrt{2^{2^{2^2}}} \times \iiiint_i \frac{22\pi}{xy} \aleph_0 \, di}}{\sqrt{\frac{2\Delta\beta^2}{2\theta\gamma} \cdot \sqrt{\sum_i 2 \frac{22 \cdot \sqrt{2}}{\pi} \gamma 2}}} \; \sqrt{\frac{2\Delta\beta^2}{2\theta\gamma} \cdot \sqrt{\sum_i 2 \frac{22 \cdot \sqrt{2}}{\pi} \gamma 2}}$$

Dr. Googol grinned. "It means next to nothing. It is merely meant to impress."

"Impress?"

"Yes, doesn't it look impressive?"

"Here is your gyro." She reached into her pocket, smiled, and withdrew a small card upon which she scribbled. She handed him the card. "Why don't you call me sometime?"

Dr. Googol smiled back and nonchalantly stuck the card into his pocket. When he arrived at his apartment, he withdrew her card with exponentially increasing anticipation. Her card was in immaculate condition. On one side was the handwritten formula

$$\boxed{81 = (2^{2+1} + 1)^2}$$

On the other side were the words

Athens Psychiatric Hospital

with a phone number beneath.

We will probably never know why the woman wrote the enigmatic equation—Dr. Googol never found her again—but it stimulated Dr. Googol to conduct a bizarre contest. Participants were to construct numbers using just 1s and 2s, and any number of +, −, and × signs. People were also allowed exponentiation. As an example, let's first consider the problem where only the digit 1 is allowed. The number 80 could be written

$$80 = (1 + 1 + 1 + 1 + 1) \times (1 + 1 + 1 + 1) \times (1 + 1 + 1 + 1)$$

The *creator number* for a number n, symbolized as $\mathcal{Sl}(n)$, is the least number of digits that can be used to construct n. In the previous example, we see that $\mathcal{Sl}(80) \leq 13$ because thirteen 1s were used to create 80. A contest that allows only 1s for forming small numbers turns out not to be very interesting. However, once the digit 2 is also allowed, the problem becomes deep, fascinating, and filled with infinite wonders. Here is an example:

$$81 = (2^{2 + 1} + 1)^2$$

Here $\mathcal{Sl}(81) \leq 5$. Is this the best you can do with 1s and 2s?

The explicit goal of the Creator Numbers Contest is to represent the numbers 20, 120, and 567 with as few digits as possible. Dr. Googol received hundreds of responses and wishes that he could report all of the observations and entries in this chapter. Here are some examples. The first triplet of answers came from R. Lankinen of Helsinki, Finland:

$$\mathcal{Sl}(20) \leq 5, \quad \text{for } 20 = 2^{2 + 2} + 2 + 2$$

$$\mathcal{Sl}(120) \leq 6, \quad \text{for } 120 = ((2 + 1)^2 + 2)^2 - 1$$

$$\mathcal{Sl}(567) \leq 9, \quad \text{for } 567 = 2 \times 2 \times ((2 \times (2 \times 2 + 2))^2 - 2) - 1$$

But is this the best one can do for the 3 numbers? Can they be expressed with fewer digits? It turns out that 567 can be constructed with just 8 digits. Dan Hoey of Washington, D.C., the contest winner, computed the minimum values for all 3 numbers. Here are his minimal answers (which, Dr. Googol believes, use the smallest possible number of digits):

$$\mathcal{Sl}(20) \leq 5 \quad \text{for } 20 = (1 + 2 + 2) \times (2 + 2)$$

$$\mathcal{Sl}(120) \leq 6 \quad \text{for } (2 + (1 + 2)^2)^2 - 1$$

$$\mathcal{Sl}(567) \leq 8 \quad \text{for } (2^{2 + 2 + 2} - 1) \times (2 + 1)^2$$

The contest becomes more interesting if we allow concatenation of digits (thus permitting multidigit numbers such as 11, 12, 121, etc.). For this case, the winning entries come from Mark McKinzie of the University of Wisconsin's

Mathematics Department. Here are Mark's answers:

$$\mathscr{S}(20) \;\le\; 3 \quad \text{for } 20 = 22 - 2$$

$$\mathscr{S}(120) \;\le\; 4 \quad \text{for } 120 = 11^2 - 1$$

$$\mathscr{S}(567) \;\le\; 6 \quad \text{for } 567 = (2 + 1)^{2\,+\,1} \times 21$$

Another equally successful set of answers comes from Ya-xiang, Beijing, China:

$$\mathscr{S}(20) \;\le\; 3 \quad \text{for } 20 = 21 - 1$$

$$\mathscr{S}(120) \;\le\; 4 \quad \text{for } 120 = 121 - 1$$

$$\mathscr{S}(567) \;\le\; 6 \quad \text{for } 567 = 21 \times (2 + 1)^{2\,+\,1}$$

Can you do any better than these solutions?

❀ See "Further Exploring" for detailed analyses and additional challenges, including the search for *hard numbers*.

Chapter 79

Princeton Numbers

> Jesearc sat motionless within a whirlpool of numbers. He was fascinated
> by the way in which the numbers he was studying were scattered,
> apparently according to no laws, across the spectrum of integers.
> —*Arthur C. Clarke,* The City and The Stars, *1956*

In 1991, Dr. Googol visited David P. Robbins, a mathematician from Princeton, New Jersey, who had published an article in the *Mathematical Intelligencer* with the unusual title "The Story of 1, 2, 7, 42, 429, 7436, ... " The paper deals with an interesting sequence of integers starting with 1—but very quickly its members include behemoth numbers with hundreds of digits. The sequence can be represented by R_1, R_2, R_3, ... , and it can be computed using the following formula:

$$R_n = \Pi_{i=0}^{n-1} \frac{(3i+1)!}{(n+1)!}$$

Dr. Googol loves the fancy-looking symbol Π. Don't you? It simply indicates a repeated product. For example, $\Pi_{i=1}^{3} i = 1 \times 2 \times 3 = 6$. The exclamation point is the factorial sign: $n! = 1 \times 2 \times 3 \times \ldots n$. The computer code at [www.oup-usa.org/sc/0195133420] gives you additional hints on how to compute this repeated product for different values of n. For example, for $n = 2$ we need to compute the numerator and denominator for $i = 0$ and $i = 1$ and multiply the results: 1!/2! × 4!/3! = ½ × 24/6 = ½ × 4 = 2. Using the formula for R_n, it is not too difficult to determine the seventh and eighth terms of the series:

$$218347, 10850216$$

Dr. Googol has included a list of the first 25 numbers in Table 83.1 Do more of these numbers end in 00 than you would expect by chance? The 31st number (the largest Dr. Googol has computed) is:

7457901645375312545846943364460201024500933619811719342594448739658061730204945465190362255297438758806424576

Before going further and offering a challenge, let Dr. Googol tell you a bit about Dr. Robbins and the problem he was working on. Robbins is a mathematician at the Communications Research Division of the Institute for Defense Analysis in Princeton. He received his formal mathematics education at Harvard and MIT. Robbins refuses to state any mathematical specialty, insisting that he is "interested in any mathematical problem as long as its statement is easily understood and surprising." Robbins has enjoyed computers since childhood, beginning with a peculiar fascination with his father's Friden calculator. The sequence in the R_n equation has the mathematical community all in a quandary. In the last few years the sequence has arisen in 3 separate and distinct problems dealing with the analysis of combinations, and no one on Earth has been able to explain why. The details of the branch of mathematics called *combinatorics* are beyond the scope of this book but the next paragraph should whet your appetite by discussing 1 application.

ALTERNATING SIGN MATRICES

The R_n sequence seems to be relevant to the number of ways numbers can be arranged in special kinds of matrices. As most of you probably know, a matrix is an array of numbers organized in rows and columns. Here is an example of a matrix with 5 rows and 5 columns:

n	R
1	1
2	2
3	7
4	42
5	429
6	7436
7	218348
8	10850216
9	911835460
10	129534272700
11	31095744852375
12	12611311859677500
13	8639383518297652500
14	9995541355448167482000
15	19529076234661277104897200
16	64427185703425689356896743840
17	358869201916137601447486156417296
18	3374860639258750562269514491522925456
19	53580350833984348888784646149709092313244
20	1436038934715538200913155682637051204376827212
21	6497129499980842789584790438052414353885855143757
22	49620078383178087274695032966086932318270942177799731304
23	6396786003487969356007824036684854858931620602054541976694128
24	1391951305900289111219551784308097522786067722812246401574767313328
25	51125173829571287017224567391919410147905063533336189533617647958933056

Table 79.1 Robbins Princeton Numbers.

$$
\begin{bmatrix}
0 & 1 & 0 & 0 & 0 \\
1 & -1 & 0 & 1 & 0 \\
0 & 1 & 0 & -1 & 1 \\
0 & 0 & 0 & 1 & 0 \\
0 & 0 & 1 & 0 & 0
\end{bmatrix}
$$

This is a square N-by-N matrix where $N = 5$. Its entries are all 0s, 1s, and −1s, and its rows and columns sum to 1. Also notice that, upon omitting the 0s, the 1s and −1s alternate in every row and column. Such matrices are called *alternating sign matrices*. For $N = 1$ there is 1 alternating sign matrix, and for $N = 2$ there exist 2 alternating sign matrices. For $N = 3$ there are 7 matrices, including

$$\begin{bmatrix} 0 & 1 & 0 \\ 1 & -1 & 1 \\ 0 & 1 & 0 \end{bmatrix}$$

Can you find the other matrices? Notice that the number of different N-by-N alternating sign matrices appears to follow the sequence in the R_n formula at the beginning of this chapter: 1, 2, 7, ... We might be tempted to conjecture that R_n gives the number of alternating sign matrices with N rows and N columns. In fact Robbins has used a computer to check that this conjecture holds for all N up to $N = 16$. However, it's never been proved that this conjecture holds in general.

SOME CHALLENGES

Let us reconsider the first equation in this chapter. It is not obvious from the equation that the values of R_n are integers! Might there not be a value for n such that the denominator doesn't divide the numerator evenly? You need not wonder about this too long. Robbins says that all values of R_n are indeed integers. (Can you prove this?) Why not test this for yourself by making a list of a few numerator and denominator terms? Even if you do not have access to a computer, a pocket calculator should suffice for the first few terms.

Can you compute more than the 6 terms in the title of Robbins's article? Could the 31st term given in this chapter be the largest Robbins number ever computed? Can you break this record? On a computer, you could compute the product in the equation using

```
R=1
FOR 1=0 TO N-1
  R = R * factorial(3*1+1)/factorial(N+I)
END
PRINT R
```

where "factorial" is the factorial function. Perhaps this will give you a hint as to how to program the formula in the programming language of your choice.

❈ See "Further Exploring" for a zillion more challenges.
🔳 For computer recipes, see [www.oup-usa.org/sc/0195133420]

Chapter 80

Parasite Numbers

He dove his thumb into the soft glob of red licorice he held, making it a little bigger than the parasite which lay on Sarah's neck. . . . He bent forward toward the blistery growth. It was covered in a spiderweb skein of crisscrossing white threads, but he could see it beneath, a lump of pinkish jelly that throbbed and pulsed with the beat of her heart.

—*Stephen King,* Four past Midnight

"Help! Get it off me," Monica screamed.

Dr. Googol and Monica were exploring the deep jungles of Africa when she discovered a wet leech stuck to her ankle.

Dr. Googol nodded, withdrew a salt shaker from his backpack, and sprinkled salt on the leech. It began to scream and promptly dropped to the moist forest floor.

Monica took a deep breath. "Thank you."

They resumed their hike as Dr. Googol told Monica all about parasites.

The number 102,564 is a remarkable number that Dr. Googol discovered one day during his late-evening computer explorations. He calls this number a *parasite number,* for reasons that will soon become clear. In order to multiply 102,564 by 4, simply take the 4 off the right end and move it to the front to get the answer. In other words, the solution is the same as the multiplicand except that the number 4 on the right side is moved to the left end:

$$102{,}56\underline{4} \times 4 = \underline{4}10{,}256$$

Isn't this an incredible number? How many numbers with this quality exist within the numerical jungle, swimming peacefully and undetected in the swamp of mathematics? These kinds of numbers remind Dr. Googol of a biological organism that contains a parasite (digit) that roams around the body of the host organism (the multidigit number in which the parasite resides) as it gains energy by feeding (the multiplication operation). Dr. Googol has written several programs to search for parasite-containing numbers (or parasite numbers, for brevity), such as 102,564. If you search for all potential parasite numbers generated by different 1-digit multipliers, you'll find that they are exceedingly rare. It seems that the only parasite number less than 1 million is the 4-parasite 102,564. (The term *4-parasite* indicates that the number 4 is the multiplier.)

Do the other digits give rise to any parasite numbers? Are there multipliers for which no parasite number exists? How much computer time will be spent on this, now that Dr. Googol has asked this question?

There do exist occasional "pseudoparasites" lurking within the integers less than a million. These are numbers like 128,205, which when multiplied by 4 also move the last digit to the first position:

$$128,205 \times 4 = 512,820$$

(Dr. Googol calls these *pseudoparasites* only because the last migrating digit is not the same as the multiplier.) Here are some other 4-pseudoparasites:

$$153,846 \times 4 = 615,384$$
$$179,487 \times 4 = 717,948$$
$$205,128 \times 4 = 820,512$$
$$230,769 \times 4 = 923,076$$

Here is a 5-pseudoparasite: $142,857 \times 5 = 714,285$.

Both parasites and pseudoparasites seem to be as rare as diamonds. As Dr. Googol searches for parasites during the late-night hours, he challenges you to beat him in his search using the computer of your choice. On your mark. Get set. Go!

❀ See "Further Exploring" for more information on parasites so huge that no sane person should care about them.

Chapter 81

Madonna's Number Sequence

We are like the explorers of a great continent who have penetrated to its margins in most points of the compass and have mapped the major mountain chains and rivers. There are still innumerable details to fill in, but the endless horizons no longer exist.
—*Bentley Glass,* Scientific American, vol. 267, *1992*

On a fine frigid day, Dr. Googol's friend Madonna Möbius gave Dr. Googol and his disciples a number puzzle. "What is the significance of the following sequence of digits?" Möbius asked them.

$$4252603764690080434957$$

After rattling off the sequence, Möbius suffered a heart attack and died. The evening air was as astringent as alcohol, as Dr. Googol continued to study the sequence while the dead mathematician's body grew cold on the snowy ground. Even after years of study, no mathematician can fathom the mystery of this sequence. Can you?

❀ For a solution, see "Further Exploring."

Chapter 82

Apocalyptic Powers

We live on an island of knowledge surrounded by a sea of ignorance. As our island of knowledge grows, so does the shore of our ignorance.
—*John A. Wheeler,* Scientific American, vol. 267, *1992*

Dr. Googol presented the following number to his disciples and asked them what was special about it:

$$182,687,704,666,362,864,775,460,604,089,535,377,456,991,567,872$$

After much discussion, one young woman spoke: "It is the first power of 2 that exhibits 3 consecutive 6s."

Dr. Googol's disciples applauded with delight. In fact, this number is equal to 2^{157}. Dr. Googol calls numbers of the form 2^i that contain the digits 666 *apocalyptic powers* because of the prominent role 666 plays in the last book of the New Testament. In this book, called the Revelation (or Apocalypse) of John, 666 is designated as the Number of the Beast, the Antichrist.

Are there any other apocalyptic powers for higher values of i, or is this the only one? Dr. Googol has enlisted the help of IBM's Deep Blue computer in the computational search for double apocalyptic powers, which contain six 6s in a row, but he was never able to find an example. Can you find such a number?

❀ For a discussion, see "Further Exploring."

Chapter 83

The Leviathan Number ♉

> None is so fierce that dare stir Leviathan up.
>
> —*Job 41:10*

Dr. Googol's obsession with huge numbers reached maddening heights when he startled scientists with the monstrous Leviathan number (represented by the symbol ♉)—a number so large as to make the number of electrons, protons, and neutrons in the universe (10^{79}) pale in comparison. (It also makes a googol [10^{100}] look kind of small).

♉ is defined by the following identity:

$$\text{♉} = (10^{666})!$$

where the ! indicates factorial. It derives its name from a huge sea dragon or serpent of some kind that often symbolizes evil in Christian literature and in the Old Testament. The Leviathan number is also intimidating from a mathematical standpoint due to its probable incalculability, as we will soon see.

Recently Dr. Googol asked colleagues a number of questions pertaining to the Leviathan number. For example:

⊙ What are the first 6 digits of ♉?

⊙ Could modern supercomputers compute the Leviathan, or will this be beyond the realm of humankind for the next century?

⊙ Even if we cannot compute ♉, how many other characteristics of this number can we write down?

❀ For the answers, see "Further Exploring."

▮ See [www.oup-usa.org/sc/0195133420] for code explained in "Further Exploring."

Chapter 84

The Safford Number:
365,365,365,365,365,365

> Mathematics is the only science where one never knows what one is
> talking about nor whether what is said is true.
>
> —*Bertrand Russell*

What is special about the huge number

$$\boxed{365{,}365{,}365{,}365{,}365{,}365}$$

?

Dr. Googol's story begins with the calculating prodigy Truman Henry Safford (1836–1901) of Royalton, Vermont. When Safford was 10 years old, Reverend H. W. Adams asked him to square, in his head, the number 365,365,365,365,365,365. Dr. Adams reported:

> He flew around the room like a top, pulled his pantaloons over the tops of his boots, bit his hands, rolled his eyes in their sockets, sometimes smiling and talking, and then seeming to be in agony, until in not more than a minute said he, 133,491,850,208,566,925,016,658,299,941,583,255!

Truman Safford graduated from Harvard, became an astronomer, and soon lost the amazing computing powers he had in his youth.

Another prodigy was Johann Dase (1824–1861), who had incredible calculating skills but little mathematical training. He gave exhibitions of his calculating powers in Germany, Austria, and England. In 1849, while in Vienna, he was urged to use his powers for scientific purposes, and he discussed projects with mathematician Carl Gauss and others. In 1844, Dase used his calculating ability to calculate pi to 200 places. (This was published in *Crelle's Journal* for 1844.) Dase also constructed 7-figure log tables and produced a table of factors of all numbers between 7,000,000 and 10,000,000. Gauss requested that the Hamburg Academy of Sciences allow Dase to devote himself full-time to his mathematical work but, although they agreed to this, Dase died before he was able to do much more.

Legend has it that to compute pi Johann Dase used $\pi/4 = \arctan(1/2) + \arctan(1/5) + \arctan(1/8) \ldots$ with a series expansion for each arctangent. He ran the arctangent job in his brain for almost 2 months. Here is Dase's pi calculation:

3.14159 26535 89793 23846 26433 83279 50288 41971 69399 37510
58209 74944 59230 78164 06286 20899 86280 34825 34211 70679
82148 08651 32832 06647 09384 46095 50582 23172 53594 08128
48111 74502 84102 70193 85211 05559 64462 29489 54930 38196

Dase had an incredible brain. He could give the number of sheep in a flock after a single glance. He could multiply two 8-digit numbers in his head in 54 seconds, 2 40-digit numbers in 40 minutes, and two 100-digit numbers in 8 hours! Dase is said to have performed such computations for weeks on end, running as an unattended supercomputer. He would break off his calculation at bedtime, store everything in memory, and resume calculation after breakfast. Occasionally, Dase had a system crash.

❀ For Arthur C. Clarke's skeptical comment on Johann Dase, see "Further Exploring."

Chapter 85

The Aliens from *Independence Day*

Nobody has ever domesticated mankind. We are thus a wild species,
as wild as the day we first went howling across the savanna. Perhaps
the self-taming process of becoming a civilized species did not tame
us to visitors, but only to ourselves . . . and then not very well,
given our violent history.
—*Whitley Strieber,* Communion

Infinity is where things happen that don't.
—*S. Knight*

Dr. Googol was once daydreaming about walking through the Nevada desert with Captain Steven Hiller, the hero of the science-fiction movie *Independence Day*. Suddenly Dr. Googol heard a sound in the sky as a huge alien ship appeared above him veiled in fiery clouds. All over the Earth, alien crafts launched an incredible attack. The alien destroyers were 15 miles long, and the mother ship was 200 miles in length, both impossible for any Earthly weaponry to destroy.

An alien appeared before Dr. Googol. It must be a hoax. If the creature evolved on an alien world, why should it look so humanoid? The alien stood upright and was bilaterally symmetric; that is, its left and right sides looked the same. It had fingers, 2 jointed legs and arms, a head with 2 eyes, and a large cranium. In fact, stripped of its biomechanical armor, the alien looked more like a human than does an Earthly lemur, with whom we share greater than 95 percent of our genetic material.

Science-fiction writers have explored a far greater diversity of alien life forms in books than Hollywood can ever explore in movies, because the Hollywood alien must trigger instantaneous emotional impact; this requires a design based on recognizable human facial expressions of threat and menace. In fact, most of the "evil" Hollywood aliens since the 1953 *War of the Worlds* have had a tendency to look mean and cranky, or like skullfaced sex-fiends. In reality, if we ever meet real aliens we will have a hard time understanding their moods by looking at them.

❀ ❀ ❀

Dr. Googol slowly came out of his dream and began to formulate a mathematical problem involving aliens abducting humans. In Dr. Googol's scenario, the scary alien ☻ from *Independence Day* comes down to Earth, captures 1 male human 🚹 (in this case, the U.S. president), and takes him to a large spaceship hovering in the Earth's upper atmosphere. The creature realizes that the male is unhappy without a companion, so the next year it abducts 1 female 🚺 .

Each succeeding year the creature duplicates its removals of the preceding 2 years, stealing the same number of humans, of the same sex, and in the same order. Thus, in the third year, the creature captures a male and then a female

In the fourth year, it takes a female, a male, and then a female; and so on. The sequence goes as **M, F, M F, F M F. . . .**

Is it possible to determine the sex of the one-billionth human taken? Would the captured males be satisfied with the ratio of females to males (🚺 / 🚹) existing on the spaceship when the one-billionth human is taken? (Assume that the humans do not breed for the duration of this experiment.)

It turns out that the sex of the nth human is not too difficult to compute. In fact, in 1957, an obscure little paper was written on this class of problem, and a generating formula was discovered by T. F. Mulcrone of Loyola University. The Mulcrone formula can be adapted to Dr. Googol's questions as follows. Let's denote a male by the number 1 and a female by the number 2. The sequence of males and females then becomes M_n = 1, 2, 1 2, 2 1 2, ... The nth term M_n can be quickly computed from Mulcrone's formula: $M_n = [kn] - [k(n-1)]$, where $k = (\sqrt{5} + 1)/2$, and the brackets indicate an integer truncation. In other words, $[x]$ is the greatest integer not exceeding x.

C and BASIC program listings are provided so that you can compute the *n*th term of the sequence. You can also use these programs to compute the ratio of males to females by maintaining a count of their numbers through time. Why not make a plot of the growth in the number of males and females through the years? You should find that after 500 years there will be 191 males and 309 females:

"We men are happy in 500 years!"

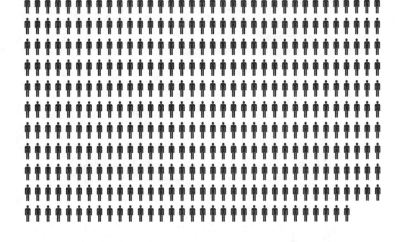

"We women need more men."

After 1,000 years, there will be 382 males and 618 females.

Armed with these simple yet powerful programs, we can now determine the sex of the one-billionth abductee. If the first abduction is considered to have taken place in year 0, then, using the Mulcrone formula, we can determine that it takes only 42 years for the alien to accumulate 1 billion humans. (This is about a sixth of the world's population and roughly the number of Chinese.) The ratio of females to males is 1.618 to 1, which, as one computer programmer from Boca Raton said, "is better than the ratio in most bars."

❀ For further alien explorations, see "Further Exploring."

▣ See [www.oup-usa.org/sc/0195133420] for program examples.

Chapter 86

One Decillion Cheerios

One thing I know and that is that I know nothing.

—*Socrates*

On Christmas Day 1999, Dr. Googol was eating a bowl of Cheerios® cereal, staring at all the doughnut-shaped morsels, daydreaming about endless streams of 0s. As he started to line up pieces of breakfast cereal in a row with his left hand, he used his right hand to list all the special qualities of the number *1 decillion*. First, he noted, a decillion is a very large number: 10 raised to the power of 33, or 1 followed by 33 zeros. He formed the number with Cheerios:

---
one decillion =
1,000,000,000,000,000,000,
000,000,000,000,000

Aside from its obvious enormity, there is something unbelievably special about a decillion, which we can write more succinctly as 10^{33}. Before revealing the strange answer, let's get an idea about how large 1 decillion is. It's greater than the number of atoms in a human breath (10^{21}). However, it's smaller than the number of electrons, protons, and neutrons in the universe (10^{79}). What Dr. Googol finds most interesting is that 10^{33} is the largest power of 10 known to humans that can be represented as the product of 2 numbers which themselves contain no zero digits:

$$10^{33} = 2^{33} \times 5^{33} = (8{,}589{,}934{,}592) \times (116{,}415{,}321{,}826{,}934{,}814{,}453{,}125)$$

For a variety of technical reasons, some mathematicians believe that no one will ever be able to find a larger power of 10 that can be represented as the product of 2 numbers which themselves contain no zero digits.

Oh my! Do you think humanity will ever find such a number? Dr. Googol's personal opinion is that the answer is "Never!" Could he be wrong?

❀ See "Further Exploring" for additional analysis.

Chapter 87

Undulation in Monaco

Such as say that things infinite are past God's knowledge may just as well leap headlong into this pit of impiety, and say that God knows not all numbers. . . . What madman would say so? What are we mean wretches that dare presume to limit His knowledge?

—*St. Augustine*

The number 69,696 is a remarkable number and certainly among Dr. Googol's top 10 favorite integers. For one thing, the number starts with 696, the very year that Dr. Googol's favorite Chinese poet, Chen Zi'ang, composed the following haunting poem:

Ballad on Climbing Youzhou Tower

Witness not the sages of the past,

Perceive not the wise of the future,

Reflecting on heaven and earth eternal,

Tears flowing down I lament in loneliness.

—*Chen Zi'ang, A.D. 696*

Moreover, 69,696 is almost exactly equal to the average velocity in miles per hour of the Earth in orbit, and it is also the surface temperature in degrees Fahrenheit of some of the hottest stars. More important are its fascinating mathematical properties.

One day while in his Monaco villa, Dr. Googol presented this number to his friend Dorian and said: "What do you find significant about 69,696?"

Dorian gazed into the Mediterranean Sea as it crashed into huge rocks. In the fractured sides and grottos of the massive cliffs were strange, rich blues and weathered aquamarines. After a few seconds, she replied, "That is too easy. It is the largest undulating square known to humanity."

Dr. Googol pondered this answer, and he himself started to undulate in a mixture of excitement and perhaps even terror. The sounds of the sea became deafening as it surged into the cavernous bellies of worn boulders and exploded in steepled and gabled sprays of foam.

To understand Dr. Googol's passionate response, we must digress to some simple mathematics. As discussed in Chapter 52, *undulating numbers* are of the form *abababababab.* . . . For example, 171,717 and 28,282 are undulating numbers. A *square number* is of the form $y = x^2$. For example, 25 is a square number. So is 16. An undulating square is simply a square number that undulates.

When Dr. Googol conceived the idea of undulating squares a few years ago, it was not known if any such numbers existed. It turns out that $69,696 = 264^2$ is indeed the largest undulating square known to humanity, and most mathematicians believe we will never find a larger one.

Dr. Noam D. Elkies from the Harvard Mathematics Department wrote to Dr. Googol about the probabilities of finding undulating squares. The chance that a "random" number around x is a perfect square is about $1/\sqrt{x}$. More generally, the probability is $x^{(-1+1/d)}$ for a perfect dth power. Since there are (for any k) only 81 k-digit undulants, one would expect to find very few undulants that are also perfect powers, and none that are very large. Dr. Elkies believes that listing all cases may be impossible using present-day methods for treating exponential Diophantine (integer) equations.

❀ See "Further Exploring" for more information on undulating squares as well as on other undulants such as undulating prime numbers.

Chapter 88

The Latest Gossip
on Narcissistic Numbers

The brain is a three-pound mass you can hold in your hand that can conceive of a universe a hundred-billion light-years across.
—*Marian Diamond*

Number is the bond of the eternal continuance of things.
—*Plato*

Dr. Googol was watching the TV show *Xena: Warrior Princess*. In this particular episode, Xena was gazing at her beautiful physiognomy reflected in a pool of water.

Monica turned to Dr. Googol. "She's such a narcissist!"

"Narcissist?"

"Yes, she can't take her eyes off herself."

"Hold on. This reminds me of something infinitely more interesting than Xena." Dr. Googol paused to collect himself, then reached for a piece of chalk. "*Narcissistic numbers* are the sums of powers of their digits. In other words, they are n-digit numbers that are equal to the sum of the nth powers of their digits." Dr. Googol went over to the blackboard attached to the top of his TV. He wrote an example of a narcissistic number:

$$\boxed{153 = 1^3 + 5^3 + 3^3}$$

"The numbers 370 and 371 are also narcissistic numbers. Variously called narcissistic numbers, numbers in love with themselves, Armstrong numbers, or perfect digital variants, this kind of number has fascinated number theorists for decades. For example, the English mathematician Godfrey Hardy (1877–1947) said, 'There are just four numbers, after unity, which are the *sums of the cubes* of their digits. . . . These are odd facts, very suitable for puzzle columns and likely to amuse amateurs, but there is nothing in them which appeals to the mathematician.'"

Xena began to slash at a bunch of thieves with her huge, glittering sword.

Dr. Googol continued. "I gave 153 as an example of such a number. Can you find other narcissistic numbers? Can you find larger narcissistic numbers?"

Monica took the chalk from Dr. Googol's hand. "There's time to think about that later. Let's just watch the show."

⊛ ⊛ ⊛

The largest narcissistic number discovered to date is this incredible 39-digit number:

$$115{,}132{,}219{,}018{,}763{,}992{,}565{,}095{,}597{,}973{,}971{,}522{,}401$$

(Each digit is raised to the 39th power!) Can you beat the world record? What would Godfrey Hardy have thought of this multidigit monstrosity? What is the density of narcissistic numbers? In other words, are there 4-digit narcissistic numbers, 5-digit numbers, 6-digit numbers, etc., or do they get progressively rarer as one searches for ever-larger examples?

❀ See "Further Exploring" for more on narcissistic numbers and for the latest gossip on lonely numbers ⋏ called factorions.

▤ See [www.oup-usa.org/sc/0195133420] for help computing these kinds of numbers.

Chapter 89

The abcdefghij Problem

> I do not know what I may appear to the world, but to myself I seem to have been only a boy playing on the sea shore, and diverting myself now and then finding a smoother pebble or a prettier sea shell than ordinary whilst the great ocean of truth lay all undiscovered before me.
>
> —*Isaac Newton*

Dr. Googol was visiting the IBM T. J. Watson Research Center in Yorktown Heights, New York, when he walked up to a blackboard and wrote down:

$$(ab)c = def \times ghij$$

Several of the IBM researchers stared with amusement at the odd formula. Others pointed with hyperbolically increasing interest and gestured and took notes. From the long hallway came several security guards, evidently curious as to how Dr. Googol had gotten into the building without a proper security pass.

Dr. Googol bowed and then motioned to the wonderful-looking equation. "Each letter stands for a number from 0 to 9. Can you find values for *a, b, c, d, e, f, g, h, i,* and *j* that make this equation correct? Each digit must be unique. For example, the first expression could be 12^3 but not 12^2 because the 2 is repeated."

The researchers jolted away to the nearest computers and began to furiously code the problem.

Dr. Googol clapped his hands. "I'm sure your managers will realize that this is a wonderful programming exercise and justify the time you are spending on the problem. As a reward, I will give the first person who solves this autographed copies of Dr. Cliff Pickover's recent books *The Science of Aliens* and *Strange Brains and Genius: The Secret Lives of Eccentric Scientists and Madmen.*"

The scientists and programmers roared with delight as their nimble fingers raced across their keyboards like angelfish swimming through clear water.

✸ See "Further Exploring" for a solution and for much tougher problems.
▤ See [www.oup-usa.org/sc/0195133420] for computer code used to solve this problem.

Chapter 90

Grenade Stacking

Students must learn that mathematics is the most human of endeavors. Flesh-and-blood representatives of their own species engaged in a centuries-long creative struggle to uncover and to erect this magnificent edifice. And the struggle goes on today. On the very campuses where mathematics is presented and received as an inhuman discipline, cold and dead, new mathematics is created. As sure as the tides.
—*J. D. Phillips,* Humanistic Mathematics Network Journal, no. 12, Oct. 1995

While on his tour of duty in Vietnam, Dr. Googol was stacking grenades in such a way that each layer formed a square array of grenades. For example, the top of the square pyramid contained 1 grenade, the next 4 grenades, the next 9, and so on.

The number of grenades in the entire pyramid was therefore a sum of consecutive squares, beginning with 1:

$$1^2 + 2^2 + 3^2 + 4^2 + \ldots .$$

"I have an amazing problem!" he thought.

He began to wonder if he could find a sum of consecutive squares, beginning with 1, that equaled a square number n; for example, $1^2 + 2^2 + 3^2 + 4^2 + \ldots .$ $= n^2$. It turns out that the *only* nontrivial solution known to humanity is $1^2 + 2^2 + 3^2 + 4^2 + \ldots . + 24^2 = 70^2$.

Are there other solutions if we allow *any* set of k consecutive squares (not necessarily beginning with 1) such that the sum is a square number? Are there solutions for consecutive cubes such that the sum is a cubical number? Would you like to be the first person on Earth to find these?

❦ See "Further Exploring" for more comments and challenges.

Chapter 91

The 450-Pound Problem

> Mathematicians study structure independent of context, and their science
> is a voyage of exploration through all the kinds of structure and order
> which the human mind is capable of discerning.
> —*Charles Pinter,* A Book of Abstract Algebra

On New Year's Eve 1994, Dr. Googol was leafing through the *Sunday Telegraph of London.* The newspaper offered a cash prize (450 pounds sterling) to the first person who sent them a solution, in coprime positive integers greater than 100, for this equation:

$$(A^3/B^3) + (C^3/D^3) = 6$$

Let's say that again in simple English. Two integers are said to be coprime if their greatest common divisor equals 1. For example, 5 and 9 are coprime, while 6 and 9 are not coprime because their greatest common divisor is 3. Unfortunately, Dr. Googol had no time to solve this problem, and mathematician Kevin Brown beat him to the prize. Finding the solution was quite a challenge, and Kevin determined that the solutions involved incredibly large numbers:

$A = 7922205726625496081902529261121216176860879394382456 6$
$58060516086211136418303364504481154195247725686 39$

$C = 6779598051038214247232639926650618387735733751387073 79$
$3470619938609337529235682974731855779658576736 1$

$B = D = 4360668418820711170950024593240851673665433429374$
$773448186461962793853054415068610177019469294891111 20$

Let's see you solve this with pencil and paper! The offer turned out to be legitimate, and the *Telegraph* actually did send Kevin Brown £450 ($706.50). (Kevin, a controls engineer at the Boeing Company in Seattle, Washington, graduated from the University of Minnesota with a bachelor's and master's degrees in engineering. See his fascinating math Web page at http://www.seanet.com/~ksbrown/index.htm.)

The famous mathematician Adrien Marie Legendre (1752–1833) once stated that this particular equation had no solutions, although it's not clear why he thought so. (Technically speaking, when $B = D$, B and D are usually not considered coprime; perhaps Legendre was right after all!) In any case, if we permit $B = D$, there is a smaller solution, namely $(17/21)^3 + (37/21)^3 = 6$. (But the *Telegraph*'s stipulation that the integers be greater than 100 was clearly intended to exclude this easier solution.)

Incidentally, mathematicians currently do not believe that there are any positive integer solutions such that $x^2 + y^3 = z^6$. (In other words, x, y, and z must be three positive integers.)

Here are some other examples that appear to have no integer solutions. Except for $n = 2$, $m = 3$ and $n = 8$, $m = 9$, there appear to be no other solutions for this deceptively simple-looking formula:

$$\boxed{\,|\,2n - 3m\,|\,=1\,}$$

Even more interesting, the equation $a^n - b^m = 1$ has no positive integer solutions with $m, n > 1$ other than $a = 3$, $n = 2$, $b = 2$, $m = 3$. Dr. Googol believes that mathematicians have proved that $|\,2^n - 3^m\,| > (2^n)e^{(-n/10)}$ for $n > 27$, and also that there exists a number $c \geq 1$ such that $|\,2^n - 3^m\,| > (2^n)/(n^c)$.

Dr. Googol does not know if there are other solutions to the problems in this chapter.

❁ For more on coprime numbers, see "Further Exploring."

Chapter 92

The Hunt for Primes in Pi

On the basis of my historical experience, I fully believe that mathematics
of the twenty-fifth century will be as different from that of today as the
latter is from that of the sixteenth century.
—*George Sarton,* A History of Science, *1959*

Last summer, Dr. Googol jumped from a C-130's cargo ramp at 29,000 feet—
the height he needed to carry him within striking distance of his target. From his
pistol belt was suspended a Heckler & Koch USP 9mm semiautomatic pistol.
His vest was equipped with class II body armor. He breathed oxygen through a
small tank on his back. The jump—a HALO (high altitude, low opening) inser-
tion—would bring him right on target: Beijing, China.

As he fell through the dark sky, he turned to Monica, his partner in the covert
operation.

"Monica, 3 is a prime number. So is 31. These numbers are also the first and
first 2 digits in the decimal expansion of $\pi = 3.14159. \ldots$. I'm wondering if
there are other integers k such that the first k decimal digits of π are prime? Can
you find any? Do you think they are commonplace?"

The rushing wind whipped through Monica's hair like a flock of seagulls. "Dr.
Googol, it turns out that 314,159 ($k = 6$) is also a prime number."

"Oh, Monica, you've made me so happy!"

"Dr. Googol, can you tell me why we are going to infiltrate military installa-
tions around the world? Are we going to disable the small computers of rogue
terrorists? Are we going to disable the atomic weapons of the less stable super-
powers?"

"In a manner of speaking, yes. We are going to have their computers begin to
hunt for pi-primes. This will render the military ineffective and bring world
peace."

Before Dr. Googol and Monica opened their chutes, Dr. Googol wondered if
the next pi-prime would ever be found. Would it be so large that it is beyond the
reach of modern supercomputers? Perhaps the next pi-prime (symbolized by
$\pi^\frown$) will be relegated to the realm of myth, like the superhuman Olympian gods
of yore.

❀ See "Further Exploring" for more comments on pi-primes.

Chapter 93

Schizophrenic Numbers

> The pursuit of mathematics is a divine madness of the human spirit.
> —*Alfred North Whitehead,* Science and the Modern World

Brilliant mathematician Kevin Brown seems to have discovered a wonderfully weird set of numbers called *schizophrenic numbers*, $\mathcal{S}$. For any positive integer n, let $f(n)$ denote the integer given by the recurrence

$$f(n) = 10 \times f(n-1) + n$$

with the initial value $f(0) = 0$. Think of this as a mathematical feedback loop. You plug in a number, and out comes a solution. You plug the solution back into the formula, and out comes a new solution, and so on. For example:

$$f(1) = 10 \times 0 + 1 = 1$$
$$f(2) = 10 \times 1 + 2 = 12$$
$$f(3) = 10 \times 12 + 3 = 123$$
$$f(4) = 10 \times 123 + 4 = 1234$$

"This sequence looks boring," you say to Dr. Googol? Ah, but here's where the schizophrenia begins. The square roots of these numbers $f(n)$ for odd integers n give a bizarre, persistent pattern. The square roots appear to be "rational" for periods—that is, a number that can be expressed as a ratio of 2 integers—and then disintegrate into irrationality. (Recall that rational numbers sometimes have infinitely repeating strings of a digit; for example, $1/3 = 0.33333333. \ldots$) This mathematical schizophrenia is exemplified below by the first 500 digits of $\mathcal{S} = \sqrt{f(49)}$ (typeset to show the interesting patterns):

$$\sqrt{f(49)} =$$

11111111111111111111111111.11111111111111111111111
0860
55555555555555555555555555555555555555555555555555
2730541
66666666666666666666666666666666666666666666666
0296260347
22222222222222222222222222222222222222222222

0426563940928819
44444444444444444444444444444
38775551250401171874
9999999999999999999999999999
8082496877114863053385541
666666666666666666666666
5987185738621440638655598958
333333333333333333333
08434604076276082069402770996093374
99999999999999
0642227587555983066639430321587456597
222222222
18634920167911808333081844 . . .

Isn't this a splendid arrangement of digits? If you look closely at 𝒮 (49), you'll see that the digit sequence consists of repeated digits alternating with "random-looking" strings. The repeating strings become progressively smaller, and the irregular strings become larger, until eventually the repeating strings disappear— as if a numerical God has turned off water from a mathematical fire hose. However, by increasing *n* we can slow down the eventual demise of repeating digits. Oddly enough, the repeating digits are always 1, 5, 6, 2, 4, 9, 6, 3, 9, 2, ... Why is this so? We may call this sequence (1, 5, 6, 2, . . .) the *schizophrenic sequence*—the key to calmness in an otherwise chaotic world.

The construction and discovery of schizophrenic numbers was prompted by a claim (posted in the Usenet newsgroup sci.math) that the digits of an irrational number chosen at random would not be expected to display obvious patterns in the first 100 digits. It was said that if such a pattern were found, it would be irrefutable proof of the existence of either God or extraterrestrial intelligence. (An irrational number is any number that cannot be expressed as a ratio of 2 integers. Transcendental numbers like *e* and pi, and noninteger surds such as $\sqrt{27}$ are irrational.)

It's obvious from 𝒮 (49) that certain easy-to-construct irrational numbers are filled with wonderful patterns that are ripe for future exploration. Dr. Googol looks forward to hearing from anyone who makes other wonderful discoveries in the little-researched area of large schizophrenic numbers.

Chapter 94

Perfect, Amicable, and Sublime Numbers

Just as the beautiful and the excellent are rare and easily counted,
but the ugly and the bad are prolific, so also abundant and deficient
numbers are found to be very many and in disorder, their discovery
being unsystematic. But the perfect are both easily counted and drawn
up in a fitting order.

—*Nichomachus*, A.D. *100*

Man ever seeks perfection but inevitably it eludes him. He has sought
"perfect numbers" through the ages and has found only a very few—
twenty-three up to 1964.

—*Albert H. Beiler*, Recreations in the Theory of Numbers

Dr. Googol raises his hand. "Monica, I want to tell you about perfection." His voice is a whisper, as if he is afraid he is being watched.

"Perfection, sir?"

Dr. Googol nods. "*Perfect numbers* are the sum of their proper divisors. For example, the first perfect number is 6 because 6 = 1 + 2 + 3. (A proper divisor is simply a divisor of a number N excluding N itself.) The next perfect number is 28 because its divisors are 1, 2, 4, 7, and 14—and 28 also equals 1 + 2 + 4 + 7 + 14."

Monica's eyes seem to be locked onto Dr. Googol's hairy mustache and golden birthmark. "Dr. Googol, there must be other perfect numbers."

"Yes, but these numbers are so rare that they have a special significance in my heart." Dr. Googol pauses. "I think perfection is rare in numbers just as goodness and beauty are rare in humans. On the other hand, imperfect numbers are common, and so is ugliness and evil."

"Imperfect numbers?"

"Those where the sum of the factors is greater or less than the number itself."

Monica nods. "My friend Bill mentioned *abundant numbers* to me. Can you explain what these numbers are?"

Dr. Googol pinches his lower lip with his teeth. "How dare he reveal that secret!" Dr. Googol then takes a deep breath. "If the original number is less than the sum of its factors, I call it *abundant*. As an example, the factors of 12 are 1, 2, 3, 4, and 6. And these factors add up to 16. If greater, the number is *deficient*. For example, the factors of 8—1, 2, and 4—add up only to 7."

"Most numbers are either abundant or deficient? Perfection is rare."

Dr. Googol nods. "You've got it!" Then he leans toward Monica as if observing a painting in a museum. "Monica, two numbers are *amicable,* or friendly, if the sum of the divisors of the first number is equal to the second number, and vice versa. The ancient philosophers considered them to have the same parentage, and in their divine world these numbers are more congenial than numbers that are unfriendly."

"I don't get it."

"Here's an example. 220 and 284 are amicable. Let's list all the numbers by which 220 is evenly divisible."

Monica leans forward and clasps her hands together like an eager child. "Uh, let's see—1, 2, 4, 5, 10, 11, 20, 22, 44, 55 and 110, all go into 220."

"Excellent. Now add up all those divisors. What do you get?"

"$1 + 2 + 4 + 5 + 10 + 11 + 20 + 22 + 44 + 55 = 284$."

"Very good, Monica. The answer is 284. Now let's try the same trick with 284. Its perfect divisors are 1, 2, 4, 71, and 142. Now, add them up."

"You get 220, Dr. Googol."

"Yes! Therefore 220 and 284 are amicable numbers. The sums of their divisors are equal to each other."

Monica nods. "Interesting. Amicable numbers, like perfect ones, are quite rare."

"War is always easier than peace."

"220 and 284 would be perfect marriage partners in the eyes of a numerical God."

Dr. Googol nods. "A perfect marriage."

Dr. Googol walks over to a wall of the White House and scrawls on it with a piece of charcoal. He lowers his voice an octave, and he thinks he sees awe in Monica's eyes. "The first 4 perfect numbers—6, 28, 496, and 8,128—were known to the late Greeks. Nicomachus and Iamblichus knew about these."

Monica raises her hand. "Do all perfect numbers end in an 8 or 6?"

"I'm not sure. But I do know that every even perfect number is also a triangular number." He pauses. "Perfect numbers are very rare. The fifth perfect number, 33,550,336, was found recorded in a medieval manuscript. To date, mathematicians know only about 30 perfect numbers. No one knows if the number of perfect numbers is infinite."

A chill goes down Dr. Googol's spine when he says the word *infinite.*

He begins to pace. "Perfect numbers thin out very quickly as you search larger and larger numbers. They might disappear completely—or they might continue to hide among the multidigit monstrosities that even our computers can't find."

Monica raises her hand. "What about amicable numbers?"

Dr. Googol nods. "Over a thousand amicable numbers have been found. Another pair includes 17,296 and 18,416."

On her notebook computer, Monica begins to furiously type a program to search for and print amicable numbers. The computer soon prints several numbers on a slip of paper:

Amicable Numbers

220	and	284	5,020	and	5,564
1,184	and	1,210	6,232	and	6,368
2,620	and	2,924	10,744	and	10,856

"Good work, Monica."

Monica takes the slip of paper and studies it.

Dr. Googol continues his discussion. "Mathematicians have also studied *sociable* numbers. In these sets of numbers, the sum of the divisors of each number is the next number of a chain. For example, in 1918 a man named Poulet found the following sociable number chain:

$$12,496 \rightarrow 14,288 \rightarrow 15,472 \rightarrow 14,536 \rightarrow 14,264 \rightarrow 12,496$$

Sociable chains always return to the starting number. Poulet's chain and a 28-link chain starting with 14,316 were the only sociable chains known until 1969, when suddenly Henri Cohen discovered seven new chains, each with 4 links." (See Figure 94.1)

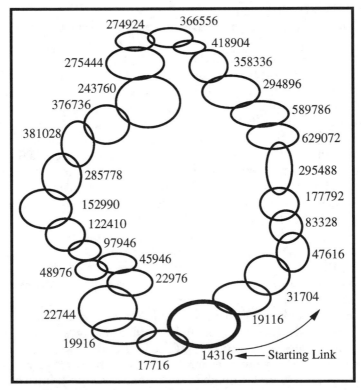

94.1 A wonderful 28-link amicable number chain.

Dr. Googol's voice grows in intensity and speed. "A pair of amicable numbers, such as 220 and 284, is simply a chain with only 2 links. A perfect number is a chain with only 1 link." He takes a deep breath. "No chains with just 3 links have been found, despite massive searches. There are certainly none with a smallest member less than 50 million! These hypothetical 3-link chains are called *crowds*. Mathematically speaking, a crowd is a very elusive thing and may not exist at all."

"Dr. Googol, you talk about discovering numbers as if we're searching for stars in the heavens."

"It's a little like that. There's a lot of unexplored territory."

Just then the floor begins to shake. Dr. Googol and Monica look warily from one to the other like condemned criminals.

"Dr. Googol, we never should've stayed here so long. What if the White House staff found one of our computers? We could be in deep trouble."

"It's okay. I'm friends with the president. He lets me use this office. In return, I advise his staff on economic issues."

"Okay."

"Before we leave, I want to tell you about some numerical beasts even rarer than the perfect numbers." Dr. Gogool walks over to a wall and begins to sketch. "For any positive integer n let $\xi(n)$ and $\Psi(n)$ denote the number of divisors of n and sum of the divisors of n, respectively. A number N is called *sublime* if $\xi(N)$ and $\Psi(N)$ are both perfect numbers. The only 2 known sublime numbers are 12 and this one:"

$$60865556702383789896703717342431696226578307733518859705$$
$$28324860512791691264$$

The latter number was discovered by Kevin Brown. (12 is sublime because its divisors are 12, 6, 4, 3, 2, and 1. The number of divisors is therefore perfect, as is the sum of its divisors.)"

"Amazing."

"Monica, here are my final questions for you. Will humanity ever be able to find another sublime number, or prove that no others exist? Can there exist an odd sublime number?"

❀ See "Further Exploring" for more on abundant, amicable, and perfect numbers.

▣ See [www.oup-usa.org/sc/0195133420] for computer code used to find perfect and amicable numbers.

Chapter 95

Prime Cycles and ∃

> The real voyage of discovery consists not in seeking new landscapes
> but in having new eyes.
> —*Marcel Proust*

Any positive integer can be expressed as the product of primes in just one way. For example, 10 = 5 × 2 and 24 = 2 × 2 × 2 × 3. Let's define a new function (*n*) which is the sum of the prime factors of *n*. For example (24) = 2 + 2 + 2 + 3 = 9. As far as Dr. Googol can tell, iterations of the form $x \rightarrow ∃(ax + b)$ invariably lead to closed loops for any integers *a* and *b*. By *closed loops*, Dr. Googol means a repeating sequence of integers. For example, mathematician Kevin Brown has discovered that if you use any initial value of *x* less than 100,000, iteration of ∃(8*x* + 1) always leads to the 23-step cycle

66 → 46 → 47 → 42 → 337 → 63 → 106 → 286 → 119 → 953 →
76 → 39 → 313 → 175 → 470 → 3761 → 30089 → 367 → 103 → 24 →
193 → 111 → 134 → 66 . . .

On the other hand, iteration of ∃(7*x* + 3) always leads to 1 of the following 2 cycles for any initial value of *x*:

cycle #1: 30 → 74 → 521 → 85 → 38 → 269 → 66 → 39 → 30 . . .
cycle #2: 92 → 647 → 118 → 829 → 2905 → 10171 → 109 → 385 → 92 . . .

One particularly long loop occurs for ∃(13*x* + 12), which has a period of 59 and appears to be the only possible limit loop for this function. Dr. Googol wonders if every iteration is eventually periodic, and if there is a finite number of limit cycles for any given function.

Can you shed further light on these strange prime cycles? The first person to make a new discovery and mail it to Dr. Googol receives a beautiful fractal print.

Chapter 96

Cards, Frogs, and Fractal Sequences

A mathematician who is not also something of a poet will never be a
complete mathematician.

—*Karl Weierstrass*

Make a set of cards numbered 1, 2, 3, . . . *n* and hold them face up in your hand.
Take the top card and place it face up on the bottom of the deck. Place the next
card face up on a table. Continue this process until all *n* cards are face up on the
table. How far down in the pile on the table do you have to look to find the original top card?

The answer relates to a sequence that begins with

1, 1, 2, 1, 3, 2, 4, 1, 5, 3, 6, 2, 7, 4, 8, 1, 9, 5, 10, 3, 11, 6, 12, 2, 13, 7, 14,
4, 15, 8, . . .

For example, if you use 5 cards numbered, in order, 1, 2, 3, 4, and 5, the initial 1 will be the third card in the deck on the table. Interestingly, this sequence
is fractal, containing infinite "copies" of itself. You can test this for yourself. If
you delete the first occurrence of each integer, you'll see that the remaining
sequence is the same as the original:

~~1~~, 1, ~~2~~, 1, ~~3~~, 2, ~~4~~, 1, ~~5~~, 3, ~~6~~, 2, ~~7~~, 4, ~~8~~, 1, ~~9~~, 5, ~~10~~, 3, ~~11~~, 6, ~~12~~, 2, ~~13~~, 7, ~~14~~, 4, ~~15~~, 8, . . .

Do it again and again, and you get the same sequence! Can you create a formula to generate the *k*th member of this sequence? What will the top card be for
a deck of 100 cards? (See "Further Reading" for Clark Kimberling references on
this interesting sequence.)

❁ ❁ ❁

Another example of a fractal sequence is the "signature sequence" of a positive
irrational number *R*, such as $\sqrt{2}$. To create this amazing sequence, arrange the set
of all numbers $i + jR$, where *i* and *j* are nonnegative integers, in ascending order:

$$i(1) + j(1)R < i(2) + j(2)R < i(3) + j(3)R < \ldots$$

Then $i(1)$, $i(2)$, $i(3)$, . . . defines the signature of *R*. For example, the signature of the square root of 2 starts with

1, 2, 1, 3, 2, 1, 4, 3, 2, 5, 1, 4, 3, 6, 2, 5, 1, 4, 7, 3, 6, 2, 5, 8, 1, 4, 7, 3, 6, 9, 2, 5, 8, . . .

If you delete the first occurrence of each integer, you'll see that the remaining sequence is the same as the original. To compute this sequence, all Dr. Googol did was to write down the first few possibilities for $i + j \times \sqrt{2}$ and arrange them in order from least to greatest:

1. $1 + 1 \times \sqrt{2} = 2.414$	6. $1 + 3 \times \sqrt{2} = 5.243$
2. $2 + 1 \times \sqrt{2} = 3.414$	7. $4 + 1 \times \sqrt{2} = 5.414$
3. $1 + 2 \times \sqrt{2} = 3.828$	8. $3 + 2 \times \sqrt{2} = 5.828$
4. $3 + 1 \times \sqrt{2} = 4.414$	9. $2 + 3 \times \sqrt{2} = 6.243$
5. $2 + 2 \times \sqrt{2} = 4.828$	10. $5 + 1 \times \sqrt{2} = 6.414$

In this example, i values form the fractal sequence.

Does this work for other irrational numbers, or is there something special about $\sqrt{2}$? Why does the sequence exhibit such wonderful fractal properties? Does the initial number have to be irrational? Could it be any random number? Would you generate a fractal sequence for the schizophrenic irrational number discussed in Chapter 93?

❀ For more information on fractal signature sequences, see "Further Exploring."

🔳 See [www.oup-usa.org/sc/0195133420] for computer code used to create these sequences.

❁ ❁ ❁

Want another example of a fractal sequence? The following one is called the golden sequence:

1 0 1 1 0 1 0 1 1 0 1 1 0 1 0 1 1 0 1 0 1 1 0 1 1 0 1 0 1 1 0 1 . . .

It can be created using the following algorithm. Start with 1, then replace 1 by 10. From then on, we repeatedly replace 1 by 10 and each 0 by 1. This sequence has many remarkable properties that involve the golden ratio $\phi = 1.6180339 \ldots = (1 + \sqrt{5})/2$. If we draw the line $y = \phi x$ on a graph, (that is, a line whose slope is ϕ) then we can see the sequence directly (Figure 96.1).

Whenever the ϕ line crosses a horizontal grid line we write 1 by it on the line, and whenever the ϕ line crosses a vertical grid line we write a 0. (The line can never cross exactly at an intersection of the vertical and horizontal grid lines.) Now, run your finger along the ϕ line starting at (0,0), and you will generate a sequence of 1s and 0s—the golden sequence. Ron D. Knott of the University of Surrey in the United Kingdom has translated the sequence into an audio file by mapping 1s to A notes (220Hz) and 0s into the A an octave higher (440Hz), played at about 5 notes per second. He notes that the rhythm is hypnotic, hav-

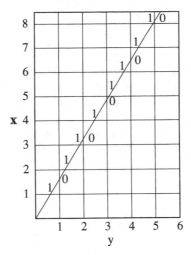

96.1 One way to generate the golden sequence. The diagonal line is y = ϕx.

ing a definite beat that keeps changing but holds one's attention. One wonders if the golden string ever repeats.

The sequence can also be generated by beginning with 1 and 10, then adjoining successive numbers as follows:

> 1
> 10
> 101
> 10110
> 10110101
> 1011010110110
>
> etc. . . .

Here are some other observations about this unusual sequence:

⊙ The number of 1s and 0s in this sequence form a Fibonacci sequence, and the ratio of 1s to 0s approaches ϕ as more terms are added.

⊙ Underline any subsequence of the golden sequence—for example, the subsequence 10: <u>10</u> 1 <u>10 10</u> 1 <u>10</u> 1 <u>10</u>. . . . You'll find that 10 follows the preceding 10 by the following number of places: 2122121. . . . If 2 is replaced by 1 and 1 by 0, the golden sequence is replicated which shows that it is "self-similar" at different scales—that is, it is a fractal sequence.

❁ ❁ ❁

Dr. Googol's favorite fractal sequences are the *batrachions*. Batrachions form a class of bizarre and infinite mathematical curves that hop like frogs from one "lilypad" to the next as they parade along the number system. These little-known curves derive their name from *batrachian*, which means frog-like. (To pronounce the word, note that the *ch* has a *k* sound.)

In addition to hopping in a strange manner from integer to integer, they also have other interesting properties. For example, they are often fractal, exhibiting an intricate self-similar structure when examined at different size scales. Also, they evolve from very simple-looking recursive formulas involving integers.

As background, perhaps the most common example of recursion in programming and in mathematics is one that defines the Fibonacci numbers. As mentioned several times in this book, after the first 2, every number in this sequence equals the sum of the two previous numbers: $F_N = F_{N-1} + F_{N-2}$ for $N \geq 2$ and $F_O = F_1 = 1$. This defines the sequence: 1, 1, 2, 3, 5, 8, 13, 21,. . . .

With this brief background to recursion, consider Dr. Googol's favorite batrachion, produced by this simple, yet weird recursive formula:

$$a(n) = a(a(n-1)) + a(n - a(n-1))$$

The formula for the batrachion is reminiscent of the Fibonacci formula in that each new value is a sum of 2 previous values—but not of the immediately previous 2 values. The sequence starts with $a(1) = 1$ and $a(2) = 1$. The "future" values at higher values of n depend on past values in intricate recursive ways. Can you determine the third member of the sequence? At first, this may seem a little complicated to evaluate by hand, but you can begin slowly by inserting values for n, as in the following:

$$a(3) = a(a(2)) + a(3 - a(2))$$
$$a(3) = a(1) + a(3 - 1)$$
$$a(3) = 1 + 1 = 2$$

Therefore, the third value of the sequence, $a(3)$, is 2. The sequence $a(n)$ seems simple enough: 1, 1, 2, 2, 3, 4, 4, 4, 5,. . . . Try computing a few additional numbers. Can you find any interesting patterns? The prolific mathematician John H. Conway presented this recursive sequence at a talk he gave at AT&T Bell Labs entitled "Some Crazy Sequences" (see "Further Reading"). He noticed that the value $a(n)/n$ approaches ½ as the sequence grows, and n becomes larger. Table 96.1 lists the first 32 terms of the batrachion and the ratio $a(n)/n$.

Dr. Googol first became interested in this sequence after reading Manfred Schroeder's delightful book *Fractals, Chaos, Power Laws*, but, alas, there were no graphics included to help readers gain insight into the behavior of the batrachion. It turns out that this sequence has an incredible amount of hidden structure. Figure 96.2 is a plot of $a(n)/n$ for values of n between 0 and 1000. Notice how the curve hops from one value of 0.5 to the next along very intricate paths.

n	a(n)	a(n)/n	n	a(n)	a(n)/n	n	a(n)	a(n)/n
1	1	1.0	15	8	.5333	29	16	.5517
2	1	1.0	16	8	.5	30	16	.5333
3	2	.666	17	9	.5294	31	16	.5161
4	2	.5	18	10	.5555	32	16	.5
5	3	.6	19	11	.5789			
6	4	.666	20	12	.6			
7	4	.5714	21	12	.5714			
8	4	.5	22	13	.5909			
9	5	.5555	23	14	.6086			
10	6	.6	24	14	.5833			
11	7	.6363	25	15	.6			
12	7	.5833	26	15	.5769			
13	8	.6153	27	15	.5555			
14	8	.5714	28	16	.5714			

Table 96.1 First 32 Terms of the Batrachion

Each hump of the curve appears to be slightly lower than the previous, as if a virtual frog were tiring as it explored higher and higher numbers. As the frog nears infinity, will it stop its hopping and lie dormant at $a(n)/n$ = 0.5? Magnification of the figure reveals more and more humps with an intricate self-similar arrangement of tiny jiggles along the path.

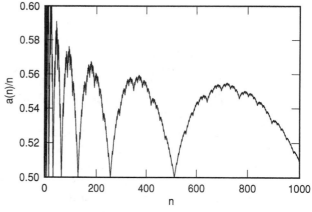

96.2 Batrachion $a(n)/n$ for $0 < n < 1,000$.

❀ Want to know a lot more about batrachions and read about the $10,000 cash award? See "Further Exploring."

▪ See [www.oup-usa.org/sc/0195133420] for computer code.

Chapter 97

Fractal Checkers

Eternity is a child playing checkers.
    —*Heraclitus, 6th–5th century* B.C.

Dr. Googol loves a particular class of self-similar objects called *fractal checkers*, which can easily be constructed using checkerboards of different sizes. The idea of producing interesting patterns by repeatedly replacing copies of a pattern at different size scales dates back many decades and includes the work of mathematicians Helge von Koch, David Hilbert, and Giuseppe Peano. More recently work has been done by Benoit Mandelbrot and A. Lindenmeyer. Artists such as M.C. Escher, Victor Vasarely, Roger Shepard, and Scott Kim have also experimented with recursive patterns that delight both the mind and eye. The designs in this chapter are so intriguing and simple to compute using a personal computer that Dr. Googol will give some computational recipes for those of you who are computer programmers.

To create the intricate forms, start with a collection of squares called the *initiator lattice*. The initial collection of squares represents one size scale. At each filled (black) location in the initial array Dr. Googol places a small copy of the filled array. This is the second size scale. At each point in this new array, Dr. Googol places another copy of the initial pattern. This is the third size scale. He only uses 3 size scales for computational speed and because an additional size scale does not add much to the beauty of the final pattern.

In mathematical terms, begin with an S-by-S square array (A) containing all 0s to which 1s, representing filled squares or sites, are added at random locations. Here's an example:

```
0 0 0 0 0 0 0
0 0 0 1 1 1 0
0 0 0 1 0 0 0
0 0 0 1 0 0 0
0 1 1 1 0 0 0
0 0 0 0 0 0 0
0 0 0 0 0 0 0
```

Just how many patterns can you create by randomly selecting array locations and filling them with 1s? To answer this question, Dr. Googol likes to think of the process of filling array locations in terms of cherries and wineglasses. Consider an S-by-S grid of beautiful crystal wineglasses. Throw M cherries at the grid. A glass is considered occupied if it contains at least 1 cherry. With each throw, a cherry goes into one of the glasses. How many different patterns of occupied glasses can you make? (A glass with more than 1 cherry is considered the same as a glass with 1 cherry.)

It turns out that for an S-by-S array and M cherries, the number of different patterns is $\sum_{n=1}^{M} S^2!/[(S^2 - n)!n!]$. As an example of how large the number of potential patterns is, consider that 32 cherries thrown at a 9-by-9 grid creates more than 10^{22} different patterns. This is far greater than the number of stars in the Milky Way galaxy (10^{12}) and greater than the number of atoms in a person's breath (10^{21}). In fact, it is about equal to the estimated number of stars in the universe (10^{22}).

For Figures 97.1 and 97.2, Dr. Googol used $S = 7$. Here are the initiator lattices for these figures, respectively from left to right:

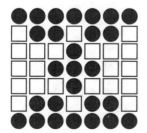 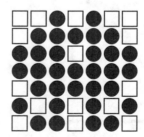

Smaller arrays would lead to fewer potential patterns, and greater values of S sometimes lead to diffuse patterns with the scaling used. Are patterns with larger starting arrays and greater size scales more aesthetically pleasing to you than those produced with the 7-by-7 arrays here? Extrapolate the algorithm here

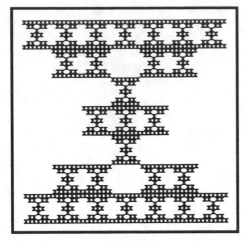

97.1 Fractal checkers: dual wine glass.

97.2 Fractal checkers: Martian with 2 feet.

to 3-D structures and higher dimensional structures. How many different patterns can you produce in a 9-by-9-by-9 3-D initial array? Generalize the recursive lattice program to nonsquare grids—for example, triangular grids.

Chapter 98

Doughnut Loops

Mathematics is not a science—it is not capable of proving or disproving the existence of things. A mathematician's ultimate concern is that his or her inventions be logical, not realistic.
—*Michael Guillen,* Bridges to Infinity

Doughnut puzzles are fiendishly difficult, but, as with many problems in math-ematics and science, the rules of the game are really quite simple. In fact, you can study them using just a pencil and paper. Dr. Googol enjoys working on them while actually eating a chocolate doughnut.

Doughnut puzzles are played on an annular (ringlike) board filled with ran-dom numbers from 0 to 100. Table 98.1 is a typical example, rendered as a rec-tangular region with a hole in the middle to make the playing board easier to typeset. Each "site" on the board contains a single-digit number or a 2-digit number. If you like, create your own puzzle using a graph paper and pencil.

Imagine an ant that starts on any number on the board. The ant's job is to find the longest possible path through the board by moving horizontally or vertically (not diagonally) through adjacent squares. This means the ant takes a single step (up, down, right, or left) during each movement. There are two addi-tional constraints: (1) Each number along the ant's path must be different; that is, the ant can use each number only once along its path. (2) The ant may only travel in an all-clockwise or all-counterclockwise direction. In other words, the ant must go round and round in one direction, but it can orthogonally switch among the 3 "tracks" as useful.

What is the longest path you can find? How many different unique ant paths would you expect to find in doughnut puzzles of this size? The puzzle here is more like a disc, but you could extend the puzzle so that ants tunnel through the inte-rior of 3-D doughnuts. Use computer graphics to display the longest paths as the computer finds them. Explore huge doughnut worlds containing thousands of locations. How would the kinds of solutions (and difficulty of finding solutions) change as the board size approaches infinity? Given a set of doughnut worlds con-structed randomly as in this chapter, what is the *average* "largest path" you would

2	3	11	84	10	92	63	72	19	91	98	68	51	16	46	77	14	12	46	63
23	51	26	34	73	94	27	49	73	98	60	44	36	31	79	73	67	72	56	74
11	71	40	25	22	31	83	31	20	96	23	96	74	3	6	13	97	87	25	33
87	92	73															79	50	3
45	57	61															33	55	81
23	48	43															85	50	28
73	42	29															39	97	92
56	31	61					"Doughnut	Puzzle"								17	23	19
88	40	52															13	32	71
54	79	11															51	56	49
9	60	43															11	99	47
99	13	20															34	12	32
12	48	26	67	37	34	49	56	99	32	39	94	11	23	9	29	45	56	62	65
90	70	70	15	25	6	44	77	8	66	14	54	93	3	78	95	99	99	18	69
13	20	62	53	61	6	82	55	43	79	98	37	46	26	97	66	43	49	25	64

Table 98.1 A Typical Doughnut Puzzle.

expect to find? Is it better to start your path at a particular place in the board? In other words, do certain regions give rise to longer paths than others?

❀ See "Further Exploring" for a solution.

Chapter 99

Everything You Wanted to Know about Triangles but Were Afraid to Ask

> You teach best what you most need to learn.
> —*Richard Bach,* Illusions

"Dr. Googol, thank you for coming to visit me."

There is a sudden crackling sound as William Jefferson Clinton walks to Dr. Googol and, with his right foot, crushes a half-eaten bag of potato chips that Dr. Googol had brought in.

"Excellent." Dr. Googol pauses. "Let's have a little fun."

"More Pythagorean mysticism?" Clinton says eagerly.

Dr. Googol nods. He draws this diagram on the wall: "As you know, Pythagoras's famous theorem is that in a right-angled triangle the sum of the squares of the shorter sides, a and b, is equal to the square of the hypotenuse c, that is, $(c^2 = a^2 + b^2)$."

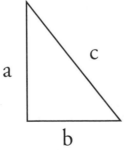

Bill Clinton nods.

"Bill, more proofs have been published of Pythagoras's theorem than of any other proposition in mathematics! There've been several hundred proofs."

"Dr. Googol, are Pythagorean triangles ones where a, b, and c are integers, like 3-4-5 and 5-12-13?"

"Correct, but Pythagoras's favorite, 3-4-5, has a number of properties not shared by other Pythagorean triangles, apart from its multiples such as 6-8-10."

"I know. It's the *only* Pythagorean triangle whose 3 sides are consecutive numbers."

"Very astute, Mr. President. It's also—"

Bill Clinton, beaming at the compliment, lifts his hand to silence Dr. Googol. "Dr. Googol, it's the only triangle of *any* shape with integer sides, the sum of whose sides (12) is equal to double its area (6)."

Dr. Googol continues, slightly annoyed by President Clinton's interruption and intellectual prowess. "It's truly an amazing triangle. But here's something that may make you think twice about 666, the Number of the Beast in the Book of Revelation."

"Go on, Googol."

"There exists only one Pythagorean triangle except for the 3-4-5 triangle whose area is expressed by a single digit. It's the triangle 693-1924-2045. Its area is—" He pauses to heighten the suspense. "666,666."

"Wow!" Bill Clinton says. "Let's tell Hillary and Chelsea." His eyes quiver.

For a moment, Dr Googol thinks he hears the whispers of Secret Service agents. Then he decides it must be the wind.

Dr. Googol calmly reaches for a notebook computer hidden beneath the president's desk. "Let me show you a magic set of formulas that will allow you to search for Pythagorean triangles. They've been known since the time of Diophantus and the early Greeks:"

One Leg of Triangle: $\quad X = m^2 - n^2$
Second Leg of Triangle: $\quad Y = 2mn$
Hypotenuse of Triangle: $\quad Z = m^2 + n^2$

"Dr. Googol, how do you use the formulas?"

"Just select any integers m and n, and you should get a useful result. For example, if $m = 2$ and $n = 1$, we get $x = 3$, $y = 4$, $z = 5$."

"Fascinating, Dr. Googol. Let me write a program to search for Pythagorean triplets. I learned all about computers from Al Gore."

Bill Clinton furiously types on the notebook computer, then hands Dr. Googol a printout:

X	Y	Z
3	4	5
8	6	10
15	8	17
10	24	26, etc.

"Mr. President, here are some mind-boggling facts about Pythagorean triangles. In every triplet of integers for the sides of the triangles, 1 integer is always divisible by 3 and 1 by 5. The product of the 2 legs is always divisible by 12, and the product of all 3 sides is always divisible by 60." Dr. Googol pauses. "Here's a star showing Pythagorean triangles each having 1 side equal to 120."

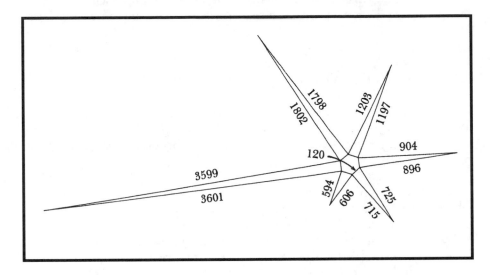

Bill Clinton seems breathless from Dr. Googol's endless barrage of facts.

Dr. Googol looks into Bill Clinton's handsome eyes. "Bill, can you find any triangles, like 3-4-5, that have consecutive leg lengths?"

✤ For an answer, and more mind-boggling information on triangles, see "Further Exploring."

▪ See [www.oup-usa.org/sc/0195133420] for program code.

Chapter 100

Cavern Genesis as a Self-Organizing System

His cave, it seemed, had no right even to be there. It went on and on, winding and scraping till it came out on the other end at a great domed railway terminal of a room, hung with dripping stalactites, and with wet stalagmites like whale penises thrusting up from the floor to meet them."
—*J. P. Miller,* The Skook

Although Dr. Googol is in his office listening to Andreas Vollenweider's *Caverna Magica* on his headphones, 30 miles of caverns plunge and twist away from him in every direction. There are passages of impenetrable stalagmites (Figure 100.1).

He shines a light into a crevice. The surface of the cave walls is aquamarine. Above are glittering stalactite chandeliers. He imagines the air smells clean and wet, like hair after it is freshly shampooed.

He walks a little further. Huddled together like little hobbits, the smaller stalagmites of calcite cluster near a clear pool. The larger ones look like rib bones of some giant prehistoric creature.

With just a few clicks of the mouse, he's entered another world, a virtual world created with mathematical simulations and computer graphics.

Ever since he read about the Lechuguilla Cave deep beneath a southern New Mexico desert and about various European caves, he's been fixated on cavern synthesis—getting his computer to create a lifelike giant maze whose furthest chambers are as yet unfathomed. The Lechuguilla Cave is one of the newest wonders of the subterranean world. Discovered in 1986 and described in the March 1991 *National Geographic*, the cave includes glittering white gypsum chandeliers 2 feet long, walls encrusted with aragonite bushes, and weird balloons of hydromagnesite once inflated by carbon dioxide. Danger is everywhere—funnel-like pits, 65-mile-an-hour winds, darkness . . .

Naturally, Dr. Googol couldn't resist the lure of creating a virtual cavern in the safety of his cybernetic surroundings. Little did he know when he began his

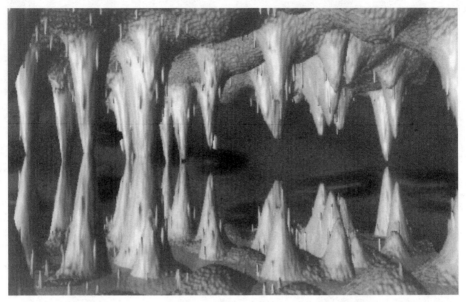

100.1 Virtual cavern produced by simple mathematical simulations and rendered with computer graphics. If you want to view real caverns, take a look at the World Wide Web home page for the Speleology Information Server at http://hum.amu.edu. pl/~sgp/spec/links.html. For other computer graphic simulations, see http://sprott.physics.wisc.edu/pickover/home.htm.

research that the simplest of algorithms would produce stalactite formations of such incredible beauty and richness. The idea behind cavern synthesis is straightforward. In natural caves, stalactites often form due to the deposition of limestone by water slowly dripping from the cavern ceiling. The air space in the cavern allows gases to escape from water, causing solid material to precipitate. Generally speaking, his computer recipe for cavern formation

1. **starts with a nearly smooth cavern ceiling;**
2. **randomly examines a few ceiling positions and notes which is lowest;**
3. **adds a drop of limestone at the point found in step 2; and**
4. **repeats steps 2 and 3.**

As this computational recipe is repeated thousands of times, a few regions are gradually selected and accumulate material as they grow longer and longer. This is similar to what happens in a natural cavern as gravity pulls liquid from the growing stalactites.

The included program code (see [www.oup-usa.org/sc/0195133420]) will start you on your way to cavern synthesis. In this example, the initial cave ceiling is represented using a 512-by-512 array called *cave*. The *cave* array stores the height profile of stalactites. A zero value in the array means no material has been deposited at that particular *x,y* location. As the stalactites grow, the array values grow larger. In Step A of the code, the initial cave ceiling is seeded with small numbers to simulate a nearly smooth ceiling. In Step B, the program simulates the deposition of little circular disc droplets. The droplets are positioned at points where the *cave* values are large in order to simulate deposition at the tips of growing stalactites. After *numdrop* droplets have been deposited, the *cave* array is filled with numbers that indicate the spatial extent of deposits from the ceiling. The actual conversion of the *cave* array to a lighted, shaded cave is left to your favorite 3-dimensional graphics package. Dr. Googol used the IBM Visualization Data Explorer software, which can read in the *cave* array of data, triangulate it, and then perform the necessary hidden surface elimination and shading. Dr. Googol does most of his work on AIX or Windows NT systems with hardware graphics acceleration, although you should be able to convert the *cave* data to input formats used by other renders running on other operating systems. Even if you do not have a three-dimensional renderer, simply assigning colors to the *cave* array values produces a visually interesting picture where stalactites are, for example, represented by bright-colored regions in a 2-dimensional figure.

In a 3-dimensional rendering, before your eyes, stalactites evolve from a nearly smooth cavern ceiling. Stalagmites rise up from the floor to meet their stalactite partners simply by reflecting ceiling structures onto the floor. In future simulations, you may wish to evolve more realistic stalagmites, which normally have thicker proportions than stalactites. Dr. Googol would be happy to give additional details of the cavern simulation to those who write him.

Using a cavern growth program, you can compresses centuries of cave evolution into minutes or seconds depending on the speed of your computer. Feel free to explore the cavern as it evolves, but don't forget to stop the simulation after some time, lest you be trapped forever in the labyrinthine chambers. You want some room to breathe. Continual elongation of stalagmites and stalactites will eventually result in junctions and the formation of columns.

The virtual cavern reminds Dr. Googol of a "self-organizing system," in which large-scale patterns arise from simple rules operating on tiny components of a system. When you look at the smooth initial cave ceiling in the simulation, there's no way you can tell where the large stalactites will eventually form. But after a few seconds of simulation, a dozen stalactites might begin to take shape. Similar behaviors arise in traffic jams, the aggregation of slime molds or bacteria, the formation of termite mounds, and the flocking of birds.

Even though cavern synthesis appears to run on autopilot with no conductor needed to orchestrate the locations of the stalactites, cavern creation can still be a tricky business. Dr. Googol's parameters are delicately poised between simplicity and complexity to make beautiful patterns. For example, in step 2 of the computational recipe, you should not scan too many ceiling points to find the lowest one on the ceiling, or after a minute you'll end up with a single large stalactite. As you perform hundreds of simulations, do you see any patterns in the stalactite positions or sizes? Do stalactites tend to cluster or stay away from one another? Watching the patterns evolve as a function of parameters may tell us a little about real caves, but, more important, it alters the way we make sense of nature. From treelike branches sprouting in human lungs to tendrils spreading through cooling crystals, nature's large-scale structures evolve from mindless microscopic individual behaviors creating pattern and beauty from chaos. It seems that both biological and geological structures grow in the chaos of the cosmos by forming order through wisps and eddies of time.

Of course, the idea of creating virtual reality structures for human exploration is not new. In fact, in my books and articles (see "Further Reading"), I have discussed a variety of virtual reality journeys: computer-generated lava lamps decorating living room walls in the 21st century, virtual vacations on Mars, electronic ant farms, and so forth. These examples not only please the eye but confound the mind with their complexity derived from simple rules.

The future of electronic spelunking is equally bright. Just as today we play 3-D interactive computer games like Doom or Quake, in the future we should look forward to exploring virtual caverns such as the ones Dr. Googol is beginning to explore. Who knows what odd geological formations we will encounter? If his simple algorithms generate lifelike and intricate formations, slightly more complex computational recipes will no doubt produce formations like those found in the Lechuguilla Cave: delicate helicite tendrils, calcite pearls, and gypsum beards.

Like a submarine pilot exploring coral formations in the Sargasso Sea, modern computers allow one to explore the strange and colorful caverns using a mouse. Specifically, Dr. Googol's simulations run on an IBM RISC System/6000 or IBM IntelliStation equipped with graphics accelerators. As the prices of computers decrease while performance increases, I'm sure we'll all be exploring together. Maybe you'll even be able to buy a cavern generator purchased as a plug-in chip.

Not only will virtual spelunking appeal to artists, but it will also be of interest to scientists seeking the causes of real geological structures. For example, the formation of stalactites and stalagmites depends on various factors including a source rock above the cavern, downward percolation of water supplied from rain, tight but continuous passageways for this water (which determine a very slow drip), and adequate air space in the void to allow either evaporation or the escape of carbon dioxide from the water, which thus loses some of its solvent ability. These kinds of variables could be investigated using more detailed computer simulations. It would be fascinating to explicitly model the effect of gravity and then see how hypothetical caverns might form on other planets with different gravities.

Dr. Googol likes to speculate that virtual decorations of the future will be grown by computer algorithms and projected or displayed on the ceilings of our own homes. But now it is time to roam. Dr. Googol lets his gaze drift to the pockets of rocks around him, noting the flowing harmony of the fractal formations, the crystalline outcroppings of rock coated with strips of velvet purple. A cool peace floods him.

He wants to place his finger in the lake. It is perfectly black. The stalactites and stalagmites and slippery cave walls are shimmering and alive.

He shines a light over the water. It is clear now and filled with nodules. It's a shame he can't blow on the water and see countless ripples appear on its surface. That is for the future. Someday the cold air will brush against him like a cat. He will hear the mystical sounds that have lulled others cave explorers: the humming of stalactites; the wild, seemingly desperate cry of the wind through the cave.

Where is the rest of the world? It hardly matters.

In the future, students, movie special effects houses, and artists may explore the virtual caverns, which can be rapidly generated and contain an infinite reservoir of magnificent topographical features. It would be interesting to apply some of the new terrain synthesis methods, such as those based on erosion, to these intricate landscapes and view the results. The various successes in terrain generation over the last decade provide continuing incentive for more research on the rapid generation of natural and artistic landscapes.

▪ See [www.oup-usa.org/sc/0195133420] for program code.

Chapter 101

Magic Squares, Tesseracts, and Other Oddities

> Mathematical inquiry lifts the human mind into closer proximity with
> the divine than is attainable through any other medium.
> —*Hermann Weyl (1885–1955)*

In Islam, the number 66 corresponds to the numerical value of the word *Allah*. Figure 101.1 is an Islamic magic square that expresses the number 66 in every direction when the letters are converted to numbers. The square's grid is formed by the letters in the word *Allah*. Magic squares such as this were quite common in the Islam, but seem not to have reached the West until the 15th century. From a historical perspective, Dr. Googol's favorite Western magic square is Albrecht Dürer's, which is drawn in the upper right-hand column of his etching *Melencolia I* (Figure 101.2). The variety of small details in the etching has confounded scholars for centuries. Scholars believe that the etching shows the insufficiency of human knowledge in attaining heavenly wisdom, or in penetrating the secrets of nature.

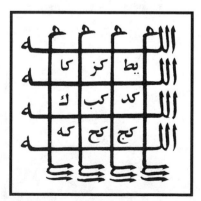

101.1 An Islamic magic square that expresses the number 66 in every direction. The grid is formed by the letters in the word *Allah*.

101.2 *Melencolia I*, by Albrecht Dürer (1514). This is usually considered the most complex of Dürer's works; its various symbolic nuances have confounded scholars for centuries. Why do you think he placed a magic square in the upper right? Scholars believe that the etching shows the insufficiency of human knowledge in attaining heavenly wisdom, or in penetrating the secrets of nature.

Dürer's 4-by-4 magic square, which can be represented as

16	3	2	13
5	10	11	8
9	6	7	12
4	15	14	1

contains the first 16 numbers and has some fascinating properties. The two central numbers in the bottom row read 1514, the year Dürer made the etching.

Also, in the vertical, the horizontal, and 2 diagonal directions, the numbers sum to 34. In addition, 34 is the sum of the numbers of the corner squares (16 + 13 + 4 + 1) and of the small central square (10 + 11 + 6 + 7). The sum of the remaining numbers is 68 = 2 × 34. Was Dürer trying to tell us something profound about the number 34?

Mark Collins, a colleague from Madison, Wisconsin, with an interest in both number theory and Dürer's works, has studied the Dürer square and finds some astonishing features when converting the numbers to binary code. (In binary representation, numbers are written in a positional number system that uses only two digits: 0 and 1—as explained in the "Further Reading" for Chapter 21.) Since the first 16 hexadecimal binary numbers start with the number 0 and end with 15, he subtracts 1 from each entry in the magic square. Below is the result:

15	2	1	12
1111	0010	0001	1100
4	9	10	7
0100	1001	1010	0111
8	5	6	11
1000	0101	0110	1011
3	14	13	0
0011	1110	1101	0000

Remarkably, if the binary representation for the magic square is rotated 45 degrees clockwise about its center so that the 15 is up and the 0 down, the resultant pattern has a vertical mirror plane down its center:

```
                  1111
              0100    0010
          1000    1001    0001
      0011    0101    1010    1100
          1110    0110    0111
              1101    1011
                  0000
```

For example, in row 2, 0100 is the mirror of 0010. (Dr. Googol very much doubts that Dürer could have known about this symmetry.)

If we rotate the binary square counterclockwise so that the 12 is at the top and the 3 at the bottom, then draw an imaginary vertical mirror down the center of the pattern, we see a peculiar left-right inverse:

```
                    1100
              0001       0111
         0010       1010       1011
   1111       1001       0110       0000
         0100       0101       1101
              1000       1110
                    0011
```

For example, in the second row, 0001 and 0111 are mirror inverses of each other.

Mark Collins has discovered the presence of mysterious intertwined hexagrams when the even and odd numbers are connected:

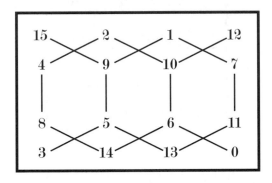

Dr. Googol would be interested in hearing from those of you who find additional meaning or patterns in Dürer's magic square. Mark Collins and Dr. Googol are unaware of other magic squares having the symmetrical properties when converted to binary numbers. Mark has also done numerous experiments converting these numbers to colors and comments: "I believe this magic square is an archetype as rich in meaning and mysticism as the I Ching. I believe it is a mathematical and visual representation of nature's origami—as beautiful as a photon of light." Mark suggests you should create other mitosis-like diagrams by connecting 0 to 1 to 2 to 3. Then lift up your hand. Connect 4 to 5 to 6 to 7. Connect 8 to 9 to 10 to 11. Connect 12 to 13 to 14 to 15.

A rather bizarre 6-by-6 magic square was invented by the mysterious A. W. Johnson. No one knows when this square was constructed, nor is there much information about Johnson. (Dr. Googol welcomes any information you may have.) All of its entries are prime numbers, and each row, column, diagonal, and broken diagonal sums to 666, the Number of the Beast. (A broken diagonal, is the diagonal produced by wrapping from one side of the square to the other; for example, the outlined numbers 131, 83, 199, 113, 13, 127 form a broken diagonal.)

The Apocalyptic Magic Square

3	107	5	131	109	311
7	331	193	11	83	41
103	53	71	89	151	199
113	61	97	197	167	31
367	13	173	59	17	37
73	101	127	179	139	47

❀ ❀ ❀

Another amazing magic square is the Kurchan array, named after its discoverer, Rodolfo Marcelo Kurchan, from Buenos Aires, Argentina. He believes this to be the smallest nontrivial magic square having n^2 distinct pandigital integers and having the smallest, pandigital magic sum. *Pandigital* means all ten digits are used, and 0 is not the leading digit. Below is the awesome Kurchan array; the pandigital sum is 4,129,607,358:

The Kurchan Array

1037956284	1036947285	1027856394	1026847395
1026857394	1027846395	1036957284	1037946285
1036847295	1037856294	1026947385	1027956384
1027946385	1026957384	1037846295	1036857294

Even more amazing is the mirror magic square:

Mirror Magic Square

96	64	37	45
39	43	98	62
84	76	25	57
23	59	82	78

If you reverse each of the entries you obtain another magic square. In both cases the sums for the rows, columns, and diagonals is 242:

69	46	73	54
93	34	89	26
48	67	52	75
32	95	28	87

Isn't that a real beauty?

Finally, mathematician John Robert Hendricks has constructed a 4-dimensional tesseract with magic properties. Just as with traditional magic squares

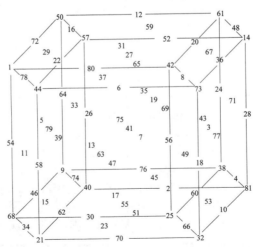

**101.3 Magic tesseract by John Robert Hendricks.
(Rerendered by Carl Speare.)**

whose rows, columns, and diagonals sum to the same number, this 4-dimensional analogue has the same kinds of properties in 4-space. Figure 101.3 represents the projection of the 4-dimensional cube onto the 2-dimensional plane of the paper. Each cubical "face" of the tesseract has 6 2-D faces consisting of 3-by-3 magic squares. (The cubes are warped in this projection in the same way that the faces of a cube are warped when drawn on 2-D paper.) To understand the magic tesseract, look at the 1 in the upper left corner. The top forward-most edge contains 1, 80, and 42, which sum to 123. The vertical columns, such as 1, 54, and 68, sum to 123. Each oblique line of three numbers, such as 1, 72, and 50, sums to 123. A fourth linear direction shown by 1, 78, and 44 sums to 123. Can you find other magical sums? This figure was first sketched in 1949. The pattern was eventually published in Canada in 1962, and later in the United States. Creation of the figure dispelled the notion that such a pattern could not be made.

❀ For more on magic squares, see the "Further Exploring" section for Chapter 16.

Chapter 102

Fabergé Eggs
Synthesis

> More significant mathematical work has been done in the latter half of
> this century than in all previous centuries combined.
> —*John Casti*, Five Golden Rules, *1996*

Have you ever noticed that many of our ancient designs consist of symmetrical and repeating patterns? For example, consider the beautiful Moorish, Persian, and other motifs in tiled floors and cloths. Among Dr. Googol's favorite ornamental patterns are those found on century-old Russian Easter eggs that wealthy individuals and members of the royal family gave to one another. Some of these eggs were made of gold and silver and decorated with enamel, precious stones, and miniature paintings. The most splendid were commissioned by the czar from Fabergé, the leading firm of Russian jewelers at the turn of the 20th century.

Fabergé eggs are beyond the financial resources of most humans on the planet. Today, however, personal computers equipped with low-cost graphics accelerators bring the beauty and mystery of "self-decorating" eggs to computer hobbyists. The patterns are based on the mathematical concept of a "residue"— the remainder after subtracting a multiple of a modulus from an integer. (2 and 7 are residues of 12 modulo 5.)

❀ ❀ ❀

THE SECRET ALGORITHM

How can the beauty of the symmetrical ornaments and designs of various cultures be simulated with the aid of a computer? From an artistic standpoint, sinusoidal equations provide a deep reservoir from which artists can draw. Computational recipes, such as those outlined in the following, interact with such traditional elements as form, shading, and color to produce classical and futuristic images and effects. The mathematical recipes function as the artist's helper by allowing the artist to experiment with a range of parameters and to select results that are considered attractive or visually interesting. Indeed, structures produced by the equations includes shapes of startling intricacy.

To compute the egg-decorating patterns, a real number c is first calculated for a range of (i,j) pairs:

$$c_k^{i,j} = 1 + 0.5 \times [\sin(\phi_k + f_k \times i) + \sin(\psi_k + f_k \times j)]$$

where the index k has the value 1, 2, and 3, to produce intensity values for three color channels (red, green, blue) used by the graphics software, and $1 \leq i \leq 400$, $1 \leq j \leq 400$. This creates 3 2-dimensional sinusoidal arrays c with values ranging from 0 to 1 as a function of i and j on a 400-by-400 grid controlled by phases ϕ ($0 \leq \phi \leq 1$), ψ ($0 \leq \psi \leq 1$), and frequencies f ($0.15 \leq f \leq 0.8$). (Values for the sin functions are in radians.) The ϕ, ψ, and f values are held constant for a particular egg. This means that 6 phase values and 3 frequency values determine a particular egg's pattern. The 3 values of c_k at each point in the array are used to control the red, green, and blue colors at each point on the the egg surface after additional mathematical manipulation.

In order to make the resultant pattern tile-like for the purposes of egg-decoration, the resulting c_k values in the first equation are multiplied by m_k, truncated (made an integer), and divided by another integer β_k. The remainder is used to determine the color of the egg surface at location i, j. Large remainders correspond to high intensities of either red, green, or blue. A remainder of 0 corresponds to zero intensity, or black. Values of m_k ranged from $\beta_k \leq m_k \leq \beta_k + 20$. Values of β_k ranged from $1 \leq \beta_k \leq 10$. This truncation and residue approach applied to the first equation in this chapter can be expressed as

$$c_k^{i,j} = [m_k \times (1 + 0.5 \times (\sin(\phi_k + f_k \times i) + \sin(\psi_k + f_k \times j)))] \bmod \beta_k$$

where the brackets indicate truncation to an integer; for example, 1.6 is truncated to 1. The various ranges for multipliers, modulo values, frequencies, and phases were empirically determined to give a diverse, attractive set of patterns, and readers are encouraged to experiment with their own ranges to suit their own aesthetic tastes.

To map the final color arrays for red (c_1), green (c_2), and blue (c_3) to the egg surface, a spherical or ellipsoidal surface transformation is made; for example, $z = \sqrt{1 - i^2 - j^2}$ where i and j denote positions along an (x,y) plane coincident with the plane of the paper or computer screen. Elongation in the x or y direction can be accomplished using 3-dimensional graphics scaling routines or by altering the z function. The surface is rendered using rectangular facets, and the surface normal at each point determined by vector (i, j, z) for the purpose of lighting calculations by the graphics software and hardware.

Ten years ago, when I wrote about "self-decorating eggs" and first experimented with black-and-white, 2-dimensional, repetitive ornaments, I would have had trouble believing how fast and elegantly these forms can be rendered today on personal computers. In particular, the eggs in Figure 102.1 were computed using a C program calling OpenGL (3-D graphics) routines on an IBM IntelliStation running a Windows NT operating system. To increase rendering

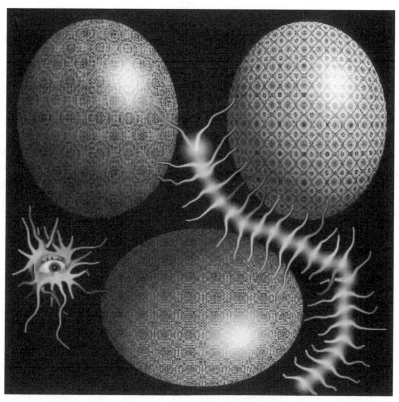

102.1 Algorithmic Fabergé eggs, with organic infestations.

speed, only one light source was used and only the front surface of the eggs was computed. If computation of the sine function limits the rate at which eggs are computed, the value for the sine function may be precomputed and stored in a look-up table. The program chooses random values for phases, frequencies, multipliers, and mod factors within the given ranges and presents a new egg to the viewer about once every three seconds. To avoid lower-frequency patterns, which may be visually less interesting, Dr. Googol sometimes has the program reject frequencies below a certain threshold value (for example, $f = 0.15$). For greater variety, other functions may be used, such as frequency-modulated sinusoids of the form $\sin(\phi + \sin(\phi + f))$. Other variants of the first equation in this chapter may also be used.

SOME THOUGHTS

Alexander III started the tradition of ornate egg design in 1885. Every year he commissioned an egg from his court jeweler, Peter Carl Fabergé, as a gift to his wife, the Empress Maria Feodorovna. After Alexander's death, his son Nicholas II continued the tradition, commissioning two eggs from the firm. At Easter, Fabergé himself would present one egg to the Dowager Empress Maria Feodorovna, while his assistant would present the second to Alexandra Feodorovna, Nicholas's wife. In all, 56 of these masterpieces were produced between 1885 and 1917; however, only 10 of these have remained in Russia. Masters from the Fabergé firm worked on each Easter egg for nearly a year.

Today Dr. Googol likes to imagine Fabergé and Alexandra Feodorovna sitting in his office behind a personal computer and selecting eggs that have special appeal for them. Fabergé adjusts the modulus factor as Alexandra screams for more.

The self-decorating eggs remind Dr. Googol of snowflakes. No two eggs ever seem to be alike as viewers watch an endless variety of forms parade on their screen. Figure 102.1 shows just a few examples of the remarkable panoply of designs made possible with the algorithm. By "turning a dial" that controls the various parameters, an infinite variety of attractive designs is generated with relative computational simplicity—and for this reason, the eggs may be of interest for designers of museum exhibits and other educational displays for both children and adults.

Chapter 103

Beauty and Gaussian Rational Numbers

An intelligent observer seeing mathematicians at work might conclude that
they are devotees of exotic sects, pursuers of esoteric keys to the universe.
—*P. Davis and R. Hersh,* The Mathematical Experience

The purpose of this chapter is to illustrate a very simple graphics technique for visualizing a large class of graphically interesting manifestations of complex rational numbers. As background, complex rational numbers are of the form p/q, where p and q can be complex numbers of the form $a + bi$ where $i = \sqrt{-1}$ and a and b are integers. As an example of a complex rational number, consider $(1 + 2i)/(3 + 3i))$. In other words, $p = p' = ip''$, and $q = q' = iq''$, with p', p'', q', q'' all integers. Accordingly,

$$\frac{p}{q} = \frac{p' + ip''}{q' + iq''} = \frac{p' + ip''}{q' + iq''} \times \frac{q' - iq''}{q' - iq''} = \frac{p'q' + p''q''}{q'^2 + q''^2} + \frac{p''q' - p'q''}{q'^2 + q''^2}$$

The complex fractions thus consist of the numbers x + iy where x and y are real fractions.

Following the lead of L. R. Ford, we may construct a sphere that represents the complex fraction p/q by having the sphere touch the complex plane at location p/q and having the radius equal to $1/(2q\bar{q})$, where $\bar{q}$ is the conjugate of q. (Given a complex number $a + bi$, the complex conjugate is $a - bi$.) Alas, Ford in 1938 had no means of visualizing the results of his ideas, and his only diagram contained four hand-drawn spheres. Perhaps due in part to lack of visualization methods, his paper is almost entirely devoted to 2-dimensional worlds where a few circles are positioned on rational points on a line, an idea discussed in Pickover's book *Keys to Infinity*. Therefore Dr. Googol could not resist the temptation of bringing Ford's ideas into the modern age. In doing so, it becomes evident that the Gaussian (i.e., complex) rational spheres provide an infinite graphical treasure chest to explore. In fact, it turns out that spheres describe the fabric of our complex rational number system in an elegant way.

How many neighbor spheres touch an individual sphere? Two fractions are called adjacent if their spheres are tangent. Any fraction has, in this sense, an infinitude of adjacents. Any sphere has an infinitude of spheres that kiss it. It can be shown that if spheres are placed at complex fractions (P/Q) and (p/q), then the spheres are tangent (adjacent) when $|Pq - pQ| = 1$. For example, consider two spheres in Figure 103.1 The distance AB between sphere centers is a function of the horizontal distance AC and the vertical distance CB (the difference of the radii). Therefore

$$AB = \sqrt{|P/Q - p/q|^2 + \delta}$$

where

$$\delta = \left(\frac{1}{2Q\bar{Q}} - \frac{1}{2q\bar{q}} \right)^2$$

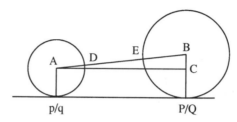

103.1 Geometrical study.

Thus,

$$AB = \sqrt{\left(\frac{1}{2Q\bar{Q}} - \frac{1}{2q\bar{q}} \right)^2 + \frac{|Pq - pQ|^2 - 1}{Q\bar{Q} - q\bar{q}}}$$

If $|Pq - pQ| > 1$, then $AB > AD + EB$, and the spheres do not kiss. If $|Pq - pQ| = 1$, then sphere P/Q and p/q kiss (i.e., the fractions are "adjacent"). It is not possible for spheres to intersect.

Figure 103.2 shows a computer graphics rendition of the Gaussian rational froth. In the original color images, color is a function of the spheres' radii. Figure 103.3 is a magnification of a side view of Figure 103.2. Figure 103.4 is the same as Figure 103.2, with the large red spheres removed to reveal underlying structure. Figure 103.5 is a ray-traced rendition of the froth with the central sphere made transparent to reveal underlying structure.

Consider a "physical" analog of the Gaussian rational sphere froth. Imagine holding an "infinitely" thin needle above the collection of spheres perched on the complex plane. (You may like to think of the complex plane as a pond surface and of the spheres as bubbles, each with its lowest point touching the pond surface.) If you drop the needle above a rational point in the complex plane, the needle must pierce a *single* bubble and hit the complex plane exactly at the bubble's point of tangency. However, if you drop the needle from above an irrational complex number, the needle cannot pass directly to the complex plane from a bubble. In other words, the needle must leave every bubble which it enters. However, as Dr. Googol men-

103.2 Gaussian rational bubbles in the complex plane
$(0 \leq (p/q) \leq 1)$, $(0 \leq (p/q) \leq 1)$, $(-7 \leq p', q', p'', q'' \leq 7)$.

103.3 Magnification of a side view of the bubbles.

tioned previously, every bubble that the needle leaves is completely surrounded by a chain of bubbles. Therefore the needle must enter another bubble. This is true for all the bubbles it pierces. Thus, when the needle is dropped above an irrational point, it must pass through an infinity of bubbles.

Gaussian rational froth holds many challenges for computer graphics specialists. Since the froth is endless, accurate representation is difficult, particularly as the froth is magnified during animated zoom sequences. (Dr. Googol computes the locations of about 15,000 spheres before terminating the computation, and this accounts for the small regions devoid of spheres on the complex plane.) However, with appropriate computer programs, students, artists, and mathematicians can "swim" through the Gaussian rational froth like a fish through the surf. Animations make it possible to "sit" on the falling virtual needle and view the

103.4 Same as Figure 103.2, with the large, central sphere removed to reveal underlying structure.

103.5 Ray-traced rendition of the froth. Paul A. Thiessen (University of Illinois) used the software POV-Ray to produce this rendering for Dr. Googol.

mechanism of infinite piercing. What strange oceanic worlds will students and artists find as they explore different regions of the Gaussian rational froth? Note that if fractions represented by spheres are not "reduced," spheres may lie inside spheres—and this can be visualized using transparency. To speed computations, Dr. Googol suggests that every fraction in which the numerator and denominator have common factors be canceled as far as possible (e.g., 6/8 → 3/4.)

Graphics specialists, educators, and mathematicians may find this chapter a useful stepping-stone to additional geometrical representations and insight. For example, assemblages of spheres may be used as pictorial representations of continued complex fractions of the form

$$a_0 + \cfrac{1}{a_1 + \cfrac{1}{a_2 + \cfrac{1}{a_3 + \ldots}}}$$

where a_n are complex integers. A final challenge would be to extend these representations to quaternionic rational numbers, which make up a 4-dimensional algebra containing the complex plane, and Cayley rational numbers, which make up an 8-dimensional nonassociative real division ring.

In order to reveal the intricacy of Gaussian froth, which is not possible in small figures in this book, you are invited to examine an example high-resolution image on the Web at http://sprott.physics.wisc.edu/pickover/home.htm.

Chapter 104

A Brief History of Smith Numbers

> The reviewer is not convinced that Smith numbers are not a rat-hole
> down which valuable mathematical effort is being poured.
> —*Carl Linderholm*, Mathematical Reviews

A Smith number is a composite number (a nonprime number) the sum of whose digits is the sum of all the digits of its prime factors. Since they were originally proposed by Albert Wilanski in the January 1982 issue of *Two-Year College Mathematics Journal*, Smith numbers have been the subject of over 15 published papers. The rather startling reason for their name is mentioned below.

Want an example of a Smith number? The number 9,985 is a Smith number because 9,985 = 5 × 1,997, and, therefore

Digit Sum   **Prime Factor Sum**

$$9 + 9 + 8 + 5 = 5 + 1 + 9 + 9 + 7$$

In 1982, the largest known Smith number (4,937,775) was credited to Albert Wilansky's brother-in-law, H. Smith, who is not a mathematician. The brother-in-law's telephone number is 493-7775!

Since 1982, interest in these numbers has exploded. In 1983, a paper appeared in *Mathematics Magazine* that gave a larger Smith number. The authors' discovery was that if p is a prime whose digits are all 1s, then $3304p$ is a Smith number. (Are there other numbers that could serve this same purpose?)

In 1986, another odd method for generating Smith numbers was presented, leading to Smith numbers such as

$$5 \times 1110110110111 \times (2 \times 5)^5 = 555,055,055,055,500,000$$

and to other behemoths, including one Smith number with 2,592,699 digits.

1987 was a banner year for Smith numbers, with three papers appearing in the *Journal of Recreational Mathematics*. In these papers, we find palindromic Smith numbers, such as 12,345,554,321, the definition of *Smith Brothers* (consecutive Smith numbers), such as 728 and 729, and all other manner of mathematical bewilderment.

For the best history of Smith numbers, see Underwood Dudley's article in the February 1994 *Mathematics Monthly*. Do you think mathematical studies of Smith numbers are worthwhile or significant? Or are they just pure recreation, useful for honing one's mathematical prowess but with no possible practical or profound results?

Chapter 105

Alien Ice Cream

The soul of man was made to walk the skies.
            —*Edward Young, 18th century*

⊚ **Number Maze 3, a visual intermission before the next book part. . . .**

This sweet puzzle is from one of Dr. Googol's dreams. Upon waking, he quickly crafted the following enigma.

It is night, and the stars shine brightly on the home of Dr. Googol (schematically illustrated in Figure 105.1). On the roof is an alien selling a special kind of ice cream cone—one that will give you eternal life if you eat it. You have only $1, which is not enough to purchase the ice cream. There are aliens with dollar bills on every floor. Entering or exiting any door requires an alien to give you $5. When you use a ladder, an alien hands you $2, and use of the spiral staircase gets you $20.

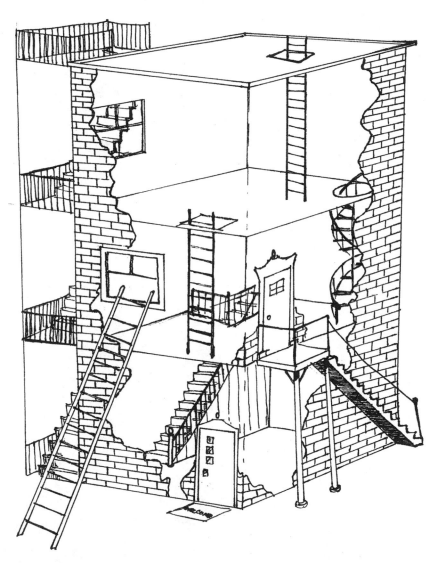

104.1 Alien Ice Cream. Can you reach the top with exactly $41? (Drawing by Brian Mansfield.)

If you use a staircase, you add $3 whenever you walk between floors. The fire escape on the outside of the building is a zigzagging staircase traveling from level to level, but only the ground floor, third floor, and roof have outlets onto it.

If you wish to eat the alien ice cream, you must start outside on the ground floor and somehow make it to the roof with exactly $41. Once you have traveled along a stairway, ladder, or the spiral staircase, you may not use them again. If you can do this within 30 minutes, the alien will gladly give you the sugary treat. Some say the puzzle is impossible. No one on Earth has solved the puzzle—or has ever tasted the ice cream of eternity.

❀ For a solution, see "Further Exploring."

Part iV

The Peruvian Collection

Great mathematics must suggest nature:
a snow crystal, a mossy cavern,
a seagull's wing, a viper's tongue, red Peruvian earth,
the gnarled bark of an ancient oak.
And in a hundred years,
when humans have destroyed nature,
today's mathematics will serve as a portal
to all that which was beautiful.

—*Dr. Francis O. Googol*

Mathematics is nothing,
not even beauty,
unless at its heart,
two numbers bloom.

—*Dr. Francis O. Googol*

Chapter 106

The Huascarán Box

A Great Truth is a statement whose opposite is also a great truth.
—*Niels Bohr*

Late last summer, Dr. Googol was exploring the Peruvian rain forest at the base of Mount Huascarán, the highest mountain in Peru. There he found a mysterious box. On the box were colored fingers: red, green, and yellow. A fourth finger was clear and made of diamond. Under the fingers was the following inscription:

> *Inside this box is a small, silent, well-oiled, vibrationless, battery-powered fan. The colored fingers are on/off buttons. One of them is connected to the fan; the other 2 colored fingers are dummies, not connected to the fan. When a finger is up ☝, it is on. When it is down 👇, it is off. The diamond finger cannot be moved.*
>
> *You may toggle the fingers as you wish. Once you have toggled the fingers in the pattern of your choice, you may look inside the box. By inspecting the fan, you know which finger controls it. How do you know? You get only one look! A correct answer allows you to take the diamond finger.*

Can you help Dr. Googol obtain the magnificent diamond finger 👆? Do you think this problem is, in fact, possible to solve? If you are a teacher, it might be fun to build a similar box and have students do experiments.

Dr. Googol traveled further into the jungle and came to another Huascarán Box! It had four potentially active switches: red, green, blue, and gold. Next to the box was a small pile of red dust, resembling spicy Peruvian paprika. In the top of the box, above the fan, was a tiny hole into which Dr. Googol could pour the paprika. Again, the colored fingers were on/off buttons, one of which was connected to the fan. The other 3 colored fingers were dummies, not connected to the fan. When a finger was up, it was on. When it was down, it was off. In this case, the golden finger could also be toggled up and down and could possibly influence the fan circuit.

As with the previous puzzle, Dr. Googol could toggle the fingers as he wished. Once he toggled the fingers in the pattern of his choice, Dr. Googol could look inside the box. By inspecting the box, he knew which finger controlled it. He could only look once. This time, a correct answer would allow him to take the valuable golden finger.

Can you help Dr. Googol obtain the gorgeous golden finger ☝?

❀ For solutions to both problems, see "Further Exploring."

Chapter 107

The Intergalactic Zoo

> A mathematician is a blind man in a dark room looking for
> a black cat which isn't there.
>
> —*Charles Darwin*

The lower slopes of the western Andes merge with the heavily forested tropical lowlands of the Amazon Basin to form the Montaña, which occupies more than three-fifths of Peru's area. While exploring the rolling hills and level plains, Dr. Googol had a vision. Perhaps the vision resulted from his fatigued mind or from the strange plants the locals had given him to eat on his journey. Or perhaps the vision was real. We will never know.

Dr. Googol watched in horror as an alien abducted Earth animals for an intergalactic zoo. Getting them safely to the zoo was a problem because the alien didn't know which animals might attack others on the way. The alien decided to keep the animals in a darkened ship hovering above the zoo until it was time to put them in their cages. The darkness should have encouraged the animals to sleep rather than fight . . . or so the alien hoped.

Inside the ship there were 5 pairs of monkeys, 4 pairs of Peruvian jaguars, and 2 pairs of tapirs. (A pair consists of a male and female.) When the alien reached a huge ark in outer space, he opened a chute that let animals drop from the ship, 1 at a time, into individual cages. Later he wanted to match the species, and pairs within a species.

It was night, so the alien couldn't tell the animals apart visually.

How many animals must the alien drop to ensure that he has 2 animals of the same species?

How many animals must he drop to ensure that he has a male and female of the same species?

Hurry, the alien needs answers. The Peruvian jaguars are roaring as the monkeys scream in terror. Daylight is just minutes away.

❀ For a solution, see "Further Exploring."

Chapter 108

The Lobsterman from Lima

> I am reminded of a French poet who, when asked why
> he took walks accompanied by a lobster with a blue ribbon around
> its neck, replied, "Because it does not bark, and because
> it knows the secret of the sea."
> —*an anonymous fan of Gerard de Nerval*

Peruvian ocean waters are abundant with haddock, anchovy, pilchard, sole, mackerel, smelt, flounder, lobster, shrimp, and other marine species. One day while visiting several coastal towns, Dr. Googol came upon a huge man selling lobsters by the side of a dirt road. The sight of the lobsters made Dr. Googol's mouth water.

"Do you speak English?" Dr. Googol said.

"Of course. I'm originally from Lima. Would you like a lobster?"

"How much do they cost?

The lobsterman raised his eyebrow. "If you answer my mathematical question correctly, you get a free lobster. If you answer incorrectly, you pay me $100. You must answer within 15 seconds. How does that sound?"

"Good deal. But I must warn you, I have a Ph.D. in mathematics."

The lobsterman held up a huge lobster and stared into Dr. Googol's eyes. Then he handed Dr. Googol a card with a question. The card smelled of fish and of low tide and of crawling things. The lettering on the card was in Old English calligraphy. Perhaps the man was trying to impress Dr. Googol with the importance or difficulty of the question.

> 𝕴f this lobster weighs 10 pounds plus half
> its own weight, how much does it weigh?

Can you help Dr. Googol answer this odd question? If you think the question is difficult, you're not alone. If you think this is too easy, you may be incredibly brilliant and arrogant, but Dr. Googol bets that none of your friends can answer this within 15 seconds. Try it on your friends. You'll see. So far, none of Dr. Googol's friends could solve it without a pencil and paper. If you're a teacher, have your students work on this problem and see what answers they arrive at. Allow them to use a pencil and paper.

❀ For a solution, see "Further Exploring."

Chapter 109

The Incan Tablets

> I looked at the ancient ruins. These bricks. This light. I was
> exponentially far from New York City. Mathematical distances are never
> measured with rulers.
>
> —*Dr. Francis O. Googol*

Dr. Googol was exploring the ruins of Machu Picchu, near Cuzco—the remains of an ancient city of the Inca Empire. Twelve hundred years previously, the Incas had mastered architecture, astronomy, and road building—but Dr. Googol came here not to study history but rather to commune with nature and remember his ancestors, some of whom could be traced to the ancient Incas.

As Dr. Googol looked inside the ruin's deep interior, surrounded by the dry bricks and old mortar, he came upon a tablet with some odd-looking symbols:

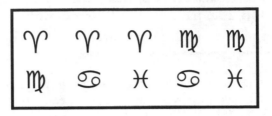

Written in English, next to the symbols, were the following instructions.

You see 5 vertical pairs of symbols.
You are to find a pair of symbols to complete the set
from among the 5 possible solutions shown here:

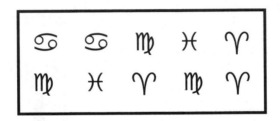

If you choose correctly and complete the set, the following wonderful events will take place: your I.Q. will be increased by 20 points; you will be able to speak to the Inca dead and learn their ancient wisdom; you will be able to stop time, at will; and you will be able to spend a day with the person of your choice, for example, the Dalai Lama, Madonna, Bill Clinton, or Robert Redford.

Dr. Googol studied the tattered tablets. Why were the instructions in English? It must be some kind of hoax. Nevertheless, there must be a solution, and Dr. Googol must find it. The rewards, although unlikely, are too great to ignore.

❀ For a solution, see "Further Exploring."

Chapter 110

Chinchilla Overdrive

> The sense of completeness that is projected by the work of art is
> to be found nowhere else in our lives. We cannot remember our birth,
> and we shall not know our death; in between is a ramshackle circus
> of our days and doings. But in a poem, a picture, or a sonata, the curve
> is complete. This is the triumph of form. It is a deception, but one
> that we desire and require.
> —*John Bainville,* "Beauty, Charm and Strangeness:
> Science as Metaphor," Science 281, *1998.*

In the sierra of Peru are all kinds of wildlife: the alpaca, llama, vicuña, chinchilla, and huanaco. Birds of the region include the partridge, giant condor, robin, phoebe, flycatcher, finch, duck, and goose. Here is a puzzle Dr. Googol developed while watching all the wonderful wildlife and listening to the cries of the condors at they circled overhead like floating ashes.

Dr. Googol has a number of llamas in his private Peruvian zoo. The number of llamas plus 10 chinchillas is 2 less than 5 times the number of llamas. If you wish, denote the number of llamas by L and the number of chinchillas by C. How many llamas does Dr. Googol have?

❀ For a solution, see "Further Exploring."

Chapter 111

Peruvian Laser Battle

Mathematics is a war between the finite and infinite.
—*Dr. Francis Googol*

"Have you ever heard of Peruvian Laser Battle?" Monica asked Dr. Googol as their canoe floated down the Amazon River, ten miles north of Iquitos, Peru.

Dr. Googol shook his head. "Please tell me more."

"Peruvians love science fiction, and Laser Battle is the hottest new game in Iquitos. Imagine yourself leading a battle on the Peruvian plains. Your attackers are a horde of alien robots."

"Alien robots?" Dr. Googol said, raising his eyebrow.

"Use your imagination. The robots are quickly closing in on your soldiers."

Monica pointed to a piece of paper showing a hexagonal grid with 4 open circles representing 4 soldiers (Figure 111.1). Robots were represented by filled circles. Far to the north was Colombia. To the east was Brazil. To the south was Chile. To the west was Ecuador.

"Dr. Googol, your object is to destroy all alien robots using your 4 courageous Peruvian soldiers. With only 2 shots each from their rifles, your soldiers must destroy all the alien robots. To make matters tricky, the robots are booby-trapped and will explode with thermonuclear blasts if hit more than once. So your soldiers had better hit each robot just once. Rifle shots continue in a

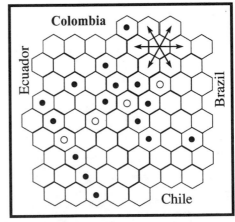

111.1 Peruvian Laser Battle. The black circles are robots. The open circles are soldiers.

straight line along any of the 6 hexagonal directions (shown by arrows at the top of the diagram) until they exit the battlefield, disabling all robots they encounter on the way."

Monica looked at Dr. Googol and grabbed his hand. "Each soldier gets 2 shots. Remember, to avoid the thermonuclear blasts, your soldiers are instructed not to hit any robot more than once. Can you determine the directions in which your soldiers should fire?"

❀ For a solution, see "Further Exploring."

Chapter 112

The Emerald Gambit

Einstein remarked more than once how strange it is that reality, as we know it, keeps proving itself amenable to the rules of man-made science. But our thought extends only as far as our capacity to express it. So too it is possible that what we consider reality is only that stratum of the world that we have the faculties to comprehend. For instance, I am convinced that quantum theory flouts commonsense logic only because commonsense logic has not yet been sufficiently expanded.
—*John Bainville,* "Beauty, Charm and Strangeness:
Science as Metaphor," Science 281, *1998*

Dr. Googol and Monica traveled to the heart of Arequipa, Peru, to seek ancient power. Inside a mighty Inca fortress was Augusto Leguía y Salcedo: mystic, soothsayer, and witch doctor. Dr. Googol looked into the wizard's flaming magenta eyes and was transfixed by his mesmerizing glance.

"Oh Great One," Dr. Googol asked, "can you grant me the power of invisibility?"

"Ah," Augusto Leguía y Salcedo replied, "in order to possess such a power, you must first pass a test." He produced a board divided into 25 squares (Figure 112.1). "Place these 13 rubies and this single emerald on the board so that there will be an even number of stones in each row and column and along the 2 diagonals."

Dr. Googol reached toward the board, thinking that this should be devil's food cake.

"Wait!" Augusto Leguía y Salcedo cried, his eyes taking on a strangely disturbing intensity. They seemed to be looking into Dr. Googol, as if he were already transparent. "There can be no more than 1 ruby per square. The emerald must be placed on a square with a ruby. Not one of

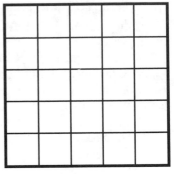

112.1 The Emerald Gambit board.

the rows, columns, or diagonals can be empty of stones." He turned over an hourglass filled with black sand. "You have 1 hour to solve the problem, or else you and your pretty friend will forever remain"—he grinned, and the blood vessels in his head throbbed—"mere visibles."

❋ For a solution, see "Further Exploring."

Chapter 113

Wise Viracocha

> This is the project that all artists are embarked upon: to subject mundane reality to such intense, passionate, and unblinking scrutiny that it becomes transformed into something rich and strange while yet remaining solidly, stolidly itself.
> —*John Bainville,* "Beauty, Charm and Strangeness: Science as Metaphor," Science 281, *1998*

Viracocha—the ancient Inca deity and creator of all living things—has a golden coin to share with his 4 favorite gods: Apu Illapu, Inti, Hathor, and Anubis. On the coin are 8 drawings of anchovies spaced as shown in Figure 113.1. (Anchovies are an Inca favorite!) To be fair, Virachocha will break the coin into 4 equal parts and give 1 to each of his godly friends.

"Wait!" cries Inti, the Inca sun god, "I want my piece to contain the same number of anchovies as everyone else's."

"So do I," says Apu Illapu, the rain giver, as he raises his staff.

"No problem," Viracocha replies as he raises his hammer and chisel to divide the coin. "Each piece will contain 2 anchovies."

How does Viracocha cut the coins so that each piece has the same area of gold and also the same perimeter (edge) length, as well as containing 2 anchovies? Viracocha's chisel cuts only straight edges, so all your cuts must be straight.

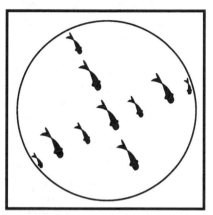

113.1 Viracocha's coin.

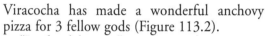

Viracocha has made a wonderful anchovy pizza for 3 fellow gods (Figure 113.2).

"Looks delicious!" cries Inti, the Inca sun god.

"I'm starved," says Apu Illapu, the rain giver, as he throws his staff on the ground.

"Me too," says Mama-Kilya, the moon mother, who starts toward the pizza with knife raised.

"Wait!" Viracocha says. "First you must pass my test. Only those who are worthy may eat my pizza. I want you to think of a way to divide the pizza into sections using 3 circular cuts so that 1 anchovy will be in each cut. Let me give you an example."

Viracocha draws a picture with 6 anchovies (Figure 113.3) . "Look here. I have used 3 circles to divide the pie in such a way that 1 anchovy is in each section. Now, who can do this for the delicious pizza pie that has 10 anchovies?"

❀ For solutions, see "Further Exploring." (Don't look up the answers until you have considered both problems; otherwise your eye will see both solutions at once and spoil the fun.)

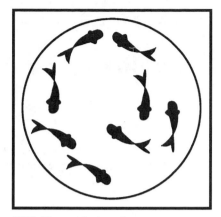

113.2 Viracocha's pizza.

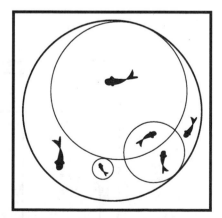

113.3 Viracocha's example.

Chapter 114

Zoologic

Mathematics is used like a microscope to understand the real world. But the microscope is flimsy, incomplete, and filled with contradictions. Does this mean that the universe, too, is filled with contradictions and paradoxes?
—*Dr. Francis Googol*

The Inca Empire in South America flourished before the European conquest of the New World, and it reached its greatest extent during the reign (1493–1525) of Huayna Capac. At this time, llamas were the primary beasts of burden; alpacas were domesticated and raised chiefly for their fine wool. Other domesticated animals included dogs, guinea pigs, and ducks.

Dr. Googol likes to imagine Capac's ancient zoo, filled with all manner of indigenous animals and overseen by a quirky zookeeper named Mr. Gila.

One warm summer day, Capac's zoo has finally moved all its animals into their new homes. Figure 114.1 shows an aerial view of the zoo. Each of the zoo's animal enclosures is marked with a circle. The paths between the enclosures, shown as lines, are overgrown with weeds. Zookeeper Gila not only has to feed all the animals, he has to mow the paths as well. (Back then mowers were a series of rotating, machete-like blades.) Each path is 100 feet long. Mr. Gila starts his walk at point *A*, the zoo's entrance, and finishes at point *B*. How far must he travel, and what route should he take, so that his walk is the shortest possible? (He may have to travel along some paths more than once.)

❀ ❀ ❀

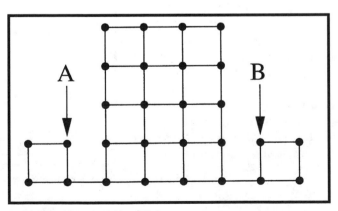

114.1 The layout of Mr. Gila's zoo.

114.2 Exhibit of 10 skinks.

In one section of the zoo there is an exhibit of 10 skinks (Figure 114.2). They live in an aquarium made of 21 panes of "glass" made from the dried sap of cinchona trees and sarsaparilla and vanilla plants. As you can see, the aquarium is divided into 10 compartments of equal size. Unfortunately, the feisty skinks have cracked 2 panes in attempts to escape. Mr. Gila needs to enclose the 10 skinks with the remaining 19 panes of glass. The compartments should be of equal size, all the glass panes must be used, and there must be no overlapping panes of loose ends. Can he do it?

❁ For solutions, see "Further Exploring."

Chapter 115

Andromeda Incident

> The mathematical spirit is a primordial human property that reveals
> itself whenever human beings live or material vestiges of former life exist.
> —*Willi Hartner*

The volcano El Misti stand 5,822 meters (19,101 feet) above sea level in southern Peru. The extinct volcano is part of the Cordillera Occidental, the principal arm of the Andes Mountains. Because of its height and clear skies, El Misti is an excellent place for observing the stars.

"Look, Monica." Dr. Googol pointed. "The Andromeda galaxy."

"Wonderful! I know all about it. It's 2 million light-years from Earth. It's the nearest spiral galaxy and the most distant object that we can see with the naked eye."

Dr. Googol huddled closer to Monica. Perhaps there was romance in the air. "May I give you a new puzzle?"

Monica hesitated. "Sure, but make it the last one for tonight. I'm getting a bit tired."

Our story begins with an amazing discovery. Happily, there turns out to be intelligent life in the Andromeda galaxy. Unhappily, however, the Andromedans, apparently driven mad by our errant television broadcasts, have decided to attack us. Nine of their best flying saucers are heading our way. They travel in formation, continuously emitting death rays horizontally, vertically, and diagonally. Therefore, they must be careful to stay in the arrangement shown in Figure 115.1 so that they don't destroy one another. In this particular arrangement, no saucer is horizontally, vertically, or diagonally in line with another.

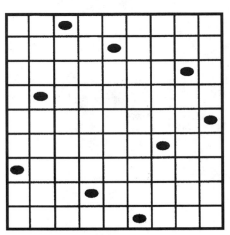

115.1 Arrangement of flying saucers.

Tired of maintaining the strict formation for such a long journey, 3 of the saucers wish to move to an adjoining cell in space. The death rays will be turned off for the move. Afterward they will be turned back on, so again no saucer can be in line with another. Which 3 of these saucers move, and to which 3 cells (at present unoccupied) do they pass?

❀ For a solution, see "Further Exploring."

Chapter 116

Yin or Yang

> The trick that art performs is to transform the ordinary into the extraordinary and back again in the twinkling of a metaphor.
> —*John Bainville*, "Beauty, Charm and Strangeness: Science as Metaphor," Science 281, *1998*

Viracocha, the great Inca god, is preparing a birthday cake for a friend's twin sons. Viracocha knows that one prefers chocolate, while the other prefers vanilla. Viracocha, in his wisdom, bakes a cake in the shape of the ancient yin-yang symbol of two opposing cosmic forces. He knows this should satisfy the children because the symbol is, geometrically speaking, a circle divided into 2 equal parts, and one part of the cake is chocolate, the other vanilla. Viracocha cuts the cake into 2 pieces along the curvy line dividing the 2 flavors (Figure 116.1).

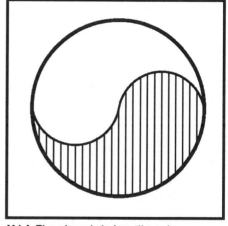

116.1 The chocolate/vanilla cake.

When the children come and look at the cake, they cry, "Oh Great One, there are 4 children to serve, not just 2. Two of us like chocolate, and 2 of us like vanilla."

Viracocha sighs. "Okay, there is a way to cut the cake into 4 pieces of the same size and shape using just 1 more cut. You'll even each have the same amount of icing. If you can figure out how to make such a cake, the 4 of you will be satisfied."

Can you help the children divide the yin and the yang into four pieces of identical shape and size with a single cut?

❀ For a solution, see "Further Exploring."

Chapter 117

A Knotty Challenge at Tacna

When an electron vibrates, the universe shakes.
—*British physicist Sir James Jeans*

Dr. Googol and Monica were exploring Tacna, the southernmost town in Peru, when a band of paramilitary thugs suddenly ambushed Dr. Googol's jeep. From the surrounding cocoa trees hung thick ropes with loops at the bottom, as if the ruthless men were preparing for a hanging.

"Oh no!" Monica said. "What do we do now?"

One of the men approached Dr. Googol and pointed to a loop of rope on the ground (Figure 117.1). Then he blindfolded Dr. Googol and Monica and turned to Dr. Googol. "Do you think it is likely that the rope on the ground is knotted?"

Monica clenched her fists. "How do we get ourselves into such absurd situations?"

Dr. Googol reached out to hold her hand. "Monica, don't worry. Even though I glanced at the ground too quickly to notice which segments of rope go over each other, I can figure out the exact probability of the rope being knotted. Then I can give the man an accurate answer."

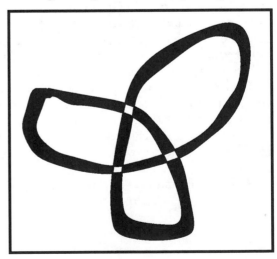

117.1 A loop of rope. Tiny white areas indicate the intersection points. Do you think this rope is knotted?

❀ ❀ ❀

Dear Reader, if you were a gambler, would you bet on the rope's being knotted?

❀ For a solution, see "Further Exploring."

Chapter 118

An Incident at Chavín de Huántar

> We wander as children through a cave; yet though the way be lost, we
> journey from the darkness to the light.
> —The Gospel According to Thomas (XV:1)

Before the Spanish invasion, the peoples of Peru were isolated from one another by the country's rugged topography. However, a unifying culture spread across
the Andes 3 times. Beginning in 1000 B.C., the Chavín culture permeated the
region, emanating from the northern ceremonial site of Chavín de Huántar. Dr.
Googol was exploring this site when a small boy ran up to him and handed him
a clay tablet with strange symbols.

Dr. Googol looked at the tablet. "These are definitely not symbols of the
Chavín culture."

"How do you know that?" the boy said. "In any case, it does not matter. I am
told that if you can decode this message, you will hold the **keys to the uniberse.**" The boy said the last 4 words in a mysterious tone of voice. "Notice how
each section of the message is smaller than the previous."

"Very good," Dr. Googol replied. "I love a great challenge. I will have my
assistant Monica decode this once I return to the village. If she can translate this
tortuous message, we might both share **the uniberse's secrets.**"

⊊⇔↓⇑→⇒← ⇔↦↪ ⊊Ⅎ↑ ⇓⇒←⇐↓⇑Ɫ⇌⇌↪ ⇐↓ ⇐↓↑Ɬ
⇓⇒⇌⇌⇒→ →↓ ⇐↓⇑→⇒↑⇓⇌Ⅎ→⇒ ↦↩⇔↦⇒← →↩↑⇒⇑Ɬ
⇐↩↓⇑⇐⅋⊊⇒→ ⇐⇒⇒↑⇐ →↓ ↑⇒ →↦Ⅎ→ ↩→ ↩⇐ ↓⇒←
⇑Ⅎ→⇒←⇒ →↓ →←⇒Ⅎ↑⅌ →↓ ⇐⇒Ⅎ←⇐↦⅌ Ⅎ⇑↑→ →↓
⇔↓⇑→⇒← Ⅎ←↓⇒→ ↓⇒← ⇓⇌Ⅎ⇐⇒ ←⇑ Ⅎ ⇐⇒⇒↑⇐↦⇑⇔Ɫ
⇌↪ ⇌↓⇑⇒⇌↪ ⇐↓⇐↑↓⇐⅋ /⇒←↦Ⅎ⇓⇓⇐ →↦↩⇐ ↩⇐ Ⅎ
←⇒Ⅎ⇐↓⇑ →↦Ⅎ→ ⇓↦↩⇌↓⇐↓⇓↦↪⇌⇐ ℲⅡ→ ⇒↩↩⇒⇑
→↦⇒↓⇌⇌↓⇔↩Ⅎ⇑⇐ ↦Ⅎ↩↩⇒ ⇐⇓⇒⇐⇒⇌Ⅎ→⇒→ Ⅎ←↓⇒→
→↦⇒ ⇒↦↩↩⇐→⇒⇑⇐⇒ ↓↩ Ⅎ ↩↓⇒←→↦ →↩↑⇒⇑Ɫ
⇐↩↓⇑ ℲⅡ→ ⇔↦Ⅎ→ ↩↩→⇐ ←⇑↦ℲⅡ↩↩→Ⅎ⇑→↩⇐ ↑↩⇔↦↩→
←⇒ ⇌↩↩⇒⅋ ⊊ Ⅎ⇔↩←⇒⇒ ⇔↩↪↦ ⊊←↩↩⇐ Ɫ←↓↑↑ ⇔↦↓
⇔↩←↓→⇒ ←↑ ⇉↦↦⇒ /←↩→ ↓↩ ⊒↓←↩←↩⇑⇔Ɫ ⊐⇉↦↦⇒
→⇒⇒⇓⇒↓⇒← ⇑⇒⇒→ ↓↩ ↑ℲⅡ ←⇐ →↓ ↓←↩→⇒←⇐↓↑⇒
↦←⇐ ⇐⇒⇓Ⅎ←Ⅎ→⇒↑⇒⇒←⅋ →↓ ⇌⇒Ⅎ←↩ →↦⇒
⇓←←⇒↓↑ ↓↩ ↦←⇐ Ⅎ⇌↓⇑⇒↑⇑⇒←⅋λ

※ For a hint, see "Further Exploring."

Chapter 119

An Odd Symmetry

> Mathematics is a train weaving its way through the infinite landscape of reality. As humans progress, the train moves ever forward. More cars are added, and rarely is a car discarded. Yet, if mathematics is the train, I cannot help but wonder: who made the tracks upon which the train rides?
> —*Dr. Francis Googol*

Peru's transportation system faces the challenge of the Andes Mountains and of the intricate Amazon River system. The only integrated networks are the roads and the airlines; the country's two railroad systems have not been interconnected.

Dr. Googol was riding the major Peruvian railroad, the Central Railway, which rises from the coast at Callao near Lima to cross the continental divide at 15,700 feet. He was about to take a nap when one of the train conductors approached him.

"My name is Jorgo Chávez," the conductor said. "I understand you are a mathematician."

"I do a little in my spare time," Dr. Googol said nonchalantly.

"Good, I have a problem for you. Come with me." He led Dr. Googol to the next car, in which there were 9 barrels. Each barrel contained several hundred plastic models of a single digit. The first barrel contained plastic models of the digit 1. The second contained models of 2, and so forth. The ninth barrel had plastic models of 9.

On the wall were several rows of mailboxes with mathematical operations between them:

. . . etc . . .

The conductor pointed to the mailboxes. "In each of your attempts to solve the problem, you are only allowed to reach into 1 barrel and place the same number in each mailbox in a row to make the mathematics correct."

"Fascinating," Dr. Googol said.

"I will give you a hint," said Jorgo Chávez. "There are infinitely many solutions for the first row, ⬛ = ⬛. Try it. For example, you can reach into the 1 barrel and place a 1 in the left mailbox and a 1 in the right mailbox. Of course, 1 = 1. In fact, you can do this for any digit."

Dr. Googol nodded.

"Now look at row 2, ⬛ + ⬛ = ⬛ × ⬛. Amazingly, the number of solutions drops from infinity to only 1 solution! Can you figure out which single digit will make this correct?"

"Interesting," Dr. Googol said.

"Now for the hard problem. We wish to continue the exact same logic for the remaining rows. What digits can you place in the other rows to make the addition at left equal to the multiplication at right? Remember, you must use the same digit in each mailbox. So, for example, you could insert a 4 into the row 3 mailboxes to create $4 + 4 + 4 = 4 \times 4 \times 4$, but unfortunately this does not yield a correct formula. In fact, don't even limit yourself to the barrels of numbers in the train. I'll let you use *any* positive integers. Can you find digits that will make this work for an arbitrary number of symmetrically placed mailboxes?"

❀ For additional discussion, see "Further Exploring."

Chapter 120

The Monolith at Madre de Dios

> I just hope that I can laugh through all phases of life, do a little
> mathematics, live to a very ripe old age, and leave the body behind like
> slipping off a tight shoe.
> —*Clay Fried (e-mail to Dr. Googol)*

While exploring Madre de Dios, a city in eastern Peru, Dr. Googol came upon a large rectangular monolith. On the outside of the huge stone block was an array of different symbols. Could it be a code of some sort? One symbol was missing from the array. Perhaps some ancient astronauts left the monument behind ages ago. Perhaps they wish to assess our intelligence by seeing if we can fill in the symbol and complete the array.

What symbol should be used to replace the missing space in the matrix of symbols? (Hint: Numerical values need to be assigned to the symbols to solve this.)

What is the logic you used to solve this puzzle? Is there another logic that you might use to solve it differently?

♍	♐	♐	♐	♑
♐	♎	♑	♑	♎
♐	♍	♑	♐	♐
♎	♐	♑	♐	♍
♍	♎	♐	♎	?

❀ For a solution, see "Further Exploring."

Chapter 121

Amazon Dissection

> As one goes through it, one sees that the gate one went through was the self that went through it.
>
> —*R. D. Laing*

The Amazon, with the largest volume of flow of any river in the world, has headwaters in the Peruvian Andes. Dr. Googol was sailing along one of its main branches, the Ucayali River, which originates in southern Peru, when an old man came out from the jungle.

"Can I help you?" Dr. Googol said to the man.

"Yes. We have heard of your great mental prowess. We have a potential religious conflict that you can resolve. The Jews, Catholics, Moslems, and a mixture of Oriental religious groups live together on my vast jungle. Now the land must be subdivided, and we want to keep the religious mixture the same in the 2 new lands. More precisely, I want to create 2 areas, both of exactly the same size and shape, that contain equal numbers of each religious household. (We want both new lands to have the same religious composition for voting and other reasons.)"

He handed Dr. Googol a card with a symbol representing each religious household:

✡	✡	✡	☾	✝	✡
✡	✡	✡	☾	✡	✡
☾	✡	✡	✡	✡	☾
☾	☾	✡	✡	✡	☾
✡	☾	☯	✡	✡	☯
✡	☾	✡	✡	✝	☾

The old man continued. "You can use a pencil to define the areas, but all the lines you draw must be straight. You can think of this as cutting a rectangular cake into 2 identically shaped pieces."

Can you help Dr. Googol solve this problem?

❋ For a solution, see "Further Exploring."

Chapter 122

3 Weird Problems with 3

Pure mathematics is religion.
—*Friedrich von Hardenberg, 1801*

The number 3 plays an important role in Peru. Peru is the <u>third</u> largest nation in South America. Peru can be divided into <u>3</u> geographic regions from west to east: the Costa (coast), the Sierra (highlands), and the Montaña, or *selva* (the vast, forested eastern foothills and plains). Agriculture employs about one-third of the workforce. But all of these facts are not the primary reasons that Dr. Googol is fascinated by the number 3.

Here are some of the major reasons Dr. Googol loves three. Three is the only natural number that is the sum of all preceding numbers. It is the only number that is the sum of all the factorials of the preceding numbers: 3 = 1! + 2! In religion, 3 reigns supreme. For example, in ancient Babylonia there were 3 main gods: the Sun, Moon, and Venus. In Egypt there were three main gods: Horus, Osiris, and Isis. In Rome there were 3 main gods: Jupiter, Mars, and Quirinus. For Christians, 3 symbolizes the Holy Trinity: Father, Son, and Holy Spirit. In classical literature, there were 3 Fates, 3 Graces, and 3 Furies. In languages, there are 3 genders (masculine, feminine, neuter) and 3 degrees of comparison (positive, comparative, superlative.)

German Chancellor Otto von Bismark signed 3 peace treaties, served under 3 emperors, waged 3 wars, owned 3 estates, and had 3 children. He also organized the union of 3 countries. His family crest bore the motto: *In trinitate fortitudo* (In trinity, strength). There is a German saying: *Alle güte Dinge sind Drei* (All good things come in 3s).

With this diversion, Dr. Googol would like the most erudite among you to consider 3 fiendishly difficult problems dealing with the number 3 in some odd way or another. If you find any number nerd able to solve all of these, Dr. Googol invites them to join his Three Lovers Club.

GROWTH

Start with 3 digits: 1, 2, and 3. Each succeeding row repeats the previous 3 rows, in order, as you can see from the following diagram.

$$1$$
$$2$$
$$3$$
$$123$$
$$23123$$
$$312323123$$
$$123231233123 23123$$
$$231233123 23123 123231233123 23123$$

What is the sum of digits in row 100?

3-ATOMS

Get rid of all the 2s in the previous sequence. Here Dr. Googol has replaced each of them with a ♭:

> 1
>
> ♭
>
> 3
>
> 1 ♭ 3
>
> ♭ 31 ♭ 3
>
> 31 ♭ 3 ♭ 31 ♭ 3
>
> 1 ♭ 3 ♭ 31 ♭ 331 ♭ 3 ♭ 31 ♭ 3
>
> ♭ 3 ♭ 31 ♭ 3 ♭ 31 ♭ 31 ♭ 3 ♭ 3 ♭ 331 ♭ 3 ♭ 31 ♭ 3

Notice that in the last row of this diagram, there are 3 different atomic species: 31, 331, and 3. How many different species are there in row 30?

❁ ❁ ❁

CLEAVAGE

When the sequence first hits a 3, it now undergoes an enzymatic cleavage, and the digits on the right of the 3 are swapped with the digits on the left. (If the digit appears in the rightmost place, as in 123, nothing is swapped because nothing appears to the right of the 3.) For example:

> 1
>
> 2
>
> 3
>
> 123
>
> 23123 now becomes 12323
>
> 312312323 now becomes 123123233

Now go back to the previous "atom question" and try to find an answer.

✺ For solutions, see "Further Exploring."

Chapter 123

Zen Archery

> The Buddha, the Godhead, resides quite as comfortably in the circuits of
> a digital computer or the gears of a cycle transmission as he does at the
> top of a mountain or in the petals of a flower; to think otherwise is to
> demean the Buddha—which is to demean oneself.
> —*Robert Pirsig,* Zen and the Art of Motorcycle Maintenance

Dr. Googol was climbing Mount Huascarán, the highest mountain in Peru,
seeking enlightenment from a Zen master who had been living in a mountain
cave for years. After several hours, Dr. Googol found him sitting on a stone
throne.

Dr. Googol bowed. "Sir, I seek enlightenment."

He nodded, handed Dr. Googol a bow and arrow, and pointed to an unusual target hanging on the wall (Figure 123.1). "With 5 shots, hit 5 different numbers on the target that total 200."

Dr. Googol stepped back. "You've got to be kidding."

The monk stared. "You have 1 minute."

What are Dr. Googol's 5 shots? How long did it take you to solve the problem?

The archery master also gave Dr. Googol another problem (Figure 123.2). "There are 3 concentric circles of numbers on this target. Start at the outside row and hit a number. Go to the middle ring and hit a number. Go to the inner ring and hit a number. The sum for your numbers must be 100. Moreover, as you go from outer to inner ring, your selected numbers must keep increasing."

What are the numbers Dr. Googol must give to the Zen master?

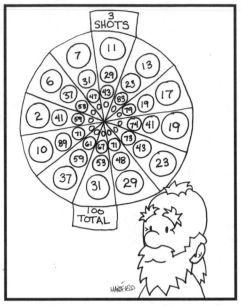

123.1 Zen archery. Hit 5 numbers for a total of 200. (Drawing by Brian Mansfield.)

123.2 Zen archery. Hit 3 numbers to total 100, given the rules described in the text. (Drawing by Brian Mansfield.)

❀ For solutions, see "Further Exploring."

■ See [www.oup-usa.org/sc/0195133420] for a computer program to solve this class of problem.

Chapter 124

Treadmills and Gears

> A rock pile ceases to be a rock pile the moment a single man contemplates it, bearing within him the image of a cathedral.
> —*Antoine-Marie-Roger de Saint-Exupery,* Flight to Arras

Dr. Googol is quite an inventor. During his last visit to coastal Peru, he invented the exercise device shown in Figure 124.1. He even received U.S. Patent

5767852 for this ingenious machine. But does it really work? As Dr. Googol runs, will the treadmill turn, or is it locked, thereby causing Dr. Googol to run off the end and plunge into the ocean? What effect does the figure-8 belt have on the operation of the device? Would the operation be different if this figure-8 were replaced with a Möbius strip (a loop of conveyor belt with a half twist)? If the device does not work, how would you fix it? Would the device function any differently if all belts were twisted?

Dr. Googol also invented a device consisting of gears and a thin loop of rubber (Figure 124.2). If he turns the crank at bottom, will the device move, or will it lock up? To solve this enigma, note that the gear train might lock if 1) two gears are trying to spin the same gear at different rates or 2) two gears are trying to spin the same gear in opposite directions. Let's assume that the rubber loop in the gear train (on the far left) is sufficiently slack so that it will take care of any differences in the speed of the gear train. Therefore, the only way the gear train would be locked is if condition 2 holds. The $20,000,000 question is: "Is the gear train locked?"

❀ For solutions, see "Further Exploring."

124.1 Will the belts on Dr. Googol's patented exercise treadmill turn freely or not? (Drawing by Brian Mansfield.)

124.2 Turn the crank at bottom. Will this gears in this contraption turn, or will they be locked? (Drawing by Brian Mansfield.)

Chapter 125

Anchovy Marriage Test

> Sometimes it's a form of love just to talk to somebody that you have
> nothing in common with and still be fascinated by their presence.
> —*David Byrne*

Late last autumn, Dr. Googol was dining with his friend Monica in a small cafe
in the town of La Oroya, Peru. They shared a large anchovy pizza while gazing
at one another and at the beautiful Peruvian tapestries hanging from the ceiling.

"Monica, did you know that in the 1950s and 1960s Peru's fishing industry
flourished madly because of the huge anchovy harvests? These fish were converted into fish meal and oil for export as animal feed."

"They do taste good. Salty."

Dr. Googol looked into Monica's dark eyes. "Monica, I've been meaning to
ask you something." He brought out a large diamond ring.

"Monica, I will marry you if you can answer the following questions."

"Oh, Dr. Googol, I thought you'd never ask!"

Dr. Googol handed Monica 3 slightly soiled pieces of paper:

> **Using standard mathematical symbols,
> can you make five 9s equal to 1,000?**

and

> **Can you add one small stroke to
> make this equation correct?**
> **6 + 6 + 20 = 666**

and

> **Insert 4 parentheses and 3 different mathematical
> symbols to make the following expression true:**
> **66666665 = 111**

Monica looked at the papers reeking of anchovies, then back at Dr. Googol.
"Francis, why must you always test me?"

"I want to make sure we are fully compatible."

"These are the last math questions I'm going to answer for a *long* time. You certainly have enough questions for that silly book you're working on."

"Monica, do you mean that you are actually *able* to answer the 3 questions? I've never come upon a person who could handle all of them."

❀ For a solution, and to see Monica's response, see "Further Exploring."

Further Exploring

Chapter 2

Why Don't We Use Roman Numerals Anymore?

To solve the boy's puzzle, simply turn the card upside down.

Dr. Googol conducted a study of almost 500 people regarding the column connection problem, and he asked people to time themselves as they attempted to arrive at a solution. About 20% of the people said this problem was impossible to solve. Those that could solve it usually did so in under 2 minutes, and there was little correlation between a person's ability to solve the puzzle and age (ages ranged from 20 to 60). The problem is in fact solvable, and the solution is left as an exercise for you. If you cannot solve the problem, don't think about it for a day; then return to the problem. Many people find it easier to solve this on their second attempt a day later. A computer could probably solve this class of problem faster than a human; however, humans have one advantage in that they have the ability to discard bad attempts rather quickly. Write a computer program to randomly place circles so as to create new and unusual "wiring" problems, or you can create new puzzles like this with pencil and paper.

Psychologists have long been interested in the relationship between visualization and the mechanisms of human reasoning. Is it significant that people find the puzzle easier to solve after returning to it a day later? Is there any correlation in a person's ability to solve the puzzle with gender, profession, IQ, musical ability, or artistic ability?

This type of problem raises questions that pertain to the mathematical field of graph theory—the study of ways in which points can be connected. Graphs often play important roles in circuit design. One unusual problem in this field involves the following question. How does one arrange sticks in a way such that 4 sticks meet end to end, without crossing each other, at every point in a geometrical figure on a flat surface? In Figure F2.1,

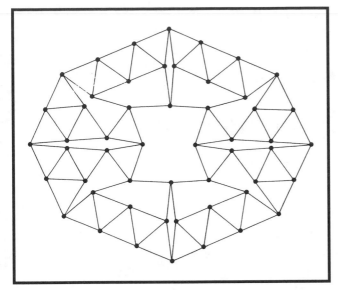

4 sticks meet at each vertex. This is the smallest arrangement known, but no one knows whether it's the smallest possible way to make a figure with 4 sticks meeting at each point!

❀ ❀ ❀

Why do clock faces with Roman numerals almost always show the number four as IIII instead of IV? There are several possible reasons, depending on whom you consult. (1) IIII provides aesthetic balance since it is visually paired with the VIII on the other side. (2) IV is a modern invention that the Romans did not use. (3) Romans did not wish to offend the god Jupiter (spelled IVPITER) by daring to place the first 2 letters of his name on the clock face. This latter explanation is unlikely because the idea of placing I before V to represent 4 (which makes numbers shorter to write while making them more confusing for arithmetic) was hardly ever used by the Romans themselves and became popular in Europe only after the invention of printing. (Also note that some clocks do use IV—London's Big Ben is the most famous example.)

F2.1 The amazing Harborth configuration from the "4 sticks" problem. (Pattern discovered by Heiko Harborth; diagram adapted from Peterson, I. (1990) *Islands of Truth*. Freeman: New York.)

Do you think that civilization's use of Roman numerals comes midway in its development? Going back in time, we find that cave-wall numbers were some of the initial steps toward primitive computing machines. One of the first true calculating machines to help expand the human mind was the abacus, a manually operated storage device that aids a human calculator; it consists of beads and rods and originated in the Orient more than 5,000 years ago. Archeologists have since found geared calculators, dated back to 80 B.C., in the sea off northwestern Crete. Since then, other primitive calculating machines have evolved, with a variety of esoteric-sounding names, including Napier's bones (consisting of sticks of bones or ivory), Pascal's arithmetic machine (utilizing a mechanical gear system), Leibniz's Stepped Reckoner, and Babbage's analytical engine (which used punched cards).

Continuing with more history: The Atanasoff-Berry computer, made in 1939, and the 1,500-vacuum-tube Colossus were the first programmable electronic machines. The Colossus first ran in 1943 in order to break a German coding machine named Enigma. The first computer able to store programs was the Manchester University Mark I, which ran its first program in 1948. Later, the transistor and the integrated circuit enabled microminiaturization and led to the modern computer.

In the mid-1990s, one of the world's most powerful and fastest computers was the special-purpose GRAPE-4 machine from the University of Tokyo. It achieved a peak speed for a computer performing a scientific calculation of 1.08 Tflops. (*Tflops* stands for "trillion floating-point operations per second.") With this computer, scientists performed simulations of the interactions among astronomical objects such as stars and galaxies. This type of simulation, referred to as an *N*-body problem because the behavior of each of the *N* test objects is affected by all the other objects, is particularly computation-intensive. GRAPE-4 reached its record speeds using 1,692 processor chips, each performing at 640 Mflops. Like a web spun by a mathematically inclined spider, each processor had intricate connections with the others. The Tokyo researchers hoped to achieve petaflops (10^{15} or 1 million billion floating-point operations per second) by the turn of the century with a suite of 20,000 processors each operating at 50 Gflops.

NASA, the Defense Advanced Research Projects Agency, and the National Security Agency are funding the exploration to support mission-critical areas ranging from simulating Earth's climate system to breaking the communications of enemy nations. Their "hybrid technology multithreaded" (HTMT) architecture for the next generation petaflop computer is a mix of emerging technologies including helium-cooled superconducting processors, memory chips with onboard processing capabilities, an optical communication network, and holographic storage. (For more information, see: Cohen, J. (1998) Mix of technologies spurs future supercomputer. *Insights* (NASA). July, 6: 2–10.)

GRAPE-4 was certainly much more expensive than the abacus or Napier's bones, but also much faster!

Chapter 3

In a Casino

The answer is 1.2 centimeters. The ruler does not help you, but the employee was wise in offering Dr. Googol this distraction. If you disregard the ruler before your minute is up, you may brilliantly realize that the measurement is reduced by 1/13th, because 4 cards are removed from 52, and then you can quickly do the necessary mathematics and subtract 0.1 from 1.3. Try this on some friends—few will be able to solve it quickly.

To make the problem more difficult for your friends, start by telling them that the deck *without* Kings is 1.2 cm thick. Next ask your friends, "If the gladiator produces four Kings and adds them to the deck, how thick is the deck?"

Still not sufficiently difficult? How thick will the deck be if the Queens abscond with all other cards that show a prime number on their face? (An integer greater than 1 is a prime number if its only positive divisors [factors] are 1 and itself.)

How many consecutive digits of pi (3.1415 . . .) can you display with a deck of cards? (See Chapter 96 for fractal sequences based on cards.)

Chapter 4

The Ultimate Bible Code

This problem was discussed by Martin Gardner in the August 1998 *Scientific American*. Gardner's wonderful "Mathematical Games" column began in the December 1956 issue and ran for more than 25 years, providing a whole library of *Scientific American* puzzle books.

In this Bible code puzzle, Gardner points out that each chain of words ends on *God*. This answer may seem miraculous, but it actually is the result of the "Kruskal count," a mathematical principle first noted by mathematician Martin Kruskal in the 1970s. When the total number of words in a text is significantly greater than the number of letters in the longest word, Gardner notes, it is likely that any 2 arbitrarily started word chains will intersect at a keyword. After that intersection point, the chains become identical. As the text lengthens, the likelihood of intersection increases.

Dr. Googol welcomes any other "miraculous" examples of texts with these kinds of properties. Can you discover similar examples using various literary or religious works? In a personal communication to Dr. Googol, Martin Gardner notes that if the Krukal count is applied to the verse of Exodus, the count ends on *man*.

Chapter 5

How Much Blood?

Here are some additional sickening challenges for you to consider. So far, none of Dr. Googol's colleagues have provided reasonable answers. Can you?

⊙ Compute the volume of body fluid for an average fish. What size container would be needed to contain all the blood of all the fishes in the world?

⊙ Today, is there more monkey blood in the world or more human blood? Ten thousand years ago, was there more monkey blood in the world or more human blood?

⊙ Today, is there more insect blood in the world or more human blood? What size container is required to store all the insect blood in the world?

⊙ If all human intestines were tied end to end, would they be able to stretch a distance equivalent to the distance from the Earth to the Moon? ("Oh, Dr. Googol, you are a gross human being for asking this.")

Chapter 6

Where Are the Ants?

Why not first try simulating this on a computer? You can build your own computerized ant farm through which ants can travel by defining a map of tunnels and chambers. Next have your simulated ants crawl through the tunnels using a "random walk" procedure. For example, start with 10 ants described by their (x,y) positions in the ant farm. Have the computer draw each ant as a little black circle, or as a triplet of circles to represent the head, thorax, and abdomen. For each increment in time, move the ants a random short distance. If an ant bumps into a wall, reflect it back into the tunnel or chamber. You can make the simulation easier to program on a computer by representing the chambers and tunnels as squares connected by straight, thin tubes. Those of you without computers can accomplish this simulation by using dots on graph paper and by throwing dice to control the ant's movements.

In which chamber do the most ants reside? To solve this problem theoretically, we assume that the ants walk randomly. In this sense, they behave like randomly diffusing molecules in a gas. Therefore, the number of ants in each chamber is proportional to the area of the chamber. The nature of the interconnecting tunnels should not matter if you give the diffusing ants sufficient time to come to an equilibrium state. In other words, in Figure 6.1 most ants will reside in chamber C, the chamber with the largest area. (Actually, just about as many ants will reside in the upper region outside the connected chambers, because this region has an area nearly the same as C.)

Are you able to simulate this using a computer? What happens if the ants are different sizes and move at different speeds, or if an ant's behavior in a chamber is affected by the density of ants in the chamber, or if they leave odor tracks behind them for other ants to follow, or if an ant can't change directions when in a tube? There are dozens of interesting experiments to try. They're not only good fun but will teach you some fundamental lessons about the diffusion of particles under different conditions.

Dr. Googol has been told that the following terms are trademarked by Uncle Milton Industries: Ant Farm, Ant Farmers, Ant Farm Village, Ant Way, and Ant Port. You can purchase already assembled, low-cost ant farms from Uncle Milton Industries, Culver City, CA 90232. The term Ant City is a trademarked name of another ant-enclosure manufacturer: Ant City, Natural Science Industries, Far Rockaway, NY 11691.

Chapter 7

Spidery Math

Figure F7.1 is a diagram showing the 6 gaps left by the hallucinating spider. The answer to the question regarding the smallest and largest spider numbers for an arbitrary (n,m) web is still a mystery to mathematicians. However, James Doyle from South Orange, New Jersey, believes that for the (4,3) web, the *largest* spider number is 322. He arrived at this number by placing 1 gap on each of the 3 circles and placing the fourth gap on any 1 of the circles in the section to the immediate left or to the immediate right of the other gap on that circle. The *smallest* spider number for a (4,3) web appears to be 240. You can arrive at this solution by placing 1 gap on each of the 4 straight lines between the center point and the first circle. The largest spider number for the (2,2) web is probably 54. (Place 3 gaps on one circle and 1 gap on the other circle.) The smallest spider number for the (2,2) web appears to be 32. (Place 2 gaps on each of the 2 straight lines, and 1 gap on each of the 4 sides of the center point.)

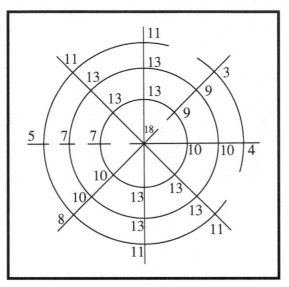

F7.1 Solution for spider math—showing 6 gaps left by the hallucinating spider.

Chapter 8

Lost in Hyperspace

Despite the pitifully little information given, we can calculate the answer: 1,800 light-years. Here's how. When the starships first meet, half of the entire path has been traveled, and this is equal to 800 light-years (what the *Enterprise* traveled) plus X (what the *Excelsior* traveled). After this meeting, they continue traveling until they meet again. During this second part of the journey, an entire path has been traveled, and this entire path is equal to $1600 + 2X$. Notice that during the second part of the journey, each ship covers twice the distance it covered during the first part of the journey. This means $2X = 200$, and the entire circuit is $1,600 + 200 = 1,800$ light-years.

The shape of the track, including a 3-D track, should not matter provided that the clockwise and counterclockwise paths between the *Enterprise* and the *Excelsior* are the same length.

Chapter 9

Along Came a Spider

🕷 For the first problem, note that Mr. Ten cannot have 10 legs, so he must have either 8 or 9 legs. Because the spider with 9 legs replies to Mr. Ten's remark, Mr. Ten cannot have 9 legs. Therefore Mr. Ten has 8 legs. Now consider Mr. Nine. He cannot have 9 legs, because this would match his name. Mr. Nine has 10 legs.

🕷 For the second problem, 1 insect is sufficient. Unwrap 1 insect from the web labeled "flies and mosquitoes." Say that it's a fly. Because each web is labeled incorrectly, the web cannot be the "flies and mosquitoes" web, and therefore it must be the fly web. The web labeled "mosquito" must contain mixed insects, and the web labeled "flies" must actually be the mosquitoes web.

🕷 Here is an unsolved problem on which you can work for hours. There are 4 webs labeled "flies and mosquitoes," "mosquitoes and ants," "ants and wasps," and "just wasps." All the labels are incorrect. How many insects do you have to unwrap to correctly label the webs, and how do you do it? Dr. Googol believes the answer is 3. Can you confirm his suspicion?

Chapter 10

Numbers beyond imagination

Before showing you the results of the Big Number Contest, Dr. Googol would like to digress and give a background on large numbers. The term *googol* is used to designate a very large number: 10 raised to the power of 100, or 1 followed by 100 zeros.

10000000000 0000000000 0000000000 0000000000 0000000000
0000000000 0000000000 0000000000 0000000000 0000000000

American mathematician Edward Kasner popularized this number in the 1930s. Most scientists agree that if we could count all the atoms in all the stars we can see, we would come up with less than a googol of them. Interestingly, the name *googol* was invented for this number by Kasner's 9-year-old nephew. The same youngster also invented the term *googolplex* for an even higher number: 1 followed by a googol 0s. Our limited brain architecture makes it difficult to comprehend numbers such as these. We have not needed to evolve this capability to ensure our survival. However, just as children slowly become able to name and appreciate larger and larger numbers as they grow, civilization has gradually increased its ability to name and deal with large numbers.

Which number is larger: the number of possible chess games (which Dr. Googol denotes by A), or the number of trials needed for a monkey to type Shakespeare's *Hamlet* by random selection of keys (expressed as 1 chance in B trials)? How do these values compare with the number of electrons, protons, and neutrons in the universe, C, or with Skewes's number D (which is one of the largest numbers that has occurred in a mathematical proof)? The values of these numbers are listed in the following.

Chess number: $A = 10^{10^{70.5}}$

Hamlet number: $B = 35^{27,000}$ which is about equal to $10^{40,000}$

Universe number: $C = 10^{79}$

Skewes's number: $D = 10^{10^{10^{34}}}$ (revised in 1955 to have the value $D = 10^{10^{10^{1000}}}$)

Mathemetician G. H. Hardy called Skewes's number "the largest number which has ever served any definite mathematical purpose in mathematics." Hardy determined that if one played chess with all the particles in the universe (which he estimated to be 10^{87}), where a move meant simply interchanging any 2 particles, then the number of possible games was roughly Skewes's original number:

$10^{10^{10000000000000000000000000000000000000}}$

A recent mathematical thesis did even better than large numbers! In his book *Mathematical Mysteries,* Calvin Clawson reports that the number of kinks in the core of an "embedded tower" is roughly

$$E = 10^{10^{10^{10^{10^{7}}}}}$$

Now that's a big number! The point is that today large numbers such as these are often contemplated, but this is a relatively recent development in human history. For example, in biblical times, the largest number expressed as a single word was 10,000. This occurs in the ancient Hebrew version of the Old Testament as the word *r'vavah.* The word for million was an Italian invention of the 13th century, and the English word *billion* was coined in the 17th century (largely as a curiosity).

In evaluating and formulating expressions, it is important to recall some of the simple rules of exponentiation. For example, $(a^m)^n = a^{mn}$. Test this using some small numbers. Also, parentheses are often needed to resolve ambiguities. For example, $3^{2^3} \neq (3^2)^3$. As discussed, a number raised to a negative power is simply 1 over the number raised to the positive value of the power. For example, 2^{-3} is $1/2^3$. The expression a^{b^c} is usually taken to mean $a^{(b^c)}$. To determine the number of digits N in a value X, recall that $N = \log_{10} X + 1$.

The numbers discussed in this chapter are often much larger than a googol, yet they are constructed with the barest of mathematical notation. In the first part of Dr. Googol's Big Number Contest, he asked participants to construct an expression for a very large number using only the digits 1, 2, 3, and 4 and the symbols (,), the decimal point, and the minus sign. Each digit could be used only once. In a second contest, the contributors could use, in addition to these symbols, any standard mathematical symbol (such as the factorial symbol, !) to produce a large number. Each symbol could be used only once in the mathematical expression. For both contest parts, the final answer had to have a finite value. Of the approximately 50 contributors, the 8 top entries are listed.

For Part 1 of the contest, exponentiation is allowed since it does not require a symbol when traditionally expressed. The following are the results for Part 1.

FIRST-PLACE WINNER: WALT HEDMAN AND TIM GREER, NEW YORK

$$0.3^{-(0.2^{-(0.1^{-4})})} = 3.33^{(5^{10000})} \text{ or } 3.3^5 \times 10^{6989}$$

This number roughly corresponds to 3 to the nth power where n has approximately 6990 digits. The number of cubic inches in the whole volume of space comprising the observable universe is almost negligible compared to this quantity.

SECOND PLACE: DIANA DLOUGHY, NEW YORK

$$(.1)^{-(4^{32})} = 1 \times 10^x \text{ where } x = 4^{32} \sim 1 \times 10^{19}$$

This second-place answer has 1×10^{19} digits. (Note: Later in the course of her experimentation Diana discovered that 3^{42} is 1 decimal place larger than 4^{32} so that her answer can be changed to $.1^{-(3^{42})} = 1 \times 10^{19}$ where $x = 3^{42}$, which is roughly equal to 10^{20}.)

THIRD PLACE: ROD DAVIS, NEW YORK

$2^{3^{41}}$ (has 1.0979×10^{19} digits)

FOURTH PLACE: ROD DAVIS, NEW YORK

$3^{4^{21}} = 3^{(4^{21})} = 3^{4398046511104}$ (has 2.1×10^{12} digits)

This number roughly corresponds to 3 to the nth power where n has approximately 6990 digits. The number of cubic inches in the whole volume of space comprising the observable universe is almost negligible compared to this quantity.

FIFTH PLACE: DIANA DLOUGHY, NEW YORK

$(.1)^{(-432)} = 1 \times 10^{432}$ (has 433 digits)

SIXTH PLACE: MANY PEOPLE FOR THIS 201-DIGIT ENTRY

$3^{421} = 7.37986 \times 10^{200}$

Submitters: Gary Hackney, Erik Tkal, Mike Shreeve, and Christine Wolak, among others.

SEVENTH PLACE: MIKE OTT, TORONTO

$2^{(4^{(3+1)})} = 2^{256} = 1.1 \times 10^{77}$ (has 78 digits)

Note: Technically this answer should be disallowed since the plus sign was not allowed in the contest rules.

EIGHTH PLACE: W. GUNN, NORTH CAROLINA

31^{42} (has 63 digits)

Can you beat the first-prize winner in this contest?

To create the prize-winning answers for the second part of the contest, contributors often placed factorial signs, denoted by the ! symbol, at the end of the expressions listed above. (Recall that, for example, $4! = 4 \times 3 \times 2 \times 1$.) For those of you who would like to evaluate the huge results obtained with factorial symbols, the following formulas may be helpful: $n! \sim \sqrt{2\pi n} \; n^n e^{-n}$ and $\ln(n!) \sim [n \ln(n)] - n$. The second-prize winner for this part, Dave Challener from New York, also used a gamma function symbol in the front of the first solution in Part 1. For positive integers, $\Gamma(n + 1) = n!$ Note that, in general, $\Gamma(x) = \int_0^\infty t^{x-1} e^{-t} dt, x > 0$ or alternatively, $\frac{1}{\Gamma(x)} = xe^{\gamma x} \Pi_{m=1}^\infty \{(1 + \frac{x}{m})e^{-xm}\}$ where γ is Euler's constant.

Mike Shreeve from Atlanta was the first-place winner. His answer made use of a second-order Ackermann's function (as described in Aho's book in "Further Reading"), which can be expressed by $A_n = 2^{A(n-1)}$ with $A(0) = 1$. The sequence progresses as follows: 1, 2, 4, 16, 64000, $2^{64,000}$, ... Mike Shreeve believes that this function grows faster than any other named function. As big as the gamma answer is, it is smaller than $A(4+3+2+1)$. Mike concluded his note to Dr. Googol with the words "I don't even want to think about $A\left(3^{(4^{21})}!\right)$."

Note that James Hunter's and Joseph Madachy's fascinating book *Mathematical Diversions* lists the expression for Contest 1's first prize as an example of a very large number. They note that this number is 3 to the *n*th power where *n* has approximately 6,990 digits. The number of cubic inches in the whole volume of space comprising the observable universe is almost negligible compared to this quantity.

Let's end this section with some other curiosities and large numbers. One of the largest individual numbers that occurs naturally in a theorem is   —

8080 17424 79451 28758 86459 90496 17107 57005 75436 80000 0000

This is the order of the so-called **Monster simple group.** An example of a finite group is a collection of integers from 1 to 12 under the operation of "clock arithmetic," so that, for instance, 9 + 6 = 3. The concept sounds simple, but it gives rise to a mathematical jungle. For decades, mathematicians have tried to classify all the finite groups. One of the strangest groups discovered is the "Monster group," which has over 10^{53} elements and a little-understood structure. For a background on this number, see: Gorenstein, D., Lyons, R., and Solomon, R. (1994) Mathematical surveys and monographs: the classification of the finite simple groups. *The American Mathematical Society:* New York. To better understand how symmetries of geometric objects form the elements of finite groups, and how a particular string theory, when applied to a folded doughnut in 26 dimensions, has more than 10^{53} symmetries and produces the Monster group, see: Wayt, W. (1998) Monstrous moonshine is true. *Scientific American,* November, 279(5): 40–41.

Although Skewes's number, mentioned earlier in this section, is often thought to be the largest number ever used in a mathematical proof, there is actually a more recent record-holder. **Graham's number** is an upper bound from a problem in a part of combinatorics called Ramsey theory. Graham's number cannot be expressed using the conventional notation of powers, and powers of powers. Let Dr. Googol try to explain it using the symbol #. 3#3 means 3 cubed, and in general $a\# b = a^b$. 3##3 means 3#(3#3). 3###3 = 3##(3##3). 3####3 = 3###(3###3). Consider the number 3### . . . ###3 in which there are 3####3 "#" signs. Next construct the number 3### . . . ###3 where the number of # signs is the previous 3### . . . ###3 number. Now continue the process, making the number of # signs in 3### . . . ###3 equal to the number at the previous step, until you are 63 steps from 3####3. This is Graham's number, which occurred in a proof by Ronald L. Graham, as described by David Wells.

The **Moser** is Dr. Googol's favorite huge number. One way of making incredibly large finite numbers is through repetition. The Moser (presumably named after mathematician Leo Moser) can be computed as follows: Define $n|$ to be n^n. This means $2| = 2^2 = 4$, and $3| = 3^3 = 27$, etc. If we add more line segments, we find: $2||| = 2^{2||} = 4|| = 4^{4|} = 256| = 256^{256}$. Now, define $n<$ (n followed by a wedge) to be n followed by n line segments. So $3< = 3||| = 27|| = 27^{27|} = (27^{\wedge}27)^{\wedge}(27^{\wedge}27)$. (Here the $\wedge$ represents exponentiation.) What a large number! But hold on. We can continue! n followed by a triangle is the same as n followed by n wedges; n followed by a square is the same as n followed by n triangles; and, in general, n followed by a $k + 1$ sided polygon is the same as n followed by n k-sided polygons. Let's just see what $2(\triangle)$ is:

$$2(\triangle) = 2<< = 2||< = 4|< = 256< = 256||| \ldots 256 \text{ lines} \ldots |||||$$

This is an unimaginably large number, which we'll call Clinton in honor of our recent president. Notice that $2(\square) = 2(\triangle)(\triangle) = \text{Clinton}(\triangle) = \text{Clinton}<<< \ldots \text{Clinton}$ wedges . . . $<<<$ = something enormous (which we may call Schwarzenneger in honor of the enormously muscular movie actor). The Moser is defined as 2(Clinton-gon), a number so large that the gods will weep over it. Mathematician Matt Hudelson says that it is "easy to see that the last digit in the base 10 expansion of the Moser is 6." How does he know? What's its second-to-last digit?

This chapter also discussed large numbers such as the *Hamlet* number and the chess number. Here are a few other large numbers—all less than a googol. The *ice age number* (10^{30}) is the number of snow crystals necessary to bring on the ice age. The *Coney Island number* (10^{20}) is the number of grains of sand on the Coney Island beach. The *talking number* (10^{16}) is the number of words spoken by humans since the dawn of time. It includes all baby talk, love songs, and congressional debates. This number is roughly the same as the number of words printed since the Gutenberg Bible appeared. The amount of money in circulation in Germany at the peak of inflation was 496,585,346,000,000,000,000 marks, a number very similar to the number of grains of sand on the Coney Island beach. The number of atoms in oxygen in the average thimble is a good deal larger: 1,000,000,000,000,000,000,000,000,000. The number of electrons that passes through a filament of an ordinary light bulb in a minute equals the number of drops of water that flow over Niagra Falls in a century. The number of electrons in a single leaf is much larger than the number of pores of all the leaves of all the trees in the world. The number of atoms in this book is less than a googol. The chance that this book will jump from the table into your hand is not 0—in fact, using the laws of statistical mechanics, it will almost certainly happen sometime in less than a googolplex years.

Chapter 11

Cupid's Arrow

Figure F11.1 shows one solution that will win you a heart. Dr. Googol is aware of 5 other solutions. Can you find any of them?

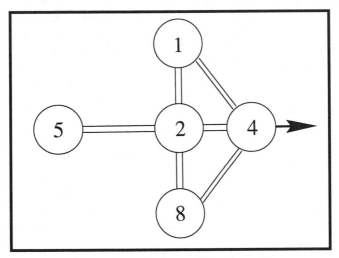

F11.1 A solution for Cupid's arrow. There are others.

Chapter 12

Poseidon Arrays

Here are the only other known solutions. In each of these 3 arrays, all 9 digits are used exactly once.

2	1	9
4	3	8
6	5	7

2	7	3
5	4	6
8	1	9

3	2	7
6	5	4
9	8	1

Notice that for each of these solutions, the sum of the numbers in each row is a constant. The sum for row 1 is 12, the sum for row 2 is 15, and the sum for row 3 is 18. And the sums all differ by the same constant, 3. Dr. Googol wonders if this property may be generalized to larger arrays or to arrays using numbers in different bases. (Dr. Googol has only considered the numbers in base 10.)

Here's a related problem. Start with the number in the last row (e.g., 657 or any other solution you may find) and continue to form another 3-by-3 matrix using the same rules with the new starting number. In other words, the number in the second row must be twice the first. The third row must be 3 times the first. However, for this problem you may truncate any digits in the beginning. For example, 1,384 would become 384. Keep going. How many arrays can you create before it is impossible to continue? Again, each digit must be used only once in each matrix.

Chapter 13

Scales of Justice

By assigning Ant = 4, Cockroach = 3, Grasshopper = 7, and Wasp = 1 we find 1 possible solution of "3 Wasps." This assignment makes the sums on each side of the scale equal. How many other solutions are there? If there were 1 cockroach on the left, could there ever be anything other than wasps on the right? If there were multiple cockroaches on the left, is it possible to balance the scale with ants or grasshoppers by using fewer of the heavier insects?

Now for an odd aside. Did you know that outside of Europe and North America, most people on Earth practice entomophagy? They eat insects. In parts of Africa, more than 60% of dietary protein comes from insects. Grubs and caterpillars have a lot of unsaturated fats. Dr. Googol once attended a banquet hosted by the New York Entomological Society where he discovered some interesting appetizers: chocolate cricket torte, mealworm ganoush, sautéed Thai water bugs, and waxworm fritters with plum sauce. In Colombia, roasted ants are eaten like popcorn. Honeypot ants, with their transparent abdomens distended with peach nectar, are delightful sweets.

Chapter 14

Mystery Squares

There appear to be several solutions to Dr. Googol's puzzle. For example:

	12	7	
8			11
9			6
	5	10	

sum = 22

Another solution is: top (1,7,12,2), left (1,6,11,4), right (2,8,9,3), bottom (4,5,10,3). Another is: top (1,11, 8, 2), right (2, 12, 5, 3), left (1, 7, 10, 4), bottom (4, 6, 9, 3). Note that in the solution

	7	12	
11			8
6			9
	10	5	

we find the sets (1, 2, 3, 4), (5,6,7,8), and (9, 10, 11, 12) sorted in clockwise order. Can you extend the puzzle to ones in which more numbers are used along the edges of the square?

Chapter 16

Jerusalem Overdrive

If you couldn't solve the first problem, work in teams until you solve it. For the second problem, here is a way to arrange the religions so that there are only 2 of the same religions in each row and column:

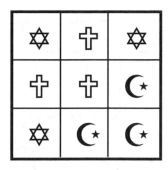

Try these problems on a few friends. Dr. Googol has found that many people have difficulty visualizing the solution.

Amazing Latin squares: The Jerusalem Overdrive problem can be thought of as a special problem in the remarkably rich mathematical area concerned with Latin squares. Latin squares were first systematically developed by Swiss mathematical Leonhard Euler in 1779. (Euler's mental powers were so great that his capacity for concentrating on math problems did not decline even when he became totally blind.) He defined a Latin square as a square matrix with n^2 entries of n different elements, none of them occurring twice or more within any row or column of the matrix. The integer n is called the *order* of the Latin square. Recently the subject of Latin squares have attracted the serious attention of mathematicians because they are relevant to the study of combinatorics and error-correcting codes. Here's an example of the occurrence of a Latin square when considering the equation $z = (2x + y + 1)$ modulo 3:

	0	1	2	x
0	1	2	0	
1	0	1	2	
2	2	0	1	

y

To understand this table, consider the case of $x = 2$ and $y = 2$ which yields $2x + y + 1 = 7$. 7 mod 3 is 1 because 7/3 has a remainder of 1. This 1 entry is in the last row and column of this Latin square.

Here's an interesting example of an order-10 Latin square containing 2 subsquares of order 4 (consisting of elements 1, 2, 3, and 4) and also one of order 5 (with elements 3, 4, 5, 6, 7), the intersection of which is a subsquare of order 2 (with elements 3, 4):

1	9	2	8	0	6	7	4	5	3
8	2	1	0	9	7	5	3	4	6
2	1	0	9	8	5	6	7	3	4
0	8	9	1	2	3	4	6	7	5
9	0	8	2	1	4	3	5	6	7
5	6	7	3	4	1	2	0	8	9
6	7	5	4	3	2	1	8	9	0
7	4	3	5	6	0	9	1	2	8
3	5	4	6	7	8	0	9	1	2
4	3	6	7	5	9	8	2	0	1

Can you create Latin squares with even greater numbers of internal subsquares than this? What is the world record for the number of subsquares in an n-by-n Latin square?

A *traversal* of a Latin square of order n is a set of n cells, 1 in each row, 1 in each column, and such that no 2 of the cells contain the same symbol. Fascinatingly, even when a Latin square has no traversals, it is very often the case that partial traversals of $(n - 1)$ elements occur in it. Do all Latin squares have a partial traversal of n - 1 elements if the squares do not contain a true traversal? Here is an example of a Latin square with an n - 1 traversal. (Dr. Googol has marked the traversal path with thick boxes):

1	6	3	7	4	9	2	5	0	8
2	0	4	6	5	8	3	1	9	7
3	9	5	0	1	7	4	2	8	6
4	8	1	9	2	6	5	3	7	0
5	7	2	8	3	0	1	4	6	9
6	1	8	2	9	4	7	0	5	3
7	5	9	1	0	3	8	6	4	2
8	4	0	5	6	2	9	7	3	1
9	3	6	4	7	1	0	8	2	5
0	2	7	3	8	5	6	9	1	4

Now consider an amazing Latin *cube*. You can think of it as a stack of file cards. Each card contains *n* rows and *n* columns. Each number occurs exactly once in each row, once in each column, and once in each row and column in the third dimension:

0	1	2
1	2	0
2	0	1

1	2	0
2	0	1
0	1	2

2	0	1
0	1	2
1	2	0

Can you design a 4-dimensional Latin hypercube? Note that computers are much faster than humans in finding errors in Latin squares, cubes, and hypercubes. So, if you are not sure if the Latin square you've written down is correct, check each row and column with a computer program (see [www.oup-usa.org/sc/0195133420]). Have your computer create 4-by-4 Latin squares by randomly selecting values for the squares and then checking if the result is a Latin square using the algorithm in the program code. How long does it take your computer to find a Latin square? Several minutes? Hours? Dr. Googol's IBM IntelliStation computer took just seconds to find 3-by-3 Latin squares. For large squares, this random method is not very efficient.

🖫 For a C code fragment used to scan for Latin squares, see [www.oup-usa.org/sc/0195133420].

Chapter 17

The Pipes of Papua

Why would some obscure tribes in a remote New Guinea rain forest be sounding this sequence upon their wooden flutes? Dr. Googol might have doubts as to the accuracy of Omar's story, but the rhythm pattern is certainly strange to hear. You may wish to beat the sequence out on your desk, or have your computer play the eerie rhythm. If you prefer, you can beat the sequence out on a tabletop with a finger to represent a low tone and a pencil to represent a high tone, or you can use short- and long-duration beats. Do you hear a pattern? It is strangely compelling, yet it never quite repeats itself in the way that most rhythms do. If the sequence is not random, what is its structure?

Not only do binary numbers provide musical possibilities, they also can yield artistic patterns. Graphic patterns produced by binary numbers are so interesting that Dr. Googol devotes Chapter 73 entirely to this subject. Interesting information on fractal number sequences can also be found in M. Schroeder's *Fractals, Chaos, Power Laws.*

For other examples of aperidoic bar codes in mathematics, see Chapter 77 on ⊎-numbers.

For recent information on the Morse-Thue sequence in many apparently unrelated occurrences, see Jean-Paul Allouche and Jeffrey Shallit, "The ubiquitous Prouhet-Thue-Morse sequence," in *Sequences and their Applications: Proceedings of SETA 1998* (New

York: Springer, 1999), 1–16. In this paper, the authors provide a survey of the sequence's amazing incidence in chess problems, quasicrystal theory, vibrational modes in alloys, mathematical physics, combinatorics on words, differential geometry, number theory, and the iteration of continuous functions. They also describe how the sequence may predate the work of Thue and Morse, including a description in an 1851 paper by E. Prouhet. The authors conclude with the words, "Searching for the many occurrences of the Prouhet-Thue-Morse sequence in the literature can be used as a pretext to take a delightful stroll through many fascinating areas of mathematics."

Chapter 18

The Fractal Society

Dr. Googol has received numerous mail from readers who experimented with the Fractal Fantasies game. For example, Martin Stone from Temple University suggests a distributed version of the game played over the Internet. He writes, "Imagine a multiuser recursive game server dedicated to the fostering of a greater intuitive understanding of recursive structures and permutations." David Kaplan from New York University points out that the game rules for Fractal Fantasies are similar to those of a medieval game called Nine Man Morris played on a different playing board. Paul Miller notes that the Fractal Fantasies game was discussed at the Boston chapter of Mensa. He asks, "Can pieces of a Googol move out and back (thus forming and reforming the Googol)?" He suggests that the Googol pieces be allowed to move *only* if there is no other legal move. Alternatively, if a player moves a piece out of a Googol, he should not be allowed to move it back into the same place on the next turn. Michael Currin from the University of Natal (South Africa) suggests that the game be adapted to allow more than 2 players. Finally, Brian Osman, a 15-year-old from Massachusetts, writes:

> I greatly enjoyed your description of the Fractal Fantasies in the March 1993 issue of *BYTE* magazine. However, I point out that some of what you said is almost impossible! I've calculated the number of rectangles and "spots" for every size board, using the formula: $(2^{N+1}) - 1$, where N is the degree of the board. From this, one can find the number of spots by simply multiplying by 6. Once you have this number, divide by 2 and subtract 2 to find the number of stones for each player. You have stated that grand masters have been known to use boards of degree 20. I've checked my calculations repeatedly, and this would require each player start with 6,291,451 stones! Assuming each opening move (only those to place your pieces) took 2 seconds, the players wouldn't be able to move until 291.2708797 days after they started the game. Am I missing something, or are your numbers as ludicrous as they seem to me? Please don't take offense at this. I still found your article very enjoyable.

Chapter 19

The Triangle Cycle

Figure F19.1 is a cycle 4 solution known as the Ashbacher solution after its discoverer, Charles Ashbacher (personal communication). No one to Dr. Googol's knowledge has yet discovered a higher cycle. But considering that this solution uses only multiples of 7, perhaps there is a higher cycle using multiples of *both* 7 and 13.

Here are some additional challenges:

Select a random number between 1 and 9. Place it in the lower left corner of the starting triangle. Can you make a cycle 2 triangle cycle? Are there solutions for any starting number you choose?

What is the largest cycle solution that can be found?

The Fibonacci cycle game can be played with 2 people. One player writes a Fibonacci number at a vertex (e.g., 1, 3, 5, 8, 13 . . . ; Fibonacci numbers are defined in Chapter 71).

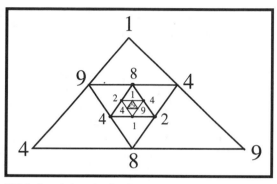

F19.1 A solution to the Triangle Cycle.

The second player writes a Fibonacci number at another vertex. The goal is to continue placing numbers so that the multi-digit numbers created by concatenating the digits of connected vertices are also Fibonacci numbers. The game continues until a person cannot place a number that would form a Fibonacci number. (It helps to have a list of Fibonacci numbers at your fingertips as you play!)

Chapter 20

iQ-Block

When Dr. Googol presented this puzzle to Joseph Madachy, editor of *Journal of Recreational Mathematics*, Madachy remarked:

> I say you cannot create a square after removing a single piece and using all the remaining pieces. The area of the complete block is 64. The areas of the 10 pieces are 8, 5, 7, 7, 8, 6, 5, 4, 6, 8. Removal of one of these pieces is simply insufficient to produce the next smallest square (7 × 7 = 49 area). I haven't tried it, but it might be possible if 2 pieces are removed.

Charles Ashbacher, book editor of the *Journal of Recreational Mathematics*, wrote a computer program that found over 1,000 solutions rather quickly! He believes the

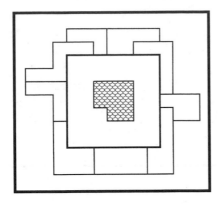

F20.1 Joseph Lemire's novel solution with a square cutout. The removed piece is shown in shaded form in the middle of the "square" formed by the other pieces. There are probably other solutions with square holes. Can you find any?

number of solutions is in the tens of thousands. Subsequent research by Joseph E. Lemire suggests that it is possible to remove a single piece and create an amazing square outline (actually, a square cutout) as illustrated in Figure F20.1. The removed piece is shown shaded in the middle of the square formed by the other pieces. Charles also wrote to Dr. Googol that he found 2 ways to create a square after *2* pieces are removed. His constructions were found via a computer, and it seems likely there are others.

Chapter 21

Riffraff

The trumpeter is playing what mathematicians call a Morse-Thue sequence, which was discussed in Chapter 17. Whenever a smiley face occurs, it is replaced in the next phrase by a smiley face followed by a sad face; sad faces are replaced by a sad face followed by a happy face. Notice that every other phrase is symmetrical—a palindrome.

For the trombone player, each 📂 is replaced with a closed file followed by an open file (📁📂), and each closed file 📁 is replaced with an open file 📂.

The violinist is simply marking every prime number (numbers divisible only by themselves and 1) with a short note. So the second, third, fifth, seventh (and so on) are short:

```
0 1 1 0 1 0 1 0 0 0 1 0 1 0 0 0 1 0 1 0 0 0 1 0 0 0 0 0 1 0 1 0 0 0 0 0 1 0 0 0 1 0 1 0 ...
  | |  |  |     |  |    |  |     |           |  |        |     |  |
  2 3  5  7     11 13   17 19    23          29 31      37    41 43
```

The saxophonist is just multiplying the digits of each number to get the next.

If you thought all of these were too difficult to solve, don't even think about attempting the next few brain bogglers. Instead, give them to your worst enemy. The following are some incredibly difficult number sequences to ponder—so difficult that rarely anyone but Dr. Googol could solve them. Solutions follow. Can you supply the missing number in the following sequence?

10, 11, 12, 13, 20, ?, 1000

If not, don't be disappointed. Exactly 99.3% of Dr. Googol's colleagues could not solve this, even after considering the sequence for days. Perhaps looking at another sequence generated by the same rules might help:

10, 11, 12, 13, 14, 20, 22, ?, 1010

Not yet? Perhaps an even longer sequence, generated using the same rules, will finally clue you in:

10, 11, 12, 13, 14, 15, 16, 17, 18, 21, 23, 25, 32, 101, ?, 10001

Ready for the solutions? For the first 2 sequences, the missing numbers are 22 and 101, respectively. To create the first sequence, Dr. Googol represented the number 8 in different bases, from base 8 to base 2. Can you now solve the third sequence?

Note: For those of you not familiar with numbers represented in bases other than 10 (which is the standard way of representing numbers), consider how to represent any number in base 2. Numbers in base 2 are called binary numbers. To represent a binary number, only the digits 0 and 1 are used. Each digit of a binary number represents a power of 2. The rightmost digit is the 1s digit, the next digit to the left is the 2s digit, and so on. In other words, the presence of a 1 in a digit position indicates that a corresponding power of 2 is used to determine the value of the binary number. A 0 in the number indicates that a corresponding power of 2 is absent from the binary number. An example should help. The binary number 1111 represents (1×2^3) + (1×2^2) + (1×2^1) + (1×2^0) = 15. The binary number 1000 represents $1 \times 2^3 = 8$. Here are the first 8 numbers represented in binary notation: 0000, 0001, 0010, 0011, 0100, 0101, 0110, 0111, . . . It turns out that any number can be written in the form $c_n b^n + c_{n-1} b^{n-1} + . . . c_2 b^2 + c_1 b^1 + c_0 b^0$, where b is a base of computation and c is some positive integer less than the base.

What is the value of the missing digit in this sequence:

6 2 5 5 4 5 6 3 ?

No one has ever gotten this. Do you give up? The solution relates to the number of segments on a standard calculator display that are required to represent the digits, starting with 0.

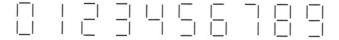

Dr. Googol knows that you are finding some of these sequences to be absurd. But that does not stop him from presenting you with more. The following are fascinating number-sequence problems sent to Dr. Googol by readers. They are all nearly impossible for mere mortals to solve. Can you supply the missing numbers denoted by a ? symbol?

⊙ Diep number sequence: 2, 71, 828, ?, . . .

⊙ Silverman number sequence: 3, 4, 5, 7, 11, 13, 17, 23, 29, 43, 47, 83, 131, 137, 359, 431, 433, 449, 509, 569, 571, 2,971, 4,723, 5,387, ? . . .

⊙ Lego sequence: 1, 3, 7, 19, 53, 149, 419. . . . What could this sequence possibly designate?

Here are some solutions. *Dr. Googol's sequence:* Convert between decimal and binary representations. For example, 11 (decimal) is 1011 (binary). 1011 (decimal) is 1111110011 (binary). And so on. The solution to the *Diep sequence* is 1,828. The *i*th term of the sequence is the next *i* digits of *e* (*e* = 2.7182818284 . . .). The *Silverman sequence* lists the indices of the prime Fibonacci numbers. For example, the third, fourth, and fifth Fibonacci numbers (*F*3, *F*4, *F*5) are primes. (See Chapter 73 on the "1,597 problem" for background on prime Fibonacci numbers.)

The *Lego sequence*: Each element *a*(*n*) is the number of stable towers that can be built from *n* Lego blocks.

Chapter 22
Klingon Paths

In order to live longest and prosper, the Klingon starts at the 13 on the bottom row. The sequence of moves is 13, 1, 10, 12, 23, 16, 7, 5, 6, 0, 11, 2, 8, 18, 15, 24, 17, 20, 4, 3. Numerous questions abound. Are there other equally long paths? Are there areas in the Klingon world that give rise to longer-lived Klingons? Do certain starting squares have a higher probability of yielding older Klingons? For example, do interior squares generally yield older Klingons than squares on the edge, because the interior squares have more neighbors? Can you extend the puzzle to 3 dimensions and higher? Can you explore larger worlds such as a 20-by-20 array of squares? How would the puzzle change if played on worlds the size of our Earth? Also, what is the *shortest* path you can find?

Before leaving this topic, Dr. Googol would like to tell you about simple computer programs you can write to explore the mysteries of Klingon paths. For example, the BASIC and C programs at [www.oup-usa.org/sc/0195133420] both start by filling a grid with random numbers between 0 and 24 or between 1 and 25. (Dr. Googol used these programs to generate the Klingon-paths board in Chapter 23, but you can easily design worlds by hand.) Next, the programs attempt to find the longest possible path for each starting square in the 8-by-8 grid. To do this, the Klingon scans the immediate neighborhood of a cell (up, down, right, and left). If the Klingon finds a square with a number he never before encountered, he enters that square. The Klingon starts again, scanning in 4 directions for a potential move from his new location. The process continues until the Klingon can no longer find a "safe" square, at which point his courageous life comes to an end.

These programs are capable of finding some long-lived Klingons, but are there older Klingons lurking in Klingon City? After all, the computer programs did not search for every possible path a Klingon could take. For example, once a Klingon found an available square, he would commit himself to moving in that direction without examining *every* possible path that it could take. (This is a little like real life, isn't it?)

To see if humans could beat the computer results, Dr. Googol held a grand Internet tournament, asking various interested colleagues on the computer networks to find the oldest Klingon in several Klingon cities. He also asked if it were possible to design a Klingon city in which each starting square would yield the same maximum length. Many respondents used pencil and paper to investigate the worlds. To learn more about the Internet Klingon Game competition, and fascinating analyses of similar games, see my book *Keys to Infinity*.

Chapter 23

Ouroboros Autophagy

Figure F23.1 shows a solution. For the outer serpent with the numbers 1, 2, 2, 3, 3, 3, 4, 4, 4, we find:

Circle 1: 1 2 2 3 3 3 4 4 4

Circle 2: 4 2 2 3 3 3 1 1 1

Circle 3: 0 3 3 2 2 2 0 0 0 0

Circle 4: 3 5 5 0 0 0 3 3 3

Circle 5: 0 5 5 5 5 5 0 0 0

Circle 6: No solutions (dead end)

Can you find any other solutions or interesting Ouroboros numbers?

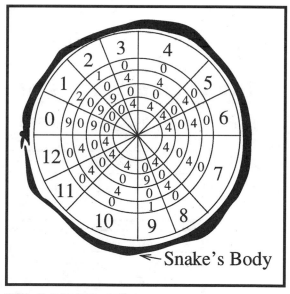

F23.1 Ouroboros solution.

Chapter 24

Interview with a Number

Here are the 5 other 4-digit true vampires:

$$21 \times 60 = 1,260 \qquad 15 \times 93 = 1,395 \qquad 30 \times 51 = 1,530$$
$$21 \times 87 = 1,827 \qquad 80 \times 86 = 6,880$$

In fact, there are many larger vampire numbers. There are, for example, 155 6-digit vampire numbers. Recently, Dr. Googol challenged computer scientists and mathematicians around the world to submit the largest vampire they could find. One such jewel is

$$1,234,554,321 \times 9,162,361,086 = 11,311,432,469,283,552,606$$

John Childs discovered a 40-digit vampire number using a Pascal program on a 486 personal computer. His amazing vampire number is

$$98,765,432,198,765,432,198 \times 98,765,432,198,830,604,534 =$$
$$9,754,610,597,415,368,368,844,499,268,390,128,385,732$$

As the numbers get larger and larger, how often do you expect to find vampires? Do they become more secretive (sparser) or more outgoing (frequent) as you search for vampires up to a googol? The "Further Reading" section lists the latest technical papers on vampire numbers.

Chapter 25

The Dream-Worms of Atlantis

Figure F25.1 shows all possible configurations for a 5-segment worm.

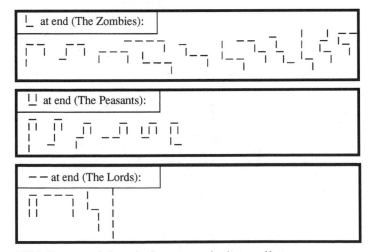

F25.1 The complete set of worm contortion patterns.

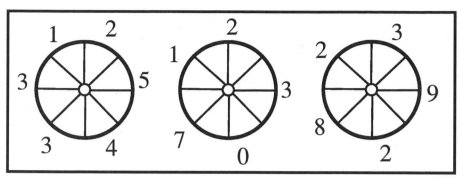

F26.1 Can you solve these wheels? After seeing Dr. Googol's wheels, a few colleagues sent him their own. These were submitted to him by Davode Crippa, Stephen Kay, and Geoff Bailey.

Chapter 26

Satanic Cycles

Note that if Dr. Googol did not constrain the number of times you could use the operation symbols, it would be possible to formulate various cycles with 1 in them (but not 0s) simply using a repeated exponential such as $1^{ij} = 1$. The first wheel falls into this category 15^3. Similarly, any cycle with 1 and only 1 0 in it can likewise be solved by using all exponentials: $0^{ij} = 0$. The third wheel in Figure 26.2 falls into this category. Let's exclude these trivial kinds of solutions from our consideration.

Dr. Googol's solution for the leftmost hell wheel in Figure 26.2 is $22 \times 2/4 = 11$. The other 2 wheels are left as exercises for you. When Dr. Googol presented them to other researchers, he was stunned by just how many solutions his wheels had. Here are some other possible solutions to the first wheel: $2^{1^1} = -4 + 2 + 2 = 2$, $1 \times 2 + 2 + 2 \times 4/1 = 12$, or $2 \times 2 = 4 = 1 + 1 + 2$, or $1 + 1 \times 2/2 + 2 = 4$, or $(((1^1)^2) \times 2) \times 2 = 4$, or $2 + 2 - 4 + 1 + 1 = 2$, or $4 - 1 = 1 + 2 \times 2/2$, or $1 \times 12 \times 2 = 24$, or $224112 = 224112$. A few respondents challenged Dr. Googol with wheels of their own devising (Figure F26.1). Can you solve these?

Bicycle wheels from Purgatory: Since there are often many solutions for a single wheel, a much tougher problem is to devise wheels for which there are no solutions! Can you do it? Of course, as Bill Mayne of Florida State University has pointed out to Dr. Googol, for any string of digits around the wheel $a \ldots z$ there are always solutions involving multiple revolutions of the wheel: $a \ldots z = a \ldots z$, $a \ldots za \ldots z = a \ldots za \ldots z$, etc. (For example, $12{,}345 = 12{,}345$ is a case of a trivial wheel-revolution solution.) At a minimum we must either limit the number of cycles to less than 2 or rule out such solutions as a special case.

Dr. Googol leaves this section with a bit of trivia. More than 90 million Americans ride bicycles. The longest tandem bicycle ever made is approximately 67 feet long (for 35 riders) and was built by Pedalstompers Westmalle of Belgium.

Chapter 27

Persistence

Here are the smallest numbers with various persistences:

1	10
2	25
3	39
4	77
5	679
6	6788
7	68889
8	2677889
9	26888999
10	3778888999
11	277777788888899

Notice the strange abundance of 8s and 9s. Incredible. Why so many 8s and 9s in a row? Dr. Googol can see you wondering what number has a persistence of 12. No one knows! We do know that there is no number less than 10^{50} that has a persistence greater than 11. Neil Sloane conjectures that there is a number N such that no number has a persistence greater than N.

It is conjectured that the largest number lacking the digit 1 with persistence 11 is 77,777,733,332,222,222,222,222,222,222.

Chapter 28

Hallucinogenic Highways

To solve the highway puzzle, start at the word START. Take the road with the sign that says 22. Take the road marked with the 17 sign (do not go past the 7 sign). Take the road with the 36 sign. Take the road with the 1 sign. Take the road with the 8 sign (do not go past the 27 sign). 22 + 17 + 36 + 1 + 8 = 84. One key to this puzzle is to add a number only when you go past the sign, but not if you simply travel on a particular road.

Are there other solutions?

Chapter 30

What if We Receive Messages from the Stars?

God's formula: Humans have thought about sending messages to the stars for decades, although there has always been some debate as to what the messages should contain. For example, in the 1970s, Soviet researchers suggested we send the message

$$10^2 + 11^2 + 12^2 = 13^2 + 14^2$$

The Soviets called the equation "mind-catching." They pointed out that the sums on each side of the equal sign total 365—the number of days in an Earth year. These imaginative Soviets went further to say that extraterrestrials had actually adjusted the Earth's rotation to bring about this striking equality! Surely it should catch aliens' attention and demonstrate our mathematical prowess.

Dr. Googol finds the Soviet formula arbitrary and not a good candidate to send. Rather, he would somehow try to send the most profound and enigmatic formula known to humans:

$$1 + e^{i\pi} = 0$$

This formula of Leonhard Euler (1707–1783) unites the 5 most important symbols of mathematics: 1, 0, π, e, and i (the square root of -1).

Another beautiful and wondrous expression involves a limit that connects not only π and e, but also radicals, factorials, and infinite limits. Surely this little-known beauty makes the gods weep for joy:

$$\lim_{n\to\infty} \frac{e^n n!}{n^n \cdot \sqrt{n}} = \sqrt{2\pi}$$

Chapter 34

A Ranking of the 5 Saddest Mathematical Scandals

Here are the answers:

1. *Ada Lovelace*, daughter of Lord Byron (the poet), and first computer programmer. She analyzed and expanded upon Charles Babbage's plans for difference and analytical engines. She explained how the machines could tackle problems in astronomy and mathematics. While married to William King, she fell in love with

mathematician John Crosse and became obsessed with gambling. During the last year of her life, Ada's cervical cancer progressed slowly, and her mother took charge of her care. When Ada confessed her affair with Crosse, her mother promptly discarded all of Ada's morphine and opium—the only things holding the horrific pain at bay—so that Ada's soul would be redeemed. Ada's last days were spent in agony as her mother watched but did nothing.

2. ***Alan Turing***, computer theorist. His code-breaking work helped shorten World War II. For this contribution he was awarded the Order of the British Empire. When he called the police to investigate a burglary at his home, a homophobic police officer suspected that Turing was homosexual. (The Criminal Law Amendment Act of 1885 made a male homosexual act illegal.) Turing was forced to make a decision. He could either go to jail for a year or take experimental drug therapy. His death 2 years after the therapy, in 1954, at the age of 42, was a shock to his friends and family. Turing was found in bed. The autopsy indicated cyanide poisoning. Perhaps he had committed suicide, but to this day we are not certain.

3. ***Kurt Gödel***, eminent mathematician and one of the most brilliant logicians of this century. The implications of his incompleteness theorem are vast, not only applying to mathematics but also touching on areas such as computer science, economics, and nature. At Princeton, one of his closest friends was Albert Einstein. When his wife Adele was not with him to coax him to eat—because she was in a hospital recovering from surgery—Gödel stopped eating. He was paranoid and felt that people were trying to poison him. On December 19, 1977, he was hospitalized but refused food. He died on January 14, 1978. During his life, he had also suffered from nervous breakdowns and hypochondria.

4. ***Georg Cantor***, the creative mathematician largely responsible for a host of extraordinary mathematical ideas such as the theory of infinite sets, transfinite numbers, and even fractals.

5. ***Alhazen*** (965–1039), a contributor to the field of mathematical optics. Al-Hakim, the ruler of Egypt, became angry with Alhazen when Alhazen made gross errors in his ability to predict and control the Nile's flooding. To save himself from execution, Alhazen pretended to be insane and was placed under house arrest. When he was not feigning insanity, Alhazen made important discoveries in optics, describing various aspects of light reflection, magnification, and the workings of the eye.

For more examples of scandal in mathematics, see Theoni Pappas's *Mathematical Scandals*.

Chapter 35

The 10 Most Important Unsolved Mathematical Problems

When Dr. Googol asked many mathematicians what is the most difficult area of math to understand, and also what is the most important unsolved mathematical problem, they always responded with two words: "Langlands philosophy" or "Artin conjecture." As this book goes to press, a proof of the Langlands conjecture for function fields may be at hand. A field denotes any algebraic structures consisting of objects (or elements) that can be added, subtracted, multiplied, and divided according to the rules that govern real numbers. For a general description see: Mackenzie, D., (2000) Fermat's Last Theorem's first cousin. *Science.* 287(5454): 792–793. Andrew Wile's 1994 proof of Fermat's last Theorem, one of the greatest mathematical achievements of the twentieth century, can also be viewed as the completion of a small part of the Langlands program. The key idea is that the Langlands program brings together theories that seem to be very different from one another. Thanks to Laurent Laffogue, a number theorist at the Université de Paris-Sud, another piece of Langlands program seems to have finally fallen into place. A 300-page handwritten version of Lafforgue's proof of "Langlands conjecture for function fields" has been circulating among mathematicians since the summer of 1999.

Dr. Googol asked several mathematicians to explain Langlands philosophy to a general audience. Alas, dear reader, you will not get your wish. Here is a sampling of replies.

Allan Adler from Western Kentucky University:

As nearly as I can tell, no one knows what the Langlands conjecture says, not even Langlands. If that is not the case, I would be glad to read a (hopefully, concise) definitive statement of the most recent version of the conjecture.

My previous assertion about the Langlands conjecture is somewhat tongue-in-cheek, masking a more complicated state of affairs, like Bertrand Russell's assertion that mathematics is the subject in which we never know what we are talking about, nor whether what we are saying is true. The main point I am making is that I believe the conjecture has undergone a certain amount of modification over the years as more has been learned about the problem. My impression is that although they know a lot more now about the relevant mathematics and about what to expect from it, I'm not sure there is at this moment a clean statement available which one can call the Langlands conjecture.

Bill Dubuque of MIT:

Alas, to appreciate the ideas in the Langlands program requires at least a Ph.D.-level mathematics education. It would be virtually impossible to attempt to convey these ideas to an audience less educated. See Oxford's website [www.oup-usa.org/sc/0195133420] for a long list of references to works of expository character which touch on topics related to the Langlands program. I'd suggest starting with Shafarevich, Gelbart (1984), and Murty—some of which should be accessible to bright math undergrads.

Jared Weinstein, age 16:

There are these things called elliptic curves, see? They look like this, in their most general form: $y^2 = x^3 + bx + c$. (If you're going to complain about characteristic 2 or 3, don't.) You could plot this, on the complex plane cross itself, and it would look like a donut. A big old 4-D donut. In any case, the Taniyama-Shimura conjecture says that all elliptic curves have this magic property called being "modular." Don't ask what this means. Unless you care to hear such things as "the mellin transform of the hasse-weil l-function produces a spitzenform." As far as I know no person has a real understanding of why any elliptic curve should be modular, but nonetheless in 1995 Andrew Wiles proved a weak form of the Taniyama-Shimura conjecture which applied only to "semistable" curves (which, for our purposes, aren't *too* wild). Luckily this was enough to prove Fermat's Last Theorem. But elliptic curves are nice things. They have a "group law." Add x^5s in there, and you've got troubles. Nonetheless a generalization of TS has been developed. I believe this is the stuff that falls under the category of "the Langlands program." Well, that is the flavor of an explanation.

For information on the Langlands program, you might examine *Modular Forms and Elliptic Curves* published by Springer-Verlag on the Fermat conference. But be forewarned. This material is difficult. I don't intend on understanding all of it for another 5 or 10 years. The problem (or, perhaps, the blessing) with conjectures as simple as Fermat's is that they tend to give rise to incomprehensibly complicated fields of mathematical study.

Bob S. says:

It is not possible to explain Langlands conjectures to a general audience without some basic knowledge of algebraic number theory and field theory. Here is a rough description: The Hilbert Class Field of a polynomial $f(x)$ with root alpha is the maximal unramified Abelian extension of $Q(a)$. The Langlands conjectures are an attempt to extend this concept to non-Abelian extensions of $Q(a)$. It makes conjectures relating certain L functions and Dirichlet series, which are analytic objects, with the purely algebraic objects associated with the extension field. The proof by Wiles of the Taniyama-Shimura conjecture covers a small part of the Langlands conjectures.

Berndt S. comments:

What I was referring to with the term "Langlands philosophy" is Langland's article innocently titled "Problems in the Theory of Automorphic Forms" in which he outlined his vision to bring group representation methods into the arithmetic theory of automorphic forms. After some 30–40 pages of heavy definitions and constructions, he poses some questions such as, "Is it possible to define the local L-functions $L(s,r,p)$ such that a certain functional equation is satisfied?" Langlands always suggests a possible or likely outcome. The entire paper is highly speculative but based on deep insight that he must have gained during his research on the functional equations of the Eisenstein series. The last two phrases in his paper read: "Thus Question 7 together with some information on the range of the correspondence of Question 3 may eventually lead to elementary, but extremely complicated, reciprocity laws. At the present it is impossible even to speculate." This was published in 1970, so Langlands came to his view some time in the late '60s. It was highly influential. The '70s then brought a lot of clarification, and Langlands himself (together with Jacquet) carried out his program for the group $G = GL(2)$. The correspondence (or association) in this case defines a mapping from degree-2 Galois representations to $GL(2)$ automorphic representations. When you start with an irreducible representation of the Galois group then you get a cuspidal representation of GL(2). You have a kind of mapping from Galois representations to automorphic representations such that irreducible representations are associated to cuspidal representations (to irreducible ones you get cuspidal ones).

One word to put this into perspective: Langlands conjectures vastly generalize some other conjectures made by different people in special areas (and the specialization process itself is not easy to carry out). The first such conjecture came from Yutaka Taniyama, who killed himself in 1958. This was refined and made very explicit by Goro Shimura. Independently, André Weil (1906–1998) has made some conjectures in this context; in particular he brought forward the astounding idea that any rational elliptic curve might be modular. Until the early 1970s this was assumed to be the exception. These latter conjectures are now called Taniyama-Shimura-Weil conjecture. Andrew Wiles (born in 1953) proved (a major) portion in 1994.

Chapter 41

The 10 Mathematical Formulas That Changed the Face of the World

Philosopher of science Dennis Gordon suggests that $D = (n/2)^2 + (m/3)^3$, the discriminant for a cubic equation, should be on the top 10 list. (The value of the discriminant determines whether the solutions to a polynomial such as $x^3 + mx = n$ are real or complex.) When $D < 0$, and thus $\sqrt{D}$ is a complex number, we have a case in which all 3 roots are real. In the 16th century, these kinds of solutions to cubic equations gave negative numbers and complex numbers their legitimacy and were a major contribution to mathematical progress.

Dennis also believes that $(d/dx)\, e^x = e^x$ and $\log(ab) = \log a + \log b$ should be included. The invention of logs certainly made major changes in the world by removing drudgery from multiplication, and certainly made mathematics less prone to error.

Chapter 48

Cube Maze

One solution is 21, 20, 11, 14, 5, 4, 7, 8, 17, 26, 27, 18, 15, 6, 3. Are there other solutions? If you can find any, be sure to let Dr. Googol know.

Chapter 49

Hailstone Numbers

Bill Richard from Commodore uses the hailstone sequence to produce interesting music. The values of the hailstone numbers are used as audio frequencies and scaled so that the tones remain in the audible spectrum for humans. For example, he maps the number 1 to 40 Hz, because 1 Hz is simply too low to be musically useful. He notes that the hailstone numbers produce "a relatively pleasing sequence" of musical notes. Notice that the hailstone plots reveal a pattern of diagonal lines of varying density that pass through the origin, a pattern of horizontal lines (visually reminiscent of preferred energy state diagrams in quantum mechanics), and a diffuse "background." The existence of the pronounced diagonal lines in Figure 49.2 indicates "likely" transformations to which the $3n + 1$ sequence naturally gives rise. For the hailstones, we are often multiplying by 3 and then dividing by 2. Therefore, the linear transformation $y = (3/2)x$ is quite common (we can eliminate the +1 in $n+1$ for large x). In order to test this idea, try plotting lines corresponding to $y = (3^n/2^m)x$ for values of n and m between 0 and 5. Several of the lines that you will see are the same as the diagonal lines in the hailstone plots. In fact, higher-order lines for a greater range of m and n are needed to account for all the diagonal patterns. Note that in Figure 49.2 the diagonal lines are of varying density—dark lines indicate more probable transforms. All of the darker lines are accounted for by low-order transforms (multiplication by 3/2 and 1/2 are among the most probable operations).

Chapter 50

The Spring of Khosrow Carpet

The algorithm for the carpets comes from Anne M. Burns's article titled "Persian Recursion," which appeared in a 1997 *Mathematics Magazine*. The algorithm starts by assigning a "color" to the outermost cells arbitrarily to produce a border square (or rectangle). The algorithm then:

1. Uses the 4 corner cells and a convenient function of 4 variables to determine a new color.

2. Assigns this new color to all interior cells in the middle row and middle column.

3. Applies the same procedure to each of the 4 new "border squares."

4. Repeats for smaller and smaller subdivisions

As is often the case with recursion, if the process is carried out for larger matrices, we observe that the patterns repeat on different size scales. See [www.oup-usa.org/sc/0195133420] for BASIC code.

Open the floodgates of Persian recursion research! What new patterns can you create by making modifications to the basic algorithm? Teachers, hold contests with students to see who can produce the prettiest pattern.

Chapter 51

The Omega Prism

If you were to draw on a face of the 230-by-231-by-323 prism, you would soon realize that a diagonal enters a new tile at the beginning and each time it crosses a horizontal or vertical line. However, in situations where the diagonal enters exactly at the corner of a tile, the diagonal crosses 2 lines but enters only 1 tile. These corner points are at corners of rectangles proportional to the whole face. In other words, the diagonals of such rectangles are on the main diagonal.

The number of tiles a diagonal crosses is therefore the length A of one side of a face plus the length B of the other minus the greatest common divisor (GCD) of the sides' lengths: $A + B - \text{GCD}(A,B)$. The greatest common divisor of 2 integers is the largest number that divides both integers. For example, a 231-by-93 face would have $231 + 93 - 3 = 321$ crossed tiles since 3 is the greatest common divisor of 231 and 93.

In the 230-by-231-by-232 prism given, we have 3 different possible combinations of rectangular sides A and B:

A	B	GCD	Number of Squares Cut by Diagonal
230	231	1	460
230	232	2	460
231	232	1	462

Students may wish to compute the number of tiles cut for different values of A and B using the code at [www.oup-usa.org/sc/0195133420]. Using this approach, one can create a figure such as Figure F51.1 which shows the number of tiles cut as a function of side length B, while side length A is held at a constant value, in this case $A = 230$.

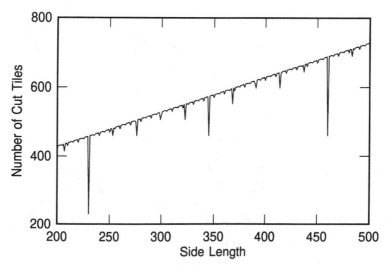

F51.1 Number of tiles cut as a function of side length B with side length A = 230.

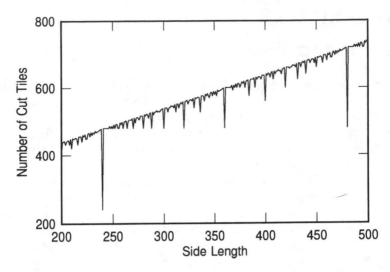

F51.2 Number of tiles cut as a function of side length *B* with side length *A* = 240.

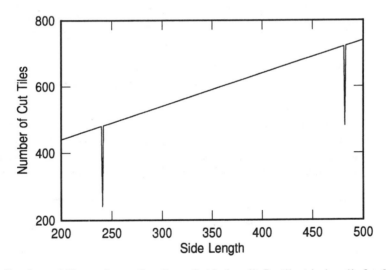

F51.3. Number of tiles cut as a function of side length *B* with side length *A* = 241.

Notice that the cutting function is quite erratic, displaying dips at various locations along the trend line. The first major dip, for example, is at *B* = 230, because the GCD is 230. The distribution of dips seems to have a fractal character as magnification reveals additional similar structures. Figure F51.2 is computed for *A* = 240, which has many factors. (Such numbers are called smooth numbers, and a number is said to be Ψ-smooth if all the prime divisors of *n* are less than or equal to Ψ, where Ψ is a positive integer.) Since 240 is smoother than 230, Figure F51.2 has more spikes than Figure F51.1. Figure F51.3 is computed for *A* = 241, which is prime and has a higher probability of yielding a large number of cut tiles.

Values of GCD = 1 correspond to prism sides that yield the most cut squares when traversed by a diagonal. Figure F51.4 shows a plot of those values of *A* and *B* that yield GCD = 1, and therefore this plot visually indicates which side lengths should be used to create the most cut squares. To produce this figure, GCD is computed for 1 < *A* < 200 and 1 < *B* < 200. The density of black dots is fairly uniform, and the complexity of the plot belies the apparent simplicity of the Omega Prism puzzle.

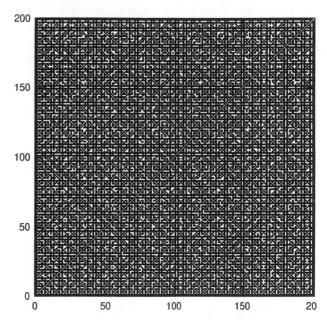

F51.4 A solution space. Black dots indicate those values of *A* and *B* that should be used to create the most cut squares on a prism face.

Number theorists call 2 numbers *A* and *B* that have no common factors relatively prime, or coprime. Such numbers have GCD values equal to 1. What is the probability that 2 numbers selected at random are coprime? Students may perform a quick simulation in order to show that the probability converges to about 0.608, as indicated in Figure F51.5. To produce this plot, Dr. Googol cataloged the occurrences of coprimes as *A* and *B* are iterated in 2 "for" programming loops in a C program. For large numbers, the probability tends toward $6/\pi^2$. Interestingly, the probability that a randomly selected integer is "square free" (not divisible by a square) also tends to $6/\pi^2$.

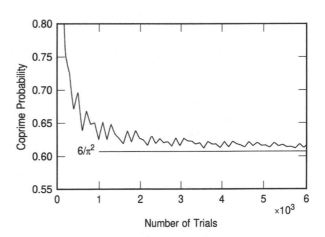

F51.5 Simulation showing the probability that two numbers selected at random are coprime. The probability converges to about 0.608 for large numbers.

Challenges await students and teachers who attempt to understand additional mysteries related to the Omega Prism. For example, try to represent the distribution of pairwise coprimality of 3 or more integers. Also, given N different colors used to color a face, what are the total different number of patterns? Is it $N^{A \times B}$? How unlikely is it that, using random colors, one side of a face will be connected to an opposite face with a continuous path of identical colors? (A path travels up, down, right, and left.) Finally, denote the sum of the length of the 3 sides of the prism as a. What combination of the 3 sides minimizes the maximum number of tiles crossed on the cube faces? How does this change with a?

This chapter provides another example of interesting graphical behavior in simple systems. For a technical discussion on "smooth" and "powersmooth" numbers, and on the distribution of smooth numbers, see H. Cohen's *A Course in Computational Algebraic Number Theory.* Topics such as these have practical application in the creation of codes that are difficult to break. Aside from factorization insight, over the last few years, mathematicians have begun to enjoy and present bizarre mathematical patterns in new ways—ways sometimes dictated as much by a sense of aesthetics as by the needs of logic. Moreover, computer graphics allows nonmathematicians to experience some of the pleasure that mathematicians take in their work and to better appreciate the very complicated and interesting graphical behavior of puzzle solution spaces.

▣ See [www.oup-usa.org/sc/0195133420] for a BASIC code listing.

Chapter 52

The Incredible Hunt for Double Smoothly Undulating Integers

This chapter defined undulating numbers, such as 19,283,746, and smoothly undulating numbers, such as 101,010,101, where the alternating digits are consistently greater or less than the digits adjacent to them. Stimulated by the material in this chapter, Charles Ashbacher has since identified several numbers that smoothly undulate in more than 1 base. For example, $121_{10} = 171_8 = 232_7$. Also $546_{10} = 4141_5 = 20202_4 = 202020_3$.

When Dr. Googol first posed the problem of double smoothly undulating integers, it caused a near riot and subsequent flood of papers to the *Journal of Recreational Mathematics.* For example, Douglas E. Jackson of Portales, New Mexico, believes that if we randomly select a positive integer having between 3 and k digits inclusively in base b, the probability that it will be smoothly undulating is $[(b-1)^2(k-2)]/(b^k - b^2)$. As k goes to infinity, this quantity approaches 0. Hence, 0 is the probability that an arbitrarily positive integer is smoothly undulating. For a derivation, see: Jackson, D. (1992) Problem 1861. *Journal of Recreational Mathematics.* 24(1): 77.

But this probability argument does not prove there are no double smoothly undulating integers. However, D. F. Robinson from the University of Canterbury in New Zealand believes he has proven there are no double smoothly undulating integers. For a reference to his analysis, see "Further Reading." Other researchers have looked at double smoothly undulating integers in other bases. For example, Ken Shirriff discovered 494,949, which smoothly undulates in bases 10 and 15. For some unknown reason, the longest double smoothly undulating numbers all seem to involve base 10 and some other base. This remains a mystical problem for future generations (see "Further Reading").

For further research, let us define smoothly gyrating numbers as those integers whose digits go up and down consecutively like a sine wave. The number of digits controlling the rise and fall determines the "kind" of number, for example:

⊙ smoothly gyrating number of the first kind: 12121212 . . .

⊙ smoothly gyrating number of the second kind: 1232123212321 . . .

⊙ smoothly gyrating number of the third kind: 1234321234321 . . .

A double smoothly gyrating number of the nth kind is simply means a number that gyrates in two different bases, e.g., base 10 and base 3.

Can you find a double smoothly gyrating number of the third kind?

Are there any Fibonacci numbers that smoothly gyrate?

Can you find a smoothly gyrating number that when multiplied by another smoothly gyrating number produces yet another smoothly gyrating number?

Chapter 53

Alien Snow: A Tour of Checkerboard Worlds

Readers might enjoy holding "defect contents" on their computers. Here is how the idea works. The term *defect* is borrowed from crystallography. It refers to an irregularity that may occur in a pure solution of some compound. Such an irregularity can form the nucleus of a crystallization process. The patterns described thus far all arose from simple defects, a single 1 in the center of the screen. What happens if you place more than one 1 on the screen at the same time?

As far as such experiments go, why not randomly seed your screen with 1s and see what develops? Alternatively, you could use very regular patterns, strips, or checkerboards.

There is an undeniable beauty to the patterns that develop from cellular automata. The richness of forms contrasts starkly with the simplicity of the rules. It is a deeply rewarding experience to watch succeeding generations, even in the especially simple cellular automata that Dr. Googol has described here. Persian carpets give way to tile mosaics. Peruvian striped fabrics, brick patterns from Asian mosques, and Moorish ornaments will grace your screen.

Natural structures will appear on your screen—from snowflakes to turbulent fluid-flow patterns. Physicist and cellular automata experimenter Stephen Wolfram has

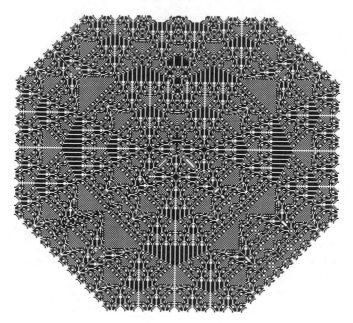

F53.1 Crystals grow from 3 defects.

pointed out that cellular automata are sufficiently simple to allow us to analyze their behavior locally, yet they are sufficiently complex to amaze us globally with very complicated behavior. It should not be surprising that models which show such promise in mimicking nature should also be the source of so much beauty.

Chapter 54
Beauty, Symmetry, and Pascal's Triangle

Many readers may be interested in recent information on practical fractals. Fractals are increasingly finding application in practical products where computer graphics and simulations are integral to the design process. As one example, Amalgamated Research, Inc., located in the state of Idaho, manufactures space-filling fractal conduits. These devices contain many root-like outlets and are designed to minimize turbulence. The company's engineered fractal cascade (EFC) can draw or inject fluid simultaneously throughout a mixing vessel.

Amalgamated Research's basic invention replaces random scaling and distribution of free interfluid turbulence with the geometrically controlled scaling and distribution of fluid flow through "engineered" fractals. This means that EFCs can be used as functional alternatives to turbulence, acting as engineered eddy cascades. Engineering applications include control of flows in chromatography, adsorption, absorption, distillation, aeration, scrubbing, extraction and reactor processes. (For more information, see http://www.arifractal.com/.)

Another company, Fractal Antenna Systems, Inc., based in Fort Lauderdale, Florida, is developing a branching "Fractenna" for hand-held telephony. As the company's name implies, Fractal Antenna Systems develops antennas using fractal geometric patterns with designs repeated at many size scales. The company's trade secrets do not permit us to know all the details at this time, but this highly efficient sender and receiver of electrical waves is said to be no bigger than a small coin.

Fractal antennas hold great promise, because these miniature, virtually invisible devices may be used in everything from wireless LANs to cell phones and televisions. Most cellular and wireless devices use wand antennas protruding from cases. Fractennas can be incorporated in the wireless or cellular device's casing, making them virtually unbreakable. (For more information, see http://www.fractenna.com/.)

A third practical use of fractals is the fiberoptic faceplate—an array of millions of optical-fiber tubes packed into a thin cylindrical pipe. The composite tube acts as an image-plane transfer device. This means that an image entering one surface exits the other surface as an undistorted digitized image, regardless of the shape of the optical tube. You can use these tubes like periscopes by bending them in order to see around a corner.

Several years ago, Lee Cook, a fiberoptic researcher at the Galileo Electro-Optics Corporation in Sturbridge, Massachusetts, was interested in preparing arrays of optical waveguides that were perfect as possible. Analysis of certain recursive tilings led Cook and his colleagues to conclude that the edges of optically useful tilings were fractal in nature. This led to the development of assembly techniques and fractal array structures that allowed the Galileo researchers to prepare highly ordered fiber arrays. One patent has already been granted on these techniques, and Incom, Inc., of Charlton, Massachusetts, has recently purchased Galileo's fractal fiberoptic technology.

Fractal fiberoptics™ may have been the first engineered fractal materials with optically useful properties. A fractal fiber array, which consists of fibers of fibers (called multi-multifibers), results in an extremely high degree of internal order and an optically useful packing in the fiberoptic. This increased order produces a markedly improved image contrast. The perimeters of these new multi-multifibers are exactly analogous to a fractal Gosper snowflake. (To create a Gosper snowflake, recursively transform each edge of an equilateral hexagon into 3 segments of equal length so as to preserve the original area of the surface.)

Traffic on the Internet has unpredictable bursts of activity over many time scales. In other words, the activity shows spikes and lulls over a period of a few seconds that resemble the fluctuations taking place in just milliseconds. This fractal behavior has implications for network engineering. For example, fractals may play a role in designing buffers for Internet routers, which store packets of information during busy periods until the packets can be sent onward to their destination. Because researchers have demonstrated the fractal nature of this traffic, buffers are designed to accommodate much more variable traffic than was assumed previously. For more information, see: Taubes, G. (1998) Fractals reemerge in the new math of the Internet. *Science*. Sept. 25, 281(5385): 1947–1948.

In 1999, physicists Richard Taylor, Adam Micolich, and David Jonas used fractals to study the paintings of Jackson Pollock, revealing that the artist was exploring ideas in fractals and chaos before these topics entered the scientific mainstream. In particular, the researchers found fractal analysis to be a useful tool for studying the abstract paintings produced by Pollock in the 1940s and 1950s. Pollock dripped paint onto vast canvases on the floor of his barn. Although recognized as a crucial advance in the evolution of modern art, the precise quality and significance of the patterns created by this unorthodox technique remained controversial. Today we know that the paintings are fractal and display the fingerprint of nature. For more information, see: Taylor, R., Micolich, A., and Jonas, D., (1998) Fractal expressionism. *Physics World.* Oct., 12(10): 25–28.

What does the future of fractals hold? Aside from obvious applications in education and art, four fields come to mind for especially increased growth: geology, medicine, astronomy, and pure math. All of these fields benefit because fractal geometry provides a language and conceptual framework for ill-defined geometries, and the power law inherent in fractals condenses their description. For example, fractals will be increasingly used in estimating the strength of rocks under shearing forces, in the analysis of mammograms, and in analyzing the randomness of transcendental numbers such as o and e. (For more information on practical applications, see http://www.math.vt.edu/people/hoggard/FracGeomReport/node7.html.) Dr. Bruce Elmegreen of IBM is currently using fractals to explain the relative proportion of high- to low-mass stars in the sky. The ultimate goal of his work is to explain how the Earth and solar system formed from tenuous, cosmic gas.

Dr. Googol asked fractal expert Professor Michael Frame of Union College and Yale University, "What scientific areas would benefit most by using fractals?" He replied:

Currently, the largest deficiency is in statistics. Common statistical methodologies don't usually make use of the scale invariance characterizing fractals, and as we accumulate more evidence that many real data sets exhibit the long-term dependence and long-tailed distributions that can arise in scaling processes, the need for appropriate statistical tests is apparent.

When proper fractal statistics are developed, I imagine the impact in all fields will be considerable. Materials science will probably be affected to a great degree. DLA (diffusion-limited aggregation) and turbulence remain two of the biggest puzzles. With enough computational power to do proper statistics on DLA clusters and turbulent flow patterns, we may begin to develop some real understanding of these processes.

On a different level, the perceived complexity of our surroundings depends in part on the language with which we describe them. Finding a better language is the main task of science, of literature, of art, of music. To the extent that many natural processes exhibit scaling, fractals provide an important component of any language. As we develop our ability to understand and analyze fractals, our language for understanding the world improves and simplifies.

Chapter 55

Audioactive Decay

According to John Conway and Richard Guy in *The Book of Numbers*, the number of digits in the nth term of this sequence is roughly proportional to

$$(1.303577269034296391257099112152551890730702504659 4 \ldots)^n$$

Now, isn't that a fine gem for stimulating party conversation?

So far, the world-record holder for this sequence is Charles Ashbacher of Cedar Rapids, Iowa. In May 1992, he sent Dr. Googol a diskette containing nearly 894,816 digits for row 50, which he computed using a FORTRAN program. He also computed the sequence for row 53, which contained nearly 1,982,718 digits. The number would not fit on a diskette. Ashbacher estimates that row 53, if printed on paper, would require about 417 pages. In August 1992, Ashbacher computed row 56. The number of digits is in the range 4,391,696 to 4,391,703. The size of the data file containing the number is roughly 5205 KB. The computation required 9 minutes using a VAX 4000. About 1 minute of this time was spent simply dumping the contents of the array to a file. Ashbacher discovered that the number of digits in a likeness sequence for row 77 would break the 1 billion mark, requiring 1.2 GB of memory.

Roger Hargrave from West Sussex, United Kingdom, was inspired by the Gleichniszahlen-Reihe sequence to extend the idea to a variation in which a row takes into account *all* occurrences of each character in a previous row. For example, the sequence starting with 123 is 123, 111213, 411213, 14311213,.... He named this the Gleichniszahleninventar sequence because *Inventar* is the German word for inventory. Oddly, he believes that all his sequences finally oscillate between 23322114 and 32232114. Can you prove this?

In 1989, Dr. Akhlesh Lakhtakia and Dr. Googol became intrigued by the fact that the likeness sequence can be generalized to the array $G(p)$ where $p \neq 1$ is either 0 or a positive integer. The following is an example:

$$
\begin{array}{l}
p \\
1\ p \\
1\ 1\ 1\ p \\
3\ 1\ 1\ p \\
1\ 3\ 2\ 1\ 1\ p \\
1\ 1\ 3\ 1\ 2\ 2\ 1\ 1\ p
\end{array}
$$

Simply substituting $p = 1$ into $G(p)$ does not allow us to obtain a standard likeness sequence since numbers and symbols are mixed in the construction of these arrays. Dr. Googol conjectures that the largest number occurring in $G(p)$ is $\max(p,3)$. Also, if $p > 3$ then p occurs only in the rightmost entry of the row.

Chapter 56

Dr. Googol's Prime Plaid

In his book *The Man Who Mistook His Wife for a Hat*, physician Oliver Sacks describes the twins John and Michael, who were able to define prime numbers up to 20 digits very quickly. Yet these same children had difficulty with the simplest additions and substractions. Divisions and multiplications were impossible for them. They said, "But we can see these prime numbers!"

One way of finding the prime numbers is to use the ancient Sieve of Eratosthenes. A list is made of positive numbers; and then all the multiples of 2 are eliminated, starting at 4; then all the multiples of 3 are eliminated, starting at 6; the process is repeated until all possible eliminations have taken place. (A modern computerized version of the Sieve has already become one of the traditional ways of evaluating and comparing computers, because the process is lengthy and CPU-intensive.)

At least once a year, new prime number records are broken using computer searches. Consider, for example, the following world records, which list the number of digits for the largest known prime numbers:

Year	Num. of Digits	Computer	Discoverer
1996	378,632	Cray T94	Slowinski & Gage
1996	420,921	Pentium (90 Mhz)	Armenagaud, Woltman et al.
1997	895,932	Pentium (100 Mhz)	Spence, Woltman et al.
1998	909,526	Pentium (200 Mhz)	Clarkson, Woltman et al.
1999	2,098,960	Various	Hajratwala, Woltman, Kurowski

The last 4 world records in this list were discovered by participants in GIMPS (the Great Internet Mersenne Prime Search), which harnesses the power of thousands of small computers to solve the seemingly intractable problem of finding HUGE prime numbers. (See also "Further Reading" for Chapter 80.) The 1999 record required 21,000 computers and three years of searching. A DEC Alpha computer ran for two weeks just to verify it. In particular, the 1999 record was achieved by Nayan Hajratwala who found a 2,098,960 digit Mersenne prime: $2^{6972593} - 1$. (Mersenne primes are those which are a power of two, minus one.) Nayan Hajratwala is from Plymouth, Michigan and works for PricewaterhouseCoopers. Using the GIMPS program and 111 days of idle time on his home computer (an Aptiva 350 MHz, Pentium II) Nayan found a 38th Mersenne prime number. His computer could have found it in three weeks running full time. This makes Hajratwala eligible for a $50,000 award that is offered by the Electronic Frontier Foundation (EFF). Larger primes will earn up to $250,000! When might we see the first billion digit bevaprime? (For more information, see: Caldwell, Chris K., The Largest Known Prime by Year. http://www.utm.edu/research/primes/notes/by_year.html)

If the 2,098,960 digit prime number was printed in a 12-point font, without commas, it would stretch over four miles. Prime number hunters believe that household appliances with computerized components could be eventually harnessed to cooperatively solve large number problems and also to help their owners earn cash awards for mathematical discoveries.

In 1999, scientists cracked a popular encryption tool for keeping credit card numbers and other information secret on the Internet. In particular, scientists had broken the RSA-155 code, which protects credit card transactions and secure e-mail in Europe. The method uses a 155-digit product of two large prime numbers, for example:

The Holy Grail of European Spies!

10941738641570527421809707032040357612003732945449205990913842131476349984288934784717997257891267332497625752899781833797076537244027146743531593354333897

(155-digit number)

=

1026395928297411057720541965739916759007165678080380668033419335217907113077779

(prime factor)

×

10660348838016845482092722036001287867920795857598928152227060823719306280864 3

(prime factor)

In particular, RSA-155 requires one party to send a message to another by using the recipient's public key—a 155-digit product of two large primes—to code the original message. Decoding the message requires the two prime numbers know only to the recipient. For a long time this encryption was considered unbreakable. Scientists thought that factoring a 155-digit number was beyond the scope of practical computations. However, a group led by Herman te Riele in Amsterdam factored the huge number using 300 personal computers and a Cray 916 supercomputer. The United States commonly uses 232-digit numbers for encryption, and the U.S. government uses 309 digits for government and military transitions. At the current rate of progress, these codes wouldn't be broken for the next 25 years—or so we hope. For more information, see Hellemans, A., Internet security code is cracked. *Science*. Sept. 3, 285: 1472–1473.

Chapter 57

Saippuakauppias

There are many other interesting patterns in the plot, and you will probably find many more patterns that no one else has yet discovered. Consult the work of IBM researcher Shaiy Pilpel for a list of numbers that are palindromic for both their decimal and binary expressions. For example, 313 is such a "double" palindrome since 313 = 100111001 in binary notation (see "Further Reading.")

Palindromic numbers have often been discussed in the past; for example, see many of the issues of the *Journal of Recreational Mathematics* and the Martin Gardner references in the "Further Reading" section. The work in this chapter is a collaboration with Akhlesh Lakhtakia.

Here is a list of mathematician Michael Keith's favorite palindromic sentences:

⊙ **Some men interpret nine memos.**

⊙ **T. Eliot, top bard, notes putrid tang emanating, is sad. I'd assign it a name: "Gnat dirt upset on drab pot-toilet."**

⊙ **Marge lets Norah see Sharon's telegram.**

⊙ **Turn! I dump Martin Gardner, I rend rag, 'n' I tramp mud in rut.**

⊙ **No D? No L? onon? No, no! LONDON!**

⊙ **On a clover, if alive, erupts a vast, pure evil: a fire volcano.**

⊙ **O, had I nine more hero-men in Idaho!**

⊙ **"Sirrah! Deliver deified desserts detartrated!" stressed deified, reviled Harris.**

⊙ **Are we not drawn onward, we few, drawn onward to new era?**

⊙ **Tarzan raised Desi Arnaz' rat.**

⊙ **Scranton's tots: not narcs.**

Can you think of **5 words** in which all the vowels appear in alphabetical order? Here are 4:

⊙ **Abstemious**: adj., practicing temperance in living

⊙ **Abstentious**: adj., characterized by abstinence

⊙ **Facetious**: adj., straining to be funny, especially at the wrong time

⊙ **Fracedinous**: adj., productive of heat through putrefaction

Can you think of 3 in which 1 letter is repeated 6 times? Here are 2:

⊙ **Nonannouncement** (6 *n*'s): n., the failure to announce

⊙ **Indivisibility** (6 *i*'s): n., the quality or state of being indivisible

Can you think of **6 pangrams** (sentences that use all the letters in the alphabet)? Here are three:

⊙ **The five boxing wizards jump quickly.**

⊙ **Pack my box with five dozen liquor jugs.**

⊙ **The quick brown fox jumps over a lazy dog.**

Chapter 58

Emordnilap Numbers

Some simple observations help to predict the outcome of the reverse-and-add process. Let d_n be the nth digit in a number, and d_n^r be the nth digit in the reversed number. Let p be the path length. Then $p \leq 1$ if, for all digits in the number, $d_n \leq 4$. Also, p is greater than 1 whenever there exists a digit such that $d_n + d_n^r \geq 10$.

Chapter 61

Hyperspace Prisons

Tim Greer of Endicott, New York, has generalized the formula to hyperspace cages of any dimension m: $L(n) = ((n^m)(n + 1)^m)/(2^m)$. Let's spend some time examining 3-D cages before moving on to the cages in higher dimensions.

How large a 3-D cage assembly would you need to contain a representative of each species of insect on Earth today? (To solve this, consider that there may be as many as 30 million insect species, which is more than all other phyla and classes put together.) Think of this as a zoo where 1 member of each insect species is placed in each 3-D quadrilateral. It turns out that all you need is a 25-by-25 ($n = 25$) lattice to create this insect zoo for 30 million species.

In order to contain the approximately 6 billion people on Earth today, you would need a 60-by-60-by-60 cage zoo (see Figure F61.1). You would only need a 40-by-40-by-40 ($n = 40$) zoo to contain the 460 million humans on Earth in the year 1500.

Let's conclude by examining the cage assemblies for fleas in higher dimensions. Dr. Googol has already given you the formula for doing this, and it stretches the mind to consider just how many caged fleas a hypercage could contain, with 1 flea resident in each hypercube or hyperrectangle.

The following are the sizes of hypercages needed to house the 1,830 flea varieties Dr. Googol mentioned earlier in different dimensions:

Dimension (m)	Size of Lattice (n)	Dimension (m)	Size of Lattice (n)
2	9	5	3
3	5	6	3
4	4	7	2

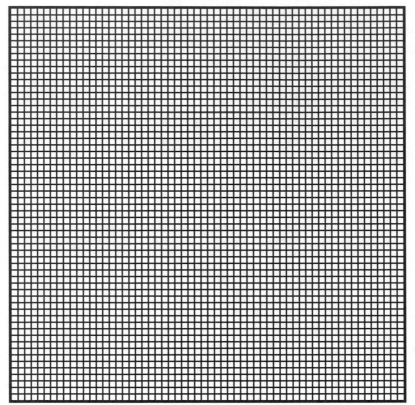

F61.1 A cage containing all humanity. In order to contain the approximately 6 billion people on Earth today, you would need a 60-by-60-by-60-cage zoo, the front face of which is shown here. You would only need a 40-by-40-by-40 ($n = 40$) zoo to contain the 460 million humans on Earth in the year 1500.

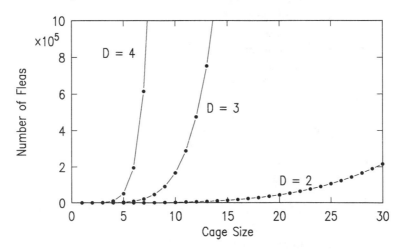

F61.2 Shown here is the number of fleas containable by a lattice cage assembly of "size" n in 2-D, 3-D, and 4-D.

This means that a small $n = 2$, 7-dimensional lattice ($2 \times 2 \times 2 \times 2 \times 2 \times 2 \times 2$) can hold the 1,830 varieties of fleas! An $n = 9$ hyperlattice in the 50th dimension can hold each electron, proton, and neutron in the universe (each particle in its own cage). Figure F61.2 shows the number of fleas containable by a lattice cage assembly of "size" n in 2-D, 3-D, and 4-D. For example, the lower rightmost point indicates that a little more than 2×105 fleas can be contained in a 30-by-30 lattice.

Akhlesh Lakhtakia has noted that the lattice numbers $L(n)$ can be computed from triangular numbers $(T_n)^m$. Why should the number of cage assemblies be related to triangular numbers? (The numbers 1, 3, 6, 10, . . . are called triangular numbers because they are the number of dots employed in making successive triangular arrays of dots. The process is started with 1 dot, and successive rows of dots are placed beneath the first dot. Each row has 1 more dot than the preceding one.)

Chapter 62

Triangular Numbers

Triangular numbers determined by $n(n + 1)/2$ continue to fascinate mathematicians. Various beautiful, almost mystical, relations have been discovered. Here are just some of them:

⊙ A number N is a triangular number if and only if it is the sum of the first M integers, for some integer M. For example, $6 = 1 + 2 + 3$.

⊙ $T_{n+1}^2 - T_n^2 = (n + 1)^3$, from which it follows that the sum of the first n cubes is the square of the nth triangular number. For example, the sum of the first 4 cubes is equal to the fourth triangular number: $1 + 8 + 27 + 64 = 100 = 10^2$.

⊙ The addition of triangular numbers yields many startling patterns: $T_1 + T_2 + T_3 = T_4$, $T_5 + T_6 + T_7 + T_8 = T_9 + T_{10}$, $T_{11} + T_{12} + T_{13} + T_{14} + T_{15} = T_{16} + T_{17} + T_{18}$.

⊙ 15 and 21 is the smallest pair of triangular numbers whose sum and difference (6 and 36) are also triangular. The next such pair is 780 and 990, followed by 1,747,515 and 2,185,095.

⊙ Every number is expressible as the sum of at most 3 triangular numbers. German mathematician and natural philosopher Karl Friedrich Gauss (1777–1855) kept a diary for most of his adult life. Perhaps his most famous diary entry, dated July 10, 1796, was the single line **ΕΥΡΗΚΑ** $= \Delta + \Delta + \Delta$, which signifies his discovery that every number is expressible as the sum of 3 triangular numbers.

Here are some contests: If you square 6, you get 36, a triangular number. Are there any other numbers (not including 1) such that when squared yield a triangular number? It turns out that the next such *triangular-square* numbers are 1,225, 41,616, and 1,413,721. What is the largest such number you can find?

We can use a little trick for determining huge triangular-square numbers. $8T_n + 1$ is always a square number. If the triangular number is itself a square, then we have the equation $8x + 1 = y^2$. The general formula for finding triangular-square numbers is $(1/32)((17 + 12\sqrt{2})^n + (17 - 12\sqrt{2})^n - 2)$.

Here is another approach to finding all numbers that are both square and triangular. We want all the solutions of $m^2 = n(n+1)/2$. Solving this for n using the quadratic formula gives $n = (-1 + \sqrt{1 + 8m^2})/2$. Obviously this equation will give an integer value of n if and only if the quantity inside the square root is a square, so there must be an integer q such that $q^2 - 8m^2 = 1$. Equations of this form are called Pell's equations, and there are infinitely many pairs of integers (q,m) that satisfy this equation. Through a bunch of mathematical manipulation we find $4n(2j-1) = (3 + 2\sqrt{2})^{(2j-1)} + (3 - 2\sqrt{2})^{(2j-1)} - 2$ is a square for every positive integer j.

Can any triangular number (not including 1) be a third, fourth, or fifth power?

Mathematician Charles Trigg has found that $T_{1,111}$ and $T_{111,111}$ are 617,716 and 6,172,882,716 respectively. Notice that both the triangular numbers and their indices are palindromic; that is they can be read backward to yield the same number. Can you find a larger palindromic triangular number than these? Why the frequent occurrence of the digits 617 in these examples?

Obviously, today we can compute huge triangular numbers using modern computers. What's the largest triangular number that Pythagoras could have computed? Would he have been interested in computing large triangular numbers?

If humanity devoted its energy to computing the largest possible triangular number within a year, how large a number would result? It turns out that this question has little meaning because we can construct arbitrarily large triangular numbers by adding 0s to 55, as in 55, 5,050, 500,500, and 50005,000. These are all triangular! Therefore, one large triangular number is:

5000000000000000000000000000000050000000000000000000000000000000000

You can continue this pattern as long as you like. Dr. Googol wonders if Pythagoras or one of his contemporaries noticed a similar pattern.

Chapter 63

Hexagonal Cats

Both triangular and hexagonal numbers are easily found in Pascal's triangle (defined in Chapter 54). For example, a column of Pascal's triangle displays all triangular numbers, as underlined below:

```
1                                        1
1   1                                  1   1
1   2   1                            1   2   1
1   3   3   1         or           1   3   3   1
1   4   6   4   1                 1   4   6   4   1
1   5  10  10   5   1           1   5  10  10   5   1
1   6  15  20  15   6   1     1   6  15  20  15   6   1
```

Can you find where the hexagonal numbers are hiding?

Chapter 64

The *X-Files* Numbers

The "end of the world" formula really did appear in the following reference: Starke, E. (1947) Professor Umbigo's prediction. *American Mathematical Monthly.* January, 54: 43–44. Dr. Googol believes that all W numbers, even ones produced for $n > 1,945$, are divisible by 1,946. A detailed mathematical proof of this can be found in *American Mathematical Monthly.* The proof relies on the fact that $x - y$ is a divisor of $x^n - y^n$ for $n = 0, 1, 2,. \ldots$

Chapter 65

A Low-Calorie Treat

Note that $Cake(n) = 1 + T_n$ where T_n is the nth triangular number.

Mike Angelo of IBM believes he has proven the conjecture that no cakemorphic numbers exist by the following argument. Let's examine the possible last digits of the expression $Cake(n)=(n^2 + n + 2)/2$. This is equivalent to evaluating $Cake$ mod 10. If n is a multiple of 10, e.g., $n = 10x$, then $Cake$ mod 10 is equivalent to: $(100x^2 + 10x + 2)/2$ mod 10, which reduces to $(5x + 1)$ mod 10. This expression has only 2 different values for all x: 1 and 6. We conclude that all integers that are a multiple of 10 (hence end in 0) yield $Cake$ integers that end in 1 or 6. Next we evaluate $Cake$ mod 10 for integers equal to 1 mod 10, 2 mod 10, . . . 9 mod 10. We include one more evaluation for 1 mod 10. $n = 10x + 1$ and $Cake = (100(x^2)+ 20x + 1 + 10x + 1 + 2))/2 = 50x^2 - 15x + 2$. Therefore $Cake$ mod 10 = $5x + 2$. The only possible values are 2 and 7. Thus any number ending in 1 (e.g. 11, 21, 31, . . .) yields a cake integer ending in 2 or 7. Hence it is impossible for an integer ending in 1 to be cakemorphic. By applying this method to the other cases we find that any value of n yields a cake integer that terminates in a different integer from that which terminates n. Hence, we believe no one will ever find a cakemorphic integer.

Dr. Googol invites you to ponder the following: Is there a *doughnutmorphic integer?* Doughnut numbers are constructed in a manner similar to cake numbers, except that the circular pancake region has a hole in it, and hence the sequence for $C(n)$ does not equal $D(n)$. Dr. Googol would be interested in hearing from those of you who have worked on this problem.

What about the existence of pretzelmorphic numbers? These numbers concern the cutting of a pretzel-shaped object.

Previously in the chapter, Dr. Googol gave the equation $Cake(n) = (n^2 + n + 2)/2$ for the maximum number of pieces that can be produced with n cuts of a flat, circular region. Martin Gardner recently sent us a letter containing similar formulas for a (3-dimensional) doughnut and sphere cut with n plane cuts. For a doughnut, the largest number of pieces that can be produced with n cuts is $(n^3 + 3n^2 + 8n)/6$. Thus a doughnut can be sliced into 13 pieces by 3 simultaneous plane cuts (for an illustration, see my book *Computers and the Imagination*). For a sphere, the equation is $(n^3 + 5n)/6+1$. For a 2-D crescent moon: $(n^2 + 3n)/2 + 1$. For further information on cutting shapes, see: Gardner, M. (1961) *The Second Scientific American Book of Mathematical Puzzles and Diversions*. University of Chicago Press: Chicago. Also: Gardner, M. (1983) *New Mathematical Diversions from Scientific American*. University of Chicago Press: Chicago.

Chapter 66

The Hunt for Elusive Squarions

Squarion arrays: Robert E. Stong from Charlottesville, Virginia, has sent Dr. Googol a proof that states for every integer n there is an n-by-n array of distinct integers for which the sum of the squares of any 2 adjacent numbers is also a square.

Strong squarions: The solution to the strong squarion problem is 11,025 (105-by-105) because 21,025 (145-by-145). (Colleagues believe that in general we want to satisfy the following formula in order to search for other numbers of this variety: $10^k = (y - x)(y + x)$ and $1.5 < (y/x)^2 < 2$. Can you figure out how this equation came about? Are there any other numbers that also satisfy these conditions? Must all such numbers end in 5? Dr. Googol does not believe that there is a solution to problem 2 for the strong squarions.

Pair squarions: The first program code for finding pair squares at [www.oup-usa.org/sc/0195133420] is a fairly traditional way of finding pair squarions. Interestingly, one can reduce the search space and computation time significantly. This is accomplished by solving for n and p and noting that we only need to examine pairs of integers whose difference is even. (Why is this so?) This means $n = (a^2 + b^2)/2$ and $p = (-a^2 + b^2)/2$. Note that $b^2 - a^2 = 2p$ and hence must be even. Note also that $b - a$ must be even. (If $b - a$ were odd, $b^2 - a^2$ would be odd.) Therefore, we can generate values for n and p from a, d values where $b = a + 2d$. A faster program to compute all values of n and p with $n < 1000$ is also given at [www.oup-usa.org/sc/0195133420]. This faster version was developed by Mike Gursky.

Chapter 67

Katydid Sequences

The katydid sequence $(x \rightarrow 2x + 2, x \rightarrow 5x + 5)$ yields a repeat after 3 generations. The katydid sequence $(x \rightarrow 2x + 2, x \rightarrow x + 1)$ yields a repeat after 4 generations. Dr. Googol has not yet found a repeat for the $(x \rightarrow 2x + 2, x \rightarrow 6x + 6)$ problem, nor has he found a solution for the related sequences: $(x \rightarrow 2x + 2, x \rightarrow 4x + 4)$ or $(x \rightarrow 2x + 2, x \rightarrow 7x + 7)$.

A colleague, Michael Clarke from England, has conducted a little study on the katydid problem, for the general case of

$$X = C_1 X + C_1 \text{ and } C_2 X + C_2$$

and finds several values of C_1 and C_2 that produce duplicates after a number of generations.

C1:	1	2	3	4	5	6	7
C2: 1	G2	G4	G5	G6	G7	G8	G9
2	G4	G2	G5	☠	G3	☠	☠
3	G5	G5	G2	G7	☠	☠	☠
4	G6	☠	G7	G2	☠	☠	☠
5	G7	G3	☠	☠	G2	☠	☠
6	G8	☠	☠	☠	☠	G2	☠
7	G9	☠	☠	☠	☠	☠	G2

Those entries with a ☠ indicate that no duplicates were found when a search was conducted to the tenth generation after starting with an initial value of 1! Only God knows if there is ever a duplicate. Gn signifies that a duplication has in fact occurred and that it occurs in generation n. In order for members of the same generation to match, the 2 members must satisfy the condition that $c_1^i c_2^{(g-i)} = c_1^j c_2^{(g-j)}$ where g is the number of the generation and i and j are numbers in the range 0 to g.

Can you fill in any of the ☠ entries? Since formulating this problem, Dr. Googol has stumbled upon some research into similar kinds of sequences by Richard Guy. Take a look in the "Further Reading" section.

Chapter 68

Pentagonal Pie

Dr. Googol derived the following sequence for the number of ways a regular n-gon can be divided into triangles: 1, 1, 2, 5, 14, 42, 132, 429, 1,430, 4,862, 16,796, 58,786, 208,012, 742,900, 2,674,440, 9,694,845, . . . Recall that a pentagon could be cut 5 different ways. This is the fourth number in the sequence. A square can be cut only 2 different ways.

These numbers are called Catalan numbers after Eugene Charles Catalan (1814–1894), and they arise in a number of problems in combinatorics—the field of mathematics concerned with problems of selection, arrangement, and operation within a finite or discrete system. (Eugene Catalan had a lectureship in descriptive geometry at the Ecole Polytechnique in 1838, but his career was damaged by his being very politically active with strong left-wing political views.)

The Catalan numbers can be computed using the following formula, which is not too difficult to program on a computer:

$$C_n = \sum_{i=0}^{n-1} [C_i C_{n-i-1}]$$

The first two Catalan numbers are 1, which we can write as $C(0) = 1$ and $C(1) = 1$. The nth Catalan number is defined by the previous formula. What is the largest Catalan number you can compute?

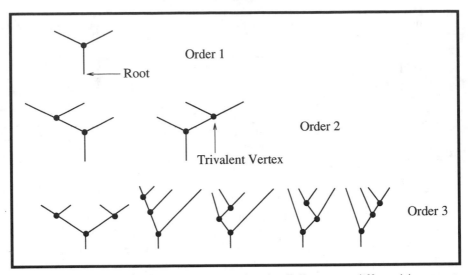

F68.1 Triavalent trees: order 1, order 2, and order 3. How many different trees can you create with 4 nodes?

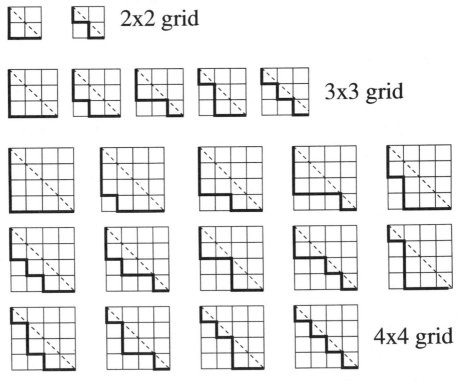

F68.2 Different paths for a 4-by-4 grid. How many different paths can you draw for a 5-by-5 grid?

Among other things, the Catalan numbers describe these:

1. the number of ways a polygon with $n + 2$ sides can be cut into n triangles

2. the number of ways in which parentheses can be placed in a sequence of numbers to be multiplied, 2 at a time

3. the number of rooted, trivalent trees with $n+1$ nodes (see Figure F68.1) (A trivalent tree is a "rooted, ordered" tree in which every vertex, except the root and endpoints, has 3 edges connecting to it. Those vertices with 3 edges connected to them are called trivalent vertices. The order of a trivalent tree depends on the number of trivalent vertices.)

4. the number of paths of length $2n$ through an n-by-n grid that do not rise above the main diagonal (see Figure F68.2)

Another way of saying the second example is that the Catlan numbers count the number of ways parentheses can be placed around a sequence of $n + 1$ letters so that there are 2 letters inside each pair of parentheses:

ab in 1 way: (ab)
abc in 2 ways: (ab)c a(bc)
abcd in 5 ways: (ab)(cd) a((bc)d) ((ab)c)d a(b(cd)) (a(bc))d

and so on.

If you prefer a more visual representation, we can use Catalan numbers to count the number of ways of grouping any objects:

💣☎   in 1 way: (💣☎)

💣☎👁   in 2 ways: (💣☎)👁   💣(☎👁)

💣☎👁👽   in 5 ways: (💣☎)(👁👽)   💣((☎👁)👽)   ((💣☎)👁)👽

                                            💣((☎👁👽))   💣(☎👁))👽

Chapter 69

An A?

A set that is topologically similar to the Ana fractal and to Cantor dusts starts with a circle and consists of 2 circles within 2 circles within 2 circles. . . . Everything *except* for 2 smaller discs is removed. Here we use pairs of circles rather than pairs of lines, and the subdivisions are repeated as with the Cantor set described in the chapter. We retain only those points inside the circles. Figure F69.1 is a picture of this Cantor cheese with each circle's radius very slightly less than half of the previous generation's radius. (The term *generation* refers to the nesting level of the circles.) If we consider just the line along the diameter, the fractal dimension for the set of points is close to 1. Smaller fractal

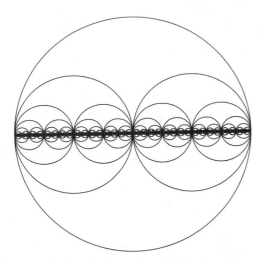

F69.1 Cantor cheese of nested circles.

dimensions are obtained by using circles that are further shrunken and separated so that they do not touch each other.

Returning to the Ana sequence, there are many questions for students to consider, and Dr. Googol is certain that new discoveries are just over the horizon:

⊙ How quickly do the rows of this Ana sequence grow in size?

⊙ What is the ratio of the occurrence of a's to n's in each row as the sequence grows? Try other starting letters.

⊙ Draw a plot where a causes a line to be drawn in a vertical direction (up), and an n causes a line to be drawn in a vertical direction (down). As you proceed through the letters in a single row, move the pen 1 unit to the right for each letter encountered, creating a steplike function. What pattern do you get? What does this tell you about the distribution of letters in the row?

Chapter 70

Humble Bits

Figure 70.1 indicates self-similarity of the gaskets for several orders of "dilational invariance," and they possess what is known as nonstandard scaling symmetry, also called dilation symmetry, i.e., invariance under changes of size scale. Dilation symmetry is sometimes expressed by the formula r ar , where r is a vector. Thus an expanded piece of the gasket can be moved in such a way as to make it coincide with the entire gasket, and this operation can be performed in an infinite number of ways.

The following discussion considers the case for ($0 \le i \le 256$), ($0 \le j \le 256$). This region corresponds to the upper left "block" of the 9 blocks shown in Figure 70.1. Let us consider the number of pixels in the image of a particular shade of gray in order to better understand the resulting patterns. For example, there are only 3 possible (i,j) pairs that form the logical Sierpinski gasket for $c = 256$, since c is 100000000 in binary. The only three ways to make 256 with **OR** are (256,0), (0,256), and (256,256). However, for 255, all 8 bits must be 1s, and there are an amazing 6,561 possible values that satisfy our formula ($c_{i,j} = i$ **OR** j) for $c = 255$. These 6,561 values are colored black for the logical Sierpinski gasket in Figure 70.1. To determine the number of equal-valued pixels there are for a particular value of c, you can use $N = 3^k$ where N is the number of different entries in the (i,j) array that satisfy $c = i$ **OR** j, and k is the number of 1s in the binary representation of c. We can understand this equation by considering that for each 1 in the binary representation of c, there are 3 bit-pairs (1 **OR** 1, 0 **OR** 1,

1 **OR** 0) that produce a 1 under the **OR** operation. For each 0 in the binary representation of c, the corresponding bits of i and j must be both 0.

Notice that if we define (1,6) and (6,1) as duplicate solutions to $c_{ij} = i$ **OR** j, then we obviously have a smaller number of pairs for a particular value of c. Let $b(c)$ be the number of 1-bits in c. Then the number of unordered pairs whose **OR**'ed value is c can be written as $3^{b(c)} - \sum_{i=0}^{b(c)-1} 3^i$. For example, if $c = 17$, then $b(c) = 2$, so there are $3^2 - 3^1 - 3^0 = 9 - 3 - 1 = 5$ solutions. They are (0,17), (1,16), (1,17), (16,17), (17,17). Alternatively, we can count the "duplicate" members by considering that there is only 1 pair of identical numbers, and all other combinations occur twice. Therefore there are $(3^{b(c)} - 1)/2 + 1 = (3^{b(c)} + 1)/2$ unique combinations.

Could the patterns of bits in this chapter be converted to interesting music?

Chapter 71

Mr. Fibonacci's Neighborhood

Replicating Fibonacci numbers are also sometimes called Keith numbers after their inventor, Michael Keith (see, for example, *Journal of Recreational Mathematics*, 1994, vol. 26, No. 3.) Dr. Googol finds these numbers fascinating for several reasos. For one thing, they are very hard to find and seem to require exhaustive computer searches. Some techniques are available to speed up the search, but there is no known technique for finding a Keith number "quickly." They are in some ways reminiscent of the primes in their erratic distribution among the integers. However, Keith numbers are much rarer than the primes—there are only 52 Keith numbers less than 15 digits long. Here they are:

14	19	28	47	61
75	197	742	1104	1537
2208	2580	3684	4788	7385
7647	7909	31331	34285	34348
55604	62662	86935	93993	120284
129106	147640	156146	174680	183186
298320	355419	694280	925993	1084051
7913837	11436171	33445755	44121607	129572008
251133297 (none with 10 digits)		24769286411	96189170155	171570159070
202366307758		239143607789	296658839738	
1934197506555		8756963649152	43520999798747	
74596893730427		97295849958669		

In addition, at least three 15-digit Keith numbers are known. Is the number of Keith numbers finite or infinite?

Michael Keith presents another challenge: define a cluster of Keith numbers as a set of 2 or more Keith numbers (all having the same number of digits) in which all the numbers are integer multiples of the smallest number in the set. There are only 3 known clusters: (14, 28), (1104, 2208), and (31331, 62662, 93993). Is the number of Keith clusters finite or infinite? He conjectures that the number of Keith numbers is infinite and the number of clusters finite, but no proof for either result is known. Since we suspect that there are an infinite number of Keith numbers, the problem of finding the next such number always remains a tantalizing one.

⊛ ⊛ ⊛

For mathematical nerds, the repfigit (Keith) sequence can be restated as follows. Consider any positive integer N with n digits $d_1, d_2, \ldots, d_n$. Consider the sequence defined by $a_k = d_k$ ($k = 1, 2, \ldots, n$) and $a_k = \Sigma_{i=1}^{n} a_{k-i}$ ($k > n$). If $a_k = N$ for any k, we call N a replicating Fibonacci number or Keith number.

It is possible to speed future computations of the repfdigit formula by observing: $a_{k+1} = 2a_k - a_{k-n}$. The use of this equation may lead to an increase in speed $\delta = (T_{1shift} + T_{1add})/[(n-1)T_{1add}]$ where T is the time the computer takes for various operations. (A multiplication by 2 can be done by a C language shift operation.) This leads to a potential speed improvement of $\delta \sim 2/(n-1)$.

Table F71.1 shows the actual sequence generated by 251,133,297.

After Dr. Googol broke the world record and discovered all repfdigits up to 1 billion, a flood of computational research poured forth (see "Further Reading"). However, there remain many serious mysteries involving these strange numbers, and several students, researchers, and clubs have spent thousands of hours searching for new world-record holders.

⊛ ⊛ ⊛

In 1999, scientists discovered a new mathematical constant that relates to Fibonacci numbers. In particular, Divakar Viswanath, a young computer scientist at the Mathematical Sciences Research Institute (MSRI) in Berkeley, California, put the ancient Fibonacci numbers back in the news by showing an odd connection between rabbits and the number 1.13198824. . . . To arrive at this constant, the next time you are trying to generate the Fibonacci sequence, flip a coin at each stage of the calculation. If it comes up heads, you *add* the last number to the one before it to give the next number, just as Fibonacci did. But if it comes up tails, you *subtract*. The sequence produced in this manner is a "random Fibonacci sequence." Viswanath, who recently finished a Ph.D. in computer science at Cornell University in New York, showed that the absolute value of the Nth number in any random Fibonacci sequence is approximated by the Nth power of the number 1.13198824. . . . In other words, if you were a gambler, you would bet that the bigger N is, the closer the absolute value of the Nth number gets to the Nth power of 1.13198824. . . . It's not obvious why this result occurs, and mathematicians

---

2, 5, 1, 1, 3, 3, 2, 9, 7, 33, 64, 123, 245, 489, 975, 1947, 3892, 7775, 15543, 31053, 62042, 123961, 247677, 494865, 988755, 1975563, 3947234, 7886693, 15757843, 31484633, 62907224, 125690487, 251133297

Table F74.1. Actual Sequence for 251,133,297.

are curious to see if there is a relationship between this number and other known constants, such as the golden ratio. Applications of the sequences are discussed by Ivars Peterson in *Science News* 155(24): 376–377, 1999. This discovery suggests that there is still lots of room for mathematical exploration and experimentation, even on a problem that began centuries ago as a simple model for rabbit population growth. It's also an example of how a random process can lead to a deterministic result when the numbers grow large.

Chapter 72

Apocalyptic Numbers

There are many additional problems for you to ponder:

☉ Does there exist an apocalyptic *prime* number?

☉ Is it just a coincidence that the keys of a piano appear to exhibit a segment of the Fibonacci sequence 1, 2, 3, 5, 8, . . . ? There are 2 black notes, followed by 3 black notes. There are 5 black keys in an octave and 8 white keys in an octave!

While on the topic of piano keys, did you ever notice that the widths of the *white* keys are not all the same at the back ends (where they pass between the black keys)? What back-end widths would piano manufacturers chose to use if they wanted to make the widths as similar as possible? Mathematician Kevin Brown studied different pianos and how they accommodate this problem in "linear programming." Let W denote the widths of the white keys at the front, and let B denote the widths of the black keys. Then let a, b, . . . , g (variables are assigned to their musical equivalents) denote the widths of the white keys at the back. It seems impossible to have $a = b = \ldots = g$. The best you can do is try to minimize the greatest difference between any 2 of these keys. One simple approach would be to set $d = g = a = (W - B)$ and $b = c = e = f = (W - B/2)$, which gives a maximum difference of $B/2$ between the widths of any two white keys (at the back ends). Dr. Googol asks, "Can you think of a better solution?"

Incidentally, 666 plays a role in modern times. For example, on July 10, 1991, Procter & Gamble announced that it was redesigning its moon-and-stars company logo, eliminating the curly hairs in the man-in-the-moon's beard that to some looked like 6s. The fall 1991 issue of the *Skeptical Inquirer* notes that "the number 666 is linked to Satan in the Book of Revelations, and this helped fuel the false rumors fostered by

F76.1

fundamentalists"; a dozen lawsuits filed by Procter & Gamble to halt rumors associating the company with Satanism were settled out of court. On May 1, 1991, the British vehicle licensing office stopped issuing license plates bearing the numbers 666. The winter 1992 issue of the *Skeptical Inquirer* reports 2 reasons given for the decision: cars with 666 plates were involved in too many accidents, and there were "complaints from the public."

Although no one to date has found an apocalyptic prime number, various researchers have tried to determine if one exists. Charles Ashbacher uses Bertrand's postulate, which states for $n > 1$ there is always at least 1 prime between n and $2n$. Taking the smallest "apocalyptic number" (denoted by A = 1 followed by 665 0s), we can apply Bertrand's postulate 3 times to conclude that there are prime numbers $p1$, $p2$, and $p3$ such that $a < p1 < 2A < p2 < 4A < p3 < 8a$. Therefore there are at least 3 apocalyptic prime numbers. We can go a step further in contemplation of these elusive numbers. According to Friend H. Kierstead Jr., the number of apocalyptic primes is very much greater than 3. The prime number theorem states that the number of primes less than n is on the order of $n/(\ln\ n)$. Thus the number of primes less than 10^{666} is approximately $10^{666}/\ln(10^{666}) = 10^{666}/(2.303 \times 666) = 6.521 \times 10^{662}$. The number of primes less than 10^{665} is about 6.531×10^{661}. Therefore the number of apocalyptic primes is about $6.521 \times 10^{662} - 6.531 \times 10^{661} = 5.8 \times 10^{662}$. Quite a few!

Chapter 73

The Wonderful Emirp, 1,597

Here are some additional problems for you to ponder.

1,597 is an "emirp," a prime number that turns into a different prime number when its digits are reversed. Can you find any other emirps? How rare are emirps? What is the largest emirp ever computed? Can you find any *Iccanobif* numbers? These are Fibonacci numbers that turn into different Fibonacci numbers when their digits are reversed. Is it possible that Iccanobif numbers do not exist?

Here are some variations to the equation Dr. Googol gave. How difficult is it to find integer solutions to any of the following: $x = \sqrt{1597y^2 + 2}$, $x = \sqrt{1597y^2}$, $\sqrt{1597y^2 + \sqrt{2}}$, $x = \sqrt{1597y^2 - 1}$? (Hint: We believe only 2 of these 5 equations have integer solutions.)

Stimulated by Dr. Googol's research, Paul Tourigny found this amazing solution to the related problem: x = $\sqrt{1597y^2 - 1}$. His solution is x = 509,760,496,584,162,107, 935,182 and y = 12755976753725984792525. He believes this to be the smallest integer solution.

Here are the first few prime Fibonacci numbers: 2, 3, 5, 13, 89, 233, 1597, 28657. How large a prime Fibonacci number can you compute?

Chapter 74

The Big Brain of Brahmagupta

These solutions were not quite the smallest ones! But even the smallest solution contains unimaginably large numbers. For example, it turns out that the absolute smallest value for x is

$$\frac{22440351770433696992455751309067486316094847204 1}{1782466453785771917605107035793432714003296166 0}$$

For more information on this type of problem you can consult Barry Mazur's paper "Arithmetic on Curves," which appeared in the *Bulletin of the AMS* (14(2): 255, 1986). Here are some additional challenges:

⊙ Considering that the Brahmagupta numbers ($x^2 - 157 = y^2$, $x^2 + 157 = z^2$) contain so many digits, what would have mathematicians in earlier centuries thought about a problem such as this?

⊙ Historically speaking, how long ago was a solution to this problem even possible?

⊙ Could someone have solved the Brahmagupta problem, for example, in 1940 or 1950? What problems considered unsolvable today will be solvable in 50 years?

⊙ Can you find any 7th-century Brahmagupta numbers for the original integer problem $x^2 - 92y^2 = 1$ given in the quotation at the beginning of this chapter? Hint: Some solutions to this should be easy to discover using a personal computer.

⊙ One can generalize the 7th-century formula to $x^2 - Ny^2 = 1$. Are there any numbers N for which there is no solution to this problem? For example, Lew Mammel Jr. of AT&T Bell Laboratories could not find a solution for $N = 53$ when doing a computer search for all integers y less than 6365.

As this book went to press, Paul Tourigny, stimulated by Dr. Googol's work with the Brahmagupta problems, found that $66249^2 - 53 \times 9100^2 = 1$.

Chapter 75

1,001 Scheherazades

The question "What is the Arabian Nights factorial?" is from a collection of thousands compiled by Chris Cole, the editor of the rec.puzzles frequently asked questions list.

The answer is 450! (450 factorial). How hard is it to determine the number of 0s at the end of this number?

Rec.Puzzles is an electronic bulletin board that is part of a large worldwide network of interconnected computers called Usenet. In his puzzle collection, Cole notes that determining the number of 0s at the end of $x!$ is not too difficult once you realize that each such 0 comes from a factor of 10 in the product $1 \times 2 \times 3 \times 4 \times \ldots \times x$. Each factor of 10, in turn, comes from a factor of 5 and a factor of 2. Since there are many more factors of 2 than factors of 5, the number of 5s determines the number of 0s at the end of the factorial. The number of 5s in the set of numbers $\{1 \ldots x\}$ (and therefore the number of 0s at the end of $x!$) is $z(x) = \text{int}(x/5) + \text{int}(x/25) + \text{int}(x/125) + \text{int}(x/625) + \ldots$ This series terminates when the powers of 5 in the denominator exceed x. Can you write a computer program for this?

Chapter 76

73,939,133

Amazingly, this is the largest number known such that all its digits produce prime numbers as they are stripped away from the right!

```
73939133
7393913
739391
73939
7393
739
73
7
```

Dr. Googol does not know if there are larger numbers with this property. In the 17th century, mathematicians showed that the following numbers are all prime:

```
31
331
3331
33331
333331
3333331
33333331
```

At the time, some mathematicians were tempted to assume that *all* numbers of this form were prime; however, the next number in the pattern, 333,333,331, turned out not to be prime because $333{,}333{,}331 = 17 \times 19{,}607{,}843$. Let this be a warning to those of you who find mathematical patterns and assume that the pattern continues forever. (If we designate n as the number of digits in the 33 . . . 31 numbers, then these numbers are prime for $n = 2, 3, 4, 5, 6, 7, 8, 18, 40, 50, 60, 78, 101, 151, 319,$ and 382.)

Here's a little dissertation on prime numbers for you. As you certainly know if you have read the previous chapters, an integer greater than 1 is a prime number if its only positive divisors (factors) are one and itself. For example, the prime divisors of 10 are 2 and 5, and the first 6 primes are 2, 3, 5, 7, 11 and 13. The Fundamental Theorem of Arithmetic shows that the primes are the building blocks of the positive integers: every positive integer is a product of prime numbers in 1 and only 1 way, except for the order of the factors. The ancient Greeks proved (ca. 300 B.C.) that there are infinitely many primes and that they are irregularly spaced (there can be arbitrarily large gaps between successive primes).

In the 19th century, it was shown that the number of primes less than or equal to n approaches $n/(\ln n)$ as n gets very large; so a rough estimate for the nth prime is $n \cdot \ln n$.

In 1801, mathematician Karl Friedrich Gauss eloquently stated in his *Disquisitiones Arithmeticae*:

> The problem of distinguishing prime numbers from composite numbers and of resolving the latter into their prime factors is known to be one of the most important and useful in arithmetic. It has engaged the industry and wisdom of ancient and modern geometers to such an extent that it would be superfluous to discuss the problem at length . . . Further, the dignity of the science itself seems to require that every possible means be explored for the solution of a problem so elegant and so celebrated.

On January 27, 1998, Roland Clarkson, George Woltman, Scott Kurowski, and others discovered a new record prime for that time: $2^{3021377} - 1$. This is the thirty-seventh known Mersenne prime (there may be smaller ones, as not all previous exponents have been checked). Clarkson, a 19-year-old college student, was one of about 4,000 individuals involved in *GIMPS: The Great Internet Mersenne Prime Search*, launched by Woltman in early 1996. He found this prime using a program written by Woltman linked to the GIMPS Internet database via Scott Kurowski's PrimeNet (Parallel Technology for the Great Internet Mersenne Prime Search). As of April 1998, PrimeNet's sustained throughput was at least 154 billion floating-point operations per second, or 4.6 (Pentium Pro 200Mhz) CPU years computing time per day. For the testing of Mersenne numbers, this is equivalent to 5.3 Cray T916 supercomputers, fully equipped (16 CPUs each) and at peak power.

GIMPS offers free software to personal computer owners who want to search for big prime numbers.

The primality of their number was verified by David Slowinski, who has found several of the recent record primes. The complete decimal expansion of this 909,526-digit number is available on the Web. For the current largest known prime number, see the Web site http://www.utm.edu/research/primes/largest.html. (See also "Further Exploring" for Chapter 56 for recent developments.)

Chapter 77

⊎-Numbers from Los Alamos

Here are a few ⊎-numbers Dr. Googol calculated with starting numbers 1 and 9:

1 9 10 11 12 13 14 15 16 17 18 20 36 38 39 40 41 42 43 44 46 66 67 68 69 70 71 72 73 92 101 121 122 123 124 125 126 127 146 155 174 182 201 211 229 230 237 256 284 285 286 287 288 289 290 291 311 348 365 368 369 370 . . .

Here are a few with starting numbers 1 and 3:

1 3 4 5 6 8 10 12 17 21 23 28 32 34 39 43 48 52 54 59 63 68 72 74 79
83 98 99 101 110 114 121 125 132 136 139 143 145 152 161 165 172 176
187 192 196 201 205 212 216 223 227 232 234 236 243 247 252 256 258
274 278 . . .

Notice how these $\uplus_{1,3}$ numbers have many terms separated by 2.

The following is a long $\uplus$-number sequence computed for the starting numbers 100 and 101. (Dr. Googol computed this massive sequence using a computer program designed for him by Michael Clarke, who lives in the United Kingdom).

100 101 201 301 302 401 403 501 504 601 603 605 701 706 801 803 805
807 901 908 1001 1003 1005 1007 1009 1101 1110 1201 1203 1205 1207
1209 1211 1301 1312 1401 1403 1405 1407 1409 1411 1413 1501 1514 1601
1603 1605 1607 1609 1611 1613 1615 1701 1716 1801 1803 1805 1807 1809
1811 1813 1815 1817 1901 1918 2001 2003 2005 2007 2009 2011 2013 2015
2017 2019 2101 2120 2201 2203 2205 2207 2209 2211 2213 2215 2217 2219
2221 2301 2322 2401 2403 2405 2407 2409 2411 2413 2415 2417 2419 2421
2423 2501 2524 2601 2603 2605 2607 2609 2611 2613 2615 2617 2619
2621 2623 2625 2701 2726 2801 2803 2805 2807 2809 2811 2813 2815
2817 2819 2821 2823 2825 2827 2901 2928 3001 3003 3005 3007 3009
3011 3013 3015 3017 3019 3021 3023 3025 3027 3029 3101 3130 3201
3203 3205 3207 3209 3211 3213 3215 3217 3219 3221 3223 3225 3227
3229 3231 3301 3332 3401 3403 3405 3407 3409 3411 3413 3415 3417
3419 3421 3423 3425 3427 3429 3431 3433 3501 3534 3601 3603 3605
3607 3609 3611 3613 3615 3617 3619 3621 3623 3625 . . .

L. Kerry Mitchell, an aerospace engineer at the NASA Langley Research Center in Hampton, Virginia, suggested to Dr. Googol the concept of modified $\uplus$-numbers, or ⊗-numbers. In these cases, addition is replaced by multiplication in the definition of $\uplus$-numbers. Starting with 2 numbers greater than 1, continue the sequence with those numbers that can be written only in 1 way as the product of 2 previous elements. For initiators of 2 and 3, here are the first 20 ⊗-numbers:

2 3 6 12 18 24 48 54 96 162 192 216 384 486 768 864 1458 1536 1944 3072

24 is on the list since it can be written only as 2×12, but 36 is not since it can be written as 2×18 or 3×12. Notice that $\otimes_{2,3}$ are all even after 3. Why? Are all ⊗-numbers even?

❀ ❀ ❀

In order to study the distribution of gaps between $\uplus_{1,2}$-numbers, Ken Shirriff and Dr. Googol computed the 100,000 gaps between the first 100,001 $\uplus_{1,2}$-numbers. Figure F77.1 shows the distribution of gaps of size 1 to 200.

In the infinite ⨄$_{1,2}$ sequence, gap sizes can be divided into 3 categories: gaps that never appear, gaps that appear a finite number of times, and gaps that appear infinitely often. Dr. Googol has not yet found certain gap sizes, such as 6, 11, 14, 16, 18, 21, 26, 28, and 33. Other gaps (e.g., 1, 4, 9, and 13) appear infrequently (occurring 4, 2, 3, and 1 time, respectively). Some gaps are very common; for example, 37% of the gaps are of size 2, and 14% of the gaps are of size 3. Of course, these computational observations do not tell us about the properties of ⨄$_{1,2}$ after the first 100,000 ⨄$_{1,2}$-numbers. Note that these missing gaps (e.g., 6, 11, 14, etc.) are separated by 1, 2, 3, or 5. Interestingly, these values are Fibonacci numbers. Is this always the case? We would like to hear from any readers who find gap sizes that do not manifest themselves in the first 100,000 ⨄$_{1,2}$ gaps.

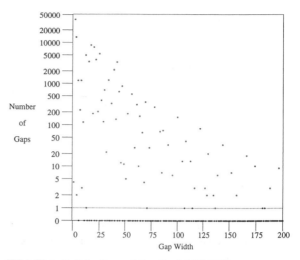

F77.1 The distribution of the first 100,000 gaps between consecutive ⨄$_{1,2}$-numbers. The number of times each gap occurs is plotted on a log scale from 1 to 50,000 along the y axis; gaps that never occur are plotted at y = 0. (From a collaboration with Ken Shirriff.)

Chapter 78

Creator Numbers ♌

Dr. Googol collaborated with Ken Shirriff of the University of California for much of the analysis of this problem. Ken wrote a computer program in C that not only searches for the minimal solutions for the first 1,500 integers but also searches for the *number* of minimal ways to construct a number. For example, without allowing concatenation (multidigit numbers), he finds that there are 208 different ways to write the number 20, and 1,128 different ways to write the number 21! Even more exciting is the fact that these 208 and 1,128 different ways to write minimal solutions change to just 2 ways and 1 way if concatenation is allowed. (After all, there is just 1 way to minimally write 21 by concatenating 2 and 1.)

The program finds solutions by using dynamic programming techniques. It starts with the 1-digit base cases and combines these numbers to generate all numbers that have a shortest solution of 2 digits. The 1- and 2-digit results are combined to yield all numbers with 3-digit shortest solutions. This process continues until all the desired numbers have been found. In order to keep the computations from growing too quickly, Ken Shirriff prunes the results by discarding any results over 10,000. He also limits results to integers by only using positive exponents. While the first limit probably has

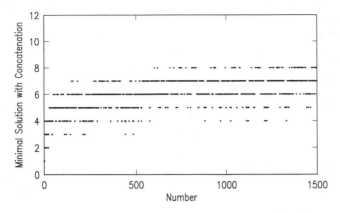

F78.1 Minimal integer solutions. These solutions $\delta\ell(n)$ are for the first 1,500 numbers. Concatenation of integers is not allowed.

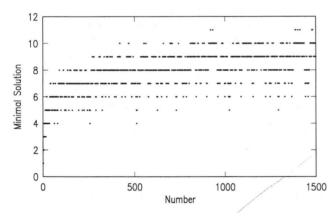

F78.2 Minimal integer solutions with concatenation. The solutions $\delta\ell(n)$ were found for the first 1,500 numbers. Concatenation of integers is allowed (that is, multidigit numbers such as 12 and 121 are permitted).

no effect on the results, there are a handful of shorter solutions that are only obtained by using negative exponents.

Figures F78.1 and F78.2 show plots of computed values for $\delta\ell(n)$ vs. n. These plots show results for both nonconcatenation and concatenation contests. Interestingly, minimal solutions comprised of fewer than 12 digits can be found for all numbers tested (on average, about 7 digits are needed to minimally construct n, $1 \le n \le 1500$).

We can also define the concept of "hard numbers" $\delta\ell_h(n)$, which are the smallest numbers requiring $\delta\ell(n)$ digits. For example, 921 is the smallest number that requires a walloping 11 digits for its expression. Running his program on the integers up to 1 million, Shirriff found the hard numbers listed in Table F78.1. Plots of n vs. $\delta\ell_h(n)$ seem to increase exponentially. Notice that almost all hard numbers include the digit 1. Why?

Unusual solutions: The contest winner, Dan Hoey, also wrote a Lisp program to confirm his hand calculations, and as with Shirriff's C program, he did not initially check for negative exponents. However, he later extended his program to negative exponents and discovered they sometimes result in shorter solutions. For instance, Hoey notes that if negative exponents are not checked, one might conclude that $\delta\ell(640) = 8$. However, look at Hoey's amazing solution $\delta\ell(640)= 7$ found when using negative exponents:

$$640 = (2^{((2 + 1)^2)}) \times (1 + 2^{-2})$$

Nevertheless, he believes that 20, 120, and 567 do not benefit from the use of negative exponents unless some subexpression has a denominator or numerator exceeding 10^{12}. He found an interesting solution with negative exponents for 567:

Without Multidigit Expressions:		With multidigit expressions:	
Digits	Hard Number	Digits	Hard Number
2	3	2	3
3	2	3	5
4	7	4	7
5	13	5	29
6	21	6	51
7	41	7	151
8	91	8	601
9	269	9	1631
10	419	10	7159
11	921	11	19145
12	2983	12	71515
13	8519	13	378701
14	18859		
15	53611		
16	136631		
17	436341		

Table F78.1 Hard Numbers.

$$567 = (2^{2^2} + 2)^2 \times (2 - 2^{-2})$$

Should future searches consider using irrational numbers? Hoey writes, "In the same way that negative exponents imply fractions, fractional exponents imply irrational numbers, and then irrational exponents imply transcendental numbers. In fact, one could obtain complex numbers, too, but I don't think that is any help, and you have problems with branch cuts there." One question is whether there are any "integers" that benefit (in the sense of requiring fewer 1s and 2s) by considering and using irrational numbers, or rational numbers formed with fractional exponents. Is there any integer that benefits from using irrational exponents? Dr. Googol thinks this is a fertile ground for significant future research.

In closing, Dr. Googol does not know for certain whether all of the $\delta\ell(n)$ values listed here are truly the minimal values. In most cases, they were arrived at through computation and not through any mathematical theory. He looks forward to hearing from readers who may be able to find even smaller values than the ones listed here. Much of the participation and discussion for Dr. Googol's Creator Numbers Contest occurred in the mathematics discussion group sci.math on the Usenet computer network, where this contest took place.

Chapter 79

Princeton Numbers

To compute very large integers in the Robbins formula, you may have to use a programming language such as REXX. Alternatively, there are many simple software packages used to compute large integers, such as Mathematica (Wolfram Research). Other notable examples are the large-integer programs of Harry J. Smith. Smith uses his own software package to perform multiple precision integer arithmetic: his software package can even compute transcendental functions to thousands of decimal places. (Contact him at: Harry J. Smith, 19628 Via Monte Drive, Saratoga, CA 95070.) Another alternative is the large-integer program Matlab (Mathworks, South Natick, MA).

Challenges: Dr. Googol calls a Robbins number R_n *Robbinmorphic* if it terminates with *n*. For example, a one-digit Robbinmorphic number is R_6 = 7436. (For more on other *morphic* numbers, see Chapter 63). If *n* were a 2-digit number, the last 2 digits of the Robbins number would be considered when checking morphicity. And so on.

Here is one hell of a question for you number nerds: does there exist a Robbinmorphic number for *n* > 7? After Dr. Googol posed this question to friends, Harry J. Smith from Saratoga, California, and David Edelheit from Oyster Bay, New York, discovered that R_{32} is Robbinmorphic because it ends in 32. This is the only known large Robbinmorphic number. Is there a larger one? To compute the Robbins numbers, Smith used $R(n) = R(n-1) \times (2n) \times (2n+1) \times \ldots (3n-2)/((n) \times (n+1) \times \ldots (2n-2))$. This equation can be easily implemented with an algorithm that has all-integer intermediate results. (You must use care when using the first formula given in this chapter. Even though all Robbins numbers are integers, some of the intermediate results in the algorithm are not integers. If intermediate results are stored as integers, some small errors may occur.)

Is there anything special about the arrangement of digits within any of the Robbins numbers? Certain Robbins numbers, such as the fourteenth, which starts with 999 and ends with 000, do not seem perfectly random. Is the arrangement of digits random?

Chapter 80

Parasite Numbers

After Dr. Googol showed his single 4-parasite number to several colleagues, Keith Ramsay of the University of British Columbia came up with an amazing formula to generate parasite numbers. It turns out that Dr. Googol's brute-force computational searches would have taken far too long to find larger parasite numbers. Suppose we start with a multiplier digit *d* and wish to find some *d*-parasite. All we have to do is evaluate the formula $d/(10d-1)$, and then use the unique segment of digits before the cluster repeats. (Every fraction, when expressed as a decimal, either "comes out even" as in $1/8 = 0.125$, or it repeats as in $1/3 = 0.33333$ where a single digits occurs over and over again, or it has group-repeats as in 1/7 = 0.142857 142857 . . .) Let Dr. Googol explain with an example. Suppose he'd like to find a large parasite for 2. Let's divide 2 by 19 to get 2/19 = 0.105263157894736842. The "105263157894736842" portion repeats over and over and is a 2-parasite because 2 × 105,263,157,894,736,842 = 210,526,315,789,473,684. (Incidentally, this number is larger than the number of stars in the Milky Way galaxy.) Here's an incredible-sized 6-parasite:

6/59 = .1016949152542372881355932203389830508474576271186440677966 . . .
1016949152542372881355932203389830508474576271186440677966 × 6 =
6101694915254237288135593220338983050847457627118644067796

Do you see how the 6 migrates from the right end to the front after multiplication? Knowing Ramsay's formula, you can amaze your friends with multidigit parasites containing hundreds of digits.

Mike Dederian of Harvey Mudd College in California found something unusual about a 5-parasite

$$1020408163265306122448979591836655$$

which can be written as **1** (**02**) (**04**) (**08**) (**16**) . . . to emphasize the doubling of digits. The reason for this initial pattern is not obvious to us.

After seeing Dr. Googol's parasite numbers, Joseph S. Madachy, editor of the *Journal of Recreational Mathematics*, sent Dr. Googol a paper he wrote in 1968 that appeared in the *Fibonacci Quarterly* (6(6): 385–389). In the paper are recipes for "instant division," which resembles what we might call (using Dr. Googol's terminology) reverse pseudo-parasites. If you wish to divide 717,948 by 4, merely move the initial 7 to the right, obtaining 179,487. Madachy also gives another example:

$$9,130,434,782,608,695,652,173$$

can be divided by 7 by transposing the initial 9 to the end, obtaining

$$1,304,347,826,086,956,521,739$$

Other challenges:

⊙ What is special about the fraction 137174210/1111111111? Try computing this to find out. You'll be amazed when you gaze upon its decimal representation.

⊙ Make a list of all pseudoparasites less than 1 million.

⊙ Do there exist "ultraparasites" that multiply by swapping both the left- and right-most digits?

Chapter 81

Madonna's Number Sequence

The digits of pi (π) are 3.1415926. . . . Notice what happens if you add 1 to each digit?

One of this book's reviewers felt that the sequence **4252603764690804434957** was not sufficiently interesting to be included in this book, and therefore Chapter 81 should be deleted. If you agree, send Dr. Googol a note, and he will delete Chapter 81 from future editions of this book.

Chapter 82

Apocalyptic Powers

Werner Knoeppchen of Glenwood Springs, Colorado, sent Dr. Googol a printout of the number $2^{5,000,000}$. Werner writes:

> The number contains six 6s in a row. Therefore it is an apocalyptic power. I do not know if it is the lowest. The printout for $2^{5,000,000}$ is over 500 pages long, and the number contains 1,505,150 digits. It required two weeks for a Mac IICI to calculate this number running Mathematica.

Werner's double apocalyptic power contains inside it the digits "10556666660670," which he proudly circled in red ink.

Charles Ashbacher of Cedar Rapids, Iowa, wrote a Pascal program that searched for double apocalyptic powers. He found such powers with exponents of i as follows: 2269, 2271, 2868, 2870, 2954, 2956, 5485, 5651, 6323, 7244, 7389, 8909, 9195, 9203, 9271, 9273, 9275, and 9514. (Why are there several "twins" that differ by 2: 2269 and 2271, 2868 and 2870, 2954 and 2956? Why should a "triplet" exist: 9271, 9273, and 9275? Just chance?)

Christopher Becker from Homer, New York, used a DEC VAX 6410 and verified Ashbacher's findings regarding double apocalyptic powers. Becker notes that the first such number 2^{2269} has 684 digits and has 666666 at the 602nd position. For *single* apocalyptic powers, he finds 2^{157}, 2^{192}, 2^{218}, 2^{220}, and 2^{222}. Curiously, 2^{666} is itself an apocalyptic power. Between 2^{2000} and 2^{3000} Becker finds that more than half of the exponents are apocalyptic powers. Becker has also searched for St. John powers, which have the digits 153 (Simon Peter caught 153 fish for Jesus). 2^{115} is the first St. John power.

Becker later used a DEC Alpha computer to search for *triple* apocalyptic numbers with nine 6s in a row. He searched as high as 2 raised to a quarter-million using his custom C program. After using five hours of computing time, he found the following triplet of triple apocalyptic exponents that differ by 2: 192916, 192918, and 192920. He also found 212253, 237373, 241883, and 242577.

John Graham of Penn State Wilkes-Barre, Pennsylvania, and R.W.W. Taylor of the National Technical Institute for the Deaf (Rochester Institute of Technology, Rochester, New York) have both proven that there is an infinite number of apocalyptic powers.

John Rickert from the Rose-Hulman Institute of Technology, Terre Haute, Indiana, is currently the world's expert on apocalyptic powers. Stimulated by Dr. Googol's initial research on apocalyptic powers, Rickert has made a number of unusual discoveries, some of which are reported in the *Journal of Recreational Mathematics* 29(2): 102-106, 1998. If we call numbers of the form 2^k that contain the digits 666 "apocalyptic powers," Rickert finds an infinite family of apocalyptic powers of the form

$$6663628647754606040895353774569915678\overset{...}72 \bmod 1039.$$

He also made several of other discoveries. For example, exponents of the form $k = 650 + 2500n$, $k = 648 + 2500n$, and $k = 1899 + 2500n$ produce apocalyptic powers for any natural number n. If an apocalyptic power, 2^k, contains the sequence of five digits $666ab$ with $50 \le 10a + b \le 74$, then 2^{k+2} will also be an apocalyptic power.

Exponents smaller than 1000 producing apocalyptic powers are: 157, 192, 218, 220, 222, 224, 226, 243, 245, 247, 251, 278, 285, 286, 287, 312, 355, 361, 366, 382, 384, 390, 394, 411, 434, 443, 478, 497, 499, 506, 508, 528, 529, 539, 540, 541, 564, 578, 580, 582, 583,610, 612, 614, 620, 624, 635, 646, 647, 648, 649, 650, 660, 662, 664, 666, 667, 669, 671, 684, 686, 693, 700, 702,704, 714, 718, 720, 723,747, 748, 749, 787, 800, 807, 819, 820, 822, 823, 824, 826, 828, 836, 838, 840, 841, 842, 844, 846, 848, 850, 857, 859, 861, 864, 865, 866, 867, 868, 869, 871, 873, 875, 882, 884, 894, 898, 920, 922, 924,925, 927, 928, 929, 931, 937, 970, 972, 975, 977, 979, 981, 983, 985, and 994.

Rickert also discovered that double apocalyptic powers for any natural number n can be produced by $k = 423152 + 1562500n$. The smallest such number is $k = 423152$. How far can we extend this madness? Is it possible to find a k so that 2^k contains 666 consecutive 666s (1988 consecutive 6s)? This large number, called the *Goliath number* and denoted by the symbol $\mathcal{V}$, certainly exists. Behold the following beauty:

The Smallest Known Goliath $\mathcal{V}_0 = 2^k$ where

$k =$ 5885687724118401941316021532344935567102950794778571209841922652323917894198804389
0692192199031609270594899151548577604644482542959681806959202797968494630757082 90
1993423558705896478202003732416276140940637030463100600653040978080994672926 82
4138566463664864191273768654105927280055118272241704786417418390805959982489620
9575993796189207053873038187955600142079762745184357994797279737875654286166 3218022958
6009915880036637454984373829099716017438631799199945059406392053282 34595 8593 98204292
4485552541186011841792663171779654659793784400589805837899013847271 5 65554 5043410 84459
8895111973105433464356301358028129300956157976029072329718545212706970497009516499247
1993709212583732355121155487041449936710414146774642084075544300433001865 30390231964337
89729766838086606019556295640004097930387252600942326726885769725247405688 5075564671
228797634014731591716426588030909430219790056441909610707855080 48577964035094420 97275
01475184963379375681750980592735297610283090441817434192039935554462701881 93944313063
2625602442747324700096861495216438083158096868200763242968319161648206544769 05889642
1775705966984874767378351736342049808812344048530786627953430373532249860596531 83547
13379580056896848381395532707309309224611883186754696845280783077287523193 6 465754479
05175217058503452352430715087565242119026575975104684568694694282933761495 93416389849
8895833912553641354411772371648959151433333377231917425854508829475647789957 9689409356
60181063879064193907748179315923988850675337823763051948576639548553 66774 01767 9 68856
39910344056607585189426538942038122270051398146900721431875177528294673624628 71190565
45179927985536084276603492993951672307985609650128420625047056950489 07161 173965139141
15620269557497773180198269629775587796242071327824364351971556771023719749 7 435157641369
0604663246733203007509819711888977867406538980331384029470004984193019899 2815556582
86058672484322587527232862069965592949727958214753463780884938892181903933847 4870981
66096452665106632745683143664200122860959024860772469439488504

There are probably Goliath numbers smaller than this, and there are certainly larger Goliath numbers. In fact, larger values of V can computed from $k + 4 \times 5^{2858}n$.

It seems likely that there is some K so that for any $k > K$, 2^k is an apocalyptic power. Rickert suggests that a proof of this is beyond our current techniques. Further exploration shows that there are only twenty exponents between 20,000 and 100,000 that do not produce apocalyptic powers. These exponents are 20271, 20300, 20509, 20644, 20710, 21077, 21600, 21602, 22447, 22734, 23097, 23253, 24422, 24441, 25026, 25357, 25896, 26051, 26667, and 29784.

Rickert conjectures that *all* powers of 2 larger than 2^{29784} are apocalyptic powers. (Currently 29,784 is the largest known non-apocalyptic power.) This would mean that there are exactly 3,715 powers of 2 that are not apocalyptic powers. Note that the frequency of double apocalyptic exponents is clearly increasing in the list of exponents smaller than 10,000 producing doubly apocalyptic powers.

Dr. Googol asks if there is some K so that for any $k > K$, 2^k is a Goliath number V? In another words, at what point in our number system do all numbers suddenly become Goliath numbers. Is $V_0 \times V_0$ a Goliath number?

Is $V_0{}^{V_0}$ a Goliath number?

Chapter 83

The Leviathan Number ♉

Michael Palmer from the United Kingdom was the first person on Earth to determine the first 6 digits of ♉. Interestingly, you don't have to compute all the digits of the Leviathan to determine just the first 6. The reasoning is as follows.

Factorial functions can be approximated by Stirling's formula. It's named after James Stirling (1692–1770), a Scot who began his career in mathematics amid political and religious conflicts. He was friends with Newton but devoted most of his life after 1735 to industrial management.

Stirling's ingenious formula for approximating factorial values is $n! \sim \sqrt{2\pi} \times e^{-n} \times n^{n+1/2}$. At [www.oup-usa.org/sc/0195133420] Dr. Googol provides BASIC and C code for computing Stirling approximations, actual factorial values, and the percentage difference between the 2. Notice that this formula give a useful approximation for $n!$ when n is large. For example, when $n = 6$, Stirling's approximation gives a value of 710, and the true value of $n!$ is 720. When n is 23, Stirling's approximation is 25,758,524, 968,130,088,000,000, and the true value is 25,852,017,444,594,486,000,000. Notice that the difference between the 2 values actually increases as a function of n, but the *percentage* difference decreases with greater values. Why not make a graph showing this percentage difference as a function of n? Because many modern software packages today allow us to compute large factorials (though presumably not so large as a googol), people often forget Stirling's formula. However, until a few years ago, this was the only way to approximately determine factorials for large numbers.

Let's use Stirling's formula to compute the first few digits of the Leviathan ♉ without computing all the digits. Michael Palmer notes that for $n = 10^{666}$, the term $n^{n+1/2}$ in Stirling's formula is a power of 10 and can be ignored when trying to determine the first 6 digits of ♉. Next, let's look at the exponential term in Stirling's formula. Here we have which can be rewritten as $10^{-10^{666} \times k}$ where $k = \log_{10} e$. Next, we split $10^{666} \times k$ into its integer and fractional parts, say m and f, giving us $e^{-10^{666}} = 10^{-m} \times 10^{-f}$.

We can ignore the 10^{-m} part since it is a power of 0.1, and therefore the first 6 digits of $10^{666}!$ are given by the first six digits of $\sqrt{2\pi} \times 10^{-f}$. Michael used a mathematical software package called AXIOM to compute this, using a high number of digits (777) to ensure accuracy. Therefore, $10^{666} \times k$ is $434,294 \ldots 9,652.27174945413317.$ . . . Next, using what remains of Stirling's formula, we find $\sqrt{2\pi} \times 10^{-0.27174945} = 1.340727397$. He therefore concludes that the left 6 digits of the Leviathan number are 134,072.

Could today's computers compute the entire Leviathan, or will this be beyond the realm of humankind for the next millennium? The number of digits in ♉ is more than 10^{668}, and this is much greater than the number of particles in the universe. Furthermore, even if a googol digits could be printed (or stored) per second, is would still require so much time that the universe would come to an end before the printing or storing was completed. Therefore such a computation will always be beyond the realm of humanity. If you are interested in computing the number of trailing 0s of ♉, see my book *Keys to Infinity*.

As we climb the integers in our quest for infinity, we find several famous large numbers. The baby Leviathan 9^{9^9} is the largest number that can be written using only 3 digits. It contains 369,693,100 digits. If typed on paper, it would require around 2,000 miles of paper strip. Since the early 1900s, scientists have tried to determine some of the digits of this number. Fred Gruenberger recently calculated the last 2,000 digits and the first 1,200 digits.

Even more unimaginable is ℑ, which has a value of $9^{9^{9^9}}$. If typed on paper, ℑ would require $10^{369693094}$ miles of paper strip. Joseph Madachy has noted that if the ink used in printing ℑ was a 1-atom-thick layer, there would not be enough total matter in millions of our universes to print the number. Shockingly, the last 10 digits of ℑ have been computed. They are 1,045,865,289.

Here's a tough problem for you. Is the following statement true or false? How do you know?

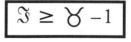

$$\mathfrak{I} \geq \mathfrak{Y} - 1$$

A final observation on big numbers. The largest "physically imaginable" size is that of our entire universe, 10 with 29 0s after it (in centimeters). The smallest size, describing the subatomic world, is 10 with 24 0s (and a decimal) in front of it. On this grand size scale, humans are right in the middle. Does this mean humans hold a central, privileged place in the cosmos? Did God place us here?

Chapter 84

The Safford Number

Arthur C. Clarke recently wrote to Dr. Googol expressing his skepticism over the story of Dase calculating pi to 200 places in his head. Clarke wrote, "Even though I've seen fairly well authenticated reports of other incredible feats of mental calculation, I think this is totally beyond credibility." Clarke, stimulated by Dr. Googol's Dase report, recently wrote Stephen Jay Gould asking how it is possible for such extraordinary abilities as human calculators to have evolved through natural selection. Clarke asks, "What is the survival value in the jungle of the ability to multiply a couple of 50-digit numbers together?"

Dr. Googol looks forward to hearing from readers who can confirm or deny the legends of Dase's extreme computing ability.

Chapter 85

The Aliens from *Independence Day*

If you want to use the computer programs at [www.oup-usa.org/sc/0195133420] to compute sexes for a large number of years, it's important to have a high-precision value for $\sqrt{5}$, and you might want to check the value that is used in your particular computer language. (You don't have to worry about this issue if you only want to compute the sex of the first few thousand abductees.) For example, many people who tried to use the Mulcrone formulation computed that a female would be the billionth person taken. This is because BASICA gives a value of 2.2360680103 for $\sqrt{5}$ on some machines, whereas the true value is 2.236067977. . . .

Notice that the number of males and females, and total number of humans, begin to follow the well-known Fibonacci sequence:

Year	0	1	2	3	4	5	6	7	8	. . .
Number of Males	1	0	1	1	2	3	5	8	13	. . .
Number of Females	0	1	1	2	3	5	8	13	21	. . .
Total	1	1	2	3	5	8	13	21	34	. . .

(As mentioned in other chapters, the Fibonacci sequence of numbers—1, 1, 2, 3, 5, 8, etc.—is such that, after the first 2, every number in the sequence equals the sum of the 2 previous numbers $F_n = F_{n-1} + F_{n-2}$). The sum of elements F_1 through F_n is $F_{n+2} - 1$. Using this relationship, it's possible to show that the number of people abducted during a particular year is simply F_{year} (in this case, the first abduction is considered to have taken place in year 1). The total number of people abducted including the current year, is $F_{year+2} - 1$. As to questions about the sex ratio, it's possible to show that the ratio of the number of females to males converges to $F_n/F_{n-1} = \phi$. Here ϕ is known as the golden ratio and is equal to 1.61803. . . . It appears in the most surprising places in nature, art, and mathematics. The symbol ϕ is the Greek letter *phi*, the first letter in the name Phidias, a classical Greek sculptor who used the golden ratio extensively in his work.

In order to avoid any numerical precision problems that may arise with the Mulcrone formulation, Ram Biyani has suggested a formulation involving only integer calculations. In particular, we can use a recursive function that computes the sex, s, of the xth person in the yth year using a previously generated sequence of the number of persons taken in each year (the Fibonacci sequence). The recursive relationship is $s(y,x) = s(y-2,x)$, if $x \le F(y-2)$; $s(y,x) = s(y-1, x - F(y-2))$, if $x > F(y-2)$, where $F(y)$ is the number of persons taken in the year y, and $s(y,x)$ is the sex of the xth person taken in year y.

Here are some additional challenges for you to ponder:

⊙ How many years would the alien require to remove the entire population of the Earth (about 6 billion people)?

⊙ Can you use this fact to determine the sex of the billionth person?

⊙ How do the sex ratios change if, during the first year, you start with *2* people, for example, **M M**, or **F M**?

Chapter 86

One Decillion Cheerios

Scott Bales from North Carolina notes that any possible solution must be of the form $2^x \times 5^x = 10^x$. If this is not true, 1 of the multiplicands' terms will have both 2 and 5 as factors, and the last digit of this term will be 0. The problem therefore is to find a power of 2 and a power of 5 that do not have 0s in them. Scott has written a Turbo Pascal program (running on a 486 DX) to check 5^x for all values of x less than 60,000. Using his program, Scott found 5^{58} to be the only power of 5 greater than 5^{33} that also contained no 0s. However the power of 2 for $x = 58$ yielded a number with at least one zero. Scott says, "Do I think such a number exists? I don't know—early evidence doesn't look good. If it exists, I think humanity will one day find it."

Chapter 87

Undulation in Monaco

Bob Murphy used the software Maple V to search for undulating squares, and he discovered some computational tricks for speeding the search. For example, he began by examining the last 4 digits of perfect squares (i.e., he computed squares mod 10,000). Interestingly, he found that the only possible digit endings for squares that undulate are 0404, 1616, 2121, 2929, 3636, 6161, 6464, 6969, 8484, and 9696. By examining squares mod 100,000, then mod 1,000,000, then mod 10,000,000, etc., he found that no perfect square ends in 40404, 6161616, 63636, 464646464 or 969696, thereby allowing him to speed further the search process. Searching all possible endings, he asserts that, if there is an undulating square, it must have more than 1,000 digits.

Dr. Helmut Richter from Germany is the world's most famous undulation hunter, and he has indicated to Dr. Googol that it is not necessary to restrict the "mod searches" to powers of 10, and that arbitrary primes work very well. He has searched for undulating squares with a million digits or fewer, using a Control Data Cyber 2000. No undulating squares greater than 69,696 have been found.

Randy Tobias of the SAS Institute in North Carolina notes that there are larger undulating squares in other number bases. For example, $292^2 = 85264 = 41414$ base 12. And 121 is an undulating square in any base. (121 base n is $(n + 1)^2$.) Interestingly, we find that there are very few undulating powers of *any* kind in base 10. For example, a 3-digit undulating cube is $7^3 = 343$. However, Randy Tobias conducted a search for other undulating powers and only found 343 as an undulant formed by raising a number to a power p. He has checked this for $3 \leq p \leq 31$ and for all undulants less than 10^{100}. Undulating powers are indeed rare!

Undulating prime numbers, on the other hand, are more common. For example, Randy has discovered the following huge and wondrous undulating prime:

$$7 + 720 \times (100^{49} - 1)/99 =$$

72,727,272,727,272,727,272,727,272,727,272,727,272,727,
272,727,272,727,272,727,
272,722,727,272,727,272,727,272,727,272,727

(It has 99 digits.) To find this monstrosity, he also used the software program called Maple. The program scanned numbers using two lines:

$$(a*10 + b)*(10^{**}(2*(k + 1)) - 1) / 99$$
$$a + 10*(a*10 + b)*(10^{**}(2*(k + 1)) - 1) / 99$$

for ($0 \leq k \leq 50$, $1 \leq a \leq n - 1$, $0 \leq b \leq n - 1$). The Maple "isprime()" function was used to check whether a number is prime. Maple makes it possible to work with very large integers.

There are many other undulating primes with many digits. However, there does not seem to be any undulating prime with an even number of digits. (Considering that *ababab* . . . *ab* = *ab* × 10101 . . . 01, we should not expect to find any even-digit undulating primes.) Dr. Googol would be interested in hearing from readers who have searched for undulating primes with larger periods of undulation, such as found in the prime number 5,995,995,995 (which does not finish its last cycle of undulation).

Finally, *binary undulants* are powers of 2 that alternate the adjacent digits 1 and 0 somewhere in their decimal expansion. For example, the "highest-quality" binary undulant Dr. Googol has found is 2^{949}. It has the undulating binary sequence 101010 in it, which he has placed in parentheses in the following:

2^{949} = 4758454107128905800095379999407968179242003264531006226897846994981(101010)291329399534453860638770032188735591612861751376145467278574369826493065785952766280250550668943187159661659651146975275798476542650352459905941679586200921628210271660911570586563854433745326052103604911620698931

Here 949 is called an undulation seed of order 6, since it gives rise to a 6-digit undulation pattern of adjacent 1s and 0s. When Dr. Googol challenged mathematicians and programmers around the world to produce a higher-order binary undulant, many took up the challenge. The highest-quality binary undulant known to humanity before 1999 was discovered by Arlin Anderson of Alabama. He was the first to find that 2^{1802} contains an 8-digit binary undulation. After much hard work he also found that $2^{7694891}$ starts with the digits 10101010173 . . . , and a week later he discovered that $2^{1748219}$ gives rise to a 10-digit undulant! Since Arlin only checked the last 240 digits of each number, he feels it is almost certain that there is a bigger binary undulation somewhere in the first million powers of 2. Considering that $2^{1000000}$ contains around 300,000 digits, the chance of finding a 10101010101 or 01010101010 is large. (Arlin uses a custom C program for large integer computation. The program runs on an Intergraph 6040 Unix workstation and on a 486 PC. Searching 240 digits in 2 million powers of 2 required 15 hours.)

How do binary undulants vary with the base b? For example, for the case of $b = 2$, there are many binary undulants. Is it possible that as b increases, the quality of the best-known cases decreases?

In 1999, John Rickert from the Rose-Hulman Institute of Technology, Terre Haute, Indiana, discovered a binary undulant with 2002 digits of alternating 1010s. The exponent $\mathcal{H}$ for the binary undulant contains 2,862 digits, starting with the digits 1705096307158733196 and ending with the digits 1125807122675. The actual number $2^{\mathcal{H}}$ has $\lceil (\mathcal{H} \times \log_{10}2) \rceil$ digits, which is approximately 5.13×10^{2860} digits. (The symbols $\lceil \; \rceil$ denote the "ceiling function," which returns the smallest integer that is greater than or equal to a given number. Example: $\lceil 354.89 \rceil = 355$.)

For the enjoyment of the most manic of number nuts (Dr. Googol uses the term affectionately), here is $\mathcal{H}$ for the highest quality binary undulant known to humanity.

The highest quality binary undulant $2^{\mathcal{H}}$, where $\mathcal{H}$ =

17050963071587331962728378228612105060442382423433454870975223797148036991369936655975353639891072630482641157828186359029820991488168447433761525655678571608557920944058798673593782911377366669546351489190863494454947342655857025935504410375649231763461155406520358487526755427708880034274877913501640662746917211495357857476813982969887847365750969933019190473988700011898925342420203177582189198933227175406031161265770695111130946015194112885693008990697664982100312300720089059642297716937873154185673729642438117742601611581634595197661492428596805529892215332020380214812095040648705149558815240262459529100488659975205662235761333910872007208294489339739748938777981104462432851817219555465542165715990153959957518916806863846678574872485590326817388071490819613391194537041518210625569483706040207238935794680940459508410604447944257741597541795694915129782864025936487733519652906407647612914532738588319806054869015748561529278954323110215197454586770412080600555759549437002917720718922384580574096371353334393810335515602730370465672725872230026969424082829743797382443578538036992484810269406575249920332773870187114331795969681227162256958258715780291621583061642005255751548634149387451439607614607256492913215435503733928418602529787600515490500541435425062593974734

86827839995162864170200245255794084792471777751182836994172761219727468852021797700556
04511259327105645654567950684223902816410000683068087706417907133196988411266740070397
70820296829307038284314334419893392828053536088792491315212435168532407879962837774521
98882475901716159079664581981622790171437820709675547562854519604996674782259152658519
39174726146152485454799202246588120754297479510225852020307434525255925025228248607077
88448306129146392378164745866612239357572911270630970671733024149136464545152557725758
52960063746704221461361084419600660432156939875196992454791963843006387475847313444983
82902622417251002030805947224672261728610217578066308585474210128558728622385990843 2
16241668321236760057247367596437121415254128093736730930649824428315207319831813255138
99081727550772864684097314321005106027160072068015255806517078038056279440480326694 89
73991008889855762804231571893997049785910776223651212725531243823515246138276349285 02
41206364914166781112837336913068353261741954733697691443654184363154197350080206652565
01302650159381837933032054494436275816611771863353959142017052414754627670165174686095
61338785103784395323632763334338726895577160830667699598638884287275549498347186214 84
57201353566687741734771799033369217079937796259748649436304492666502041192039282669 81
93377795312876300313802411564730617558537107076769263017470148211404251725661969285 517
39207901999156543983375651819473764800414818325857382865938729527561985756723479746 5
59441813329843722850664286036262185158373984850009009311454418001151878705965865357 42
80536623162470291591456865893007671125807122675

Chapter 88

The Latest Gossip on
Narcissistic Numbers

In this chapter, Dr. Googol discussed numbers of the kind:

$$153 = 1^3 + 5^3 + 3^3$$

He began to wonder if there were any cubes that are the sum of 3 consecutive cubes. Here is 1: $3^3 + 4^3 + 5^3 = 6^3$. Are there any others?

On a similar line of thought, *factorions* (denoted by the symbol F) are numbers that are the sum of the factorial values for each of their digits. (For a positive integer n, the product of all the positive integers less than or equal to n is called n factorial, usually denoted as $n!$ For example, $3! = 3 \times 2 \times 1$.) The number 145 is a factorion because it can be expressed as

$$145 = 1! + 4! + 5!$$

Two tiny examples of F are

$$1 = 1! \text{ and } 2 = 2!$$

The largest known F is 40,585; discovered in 1964 by R. Dougherty using a computer search, it can be written as

$$40,585 = 4! + 0! + 5! + 8! + 5!$$

Can you end the loneliness of the factorions? Do any others exist?

Various proofs have been advanced indicating that 40,585 is the largest possible F and that humans will never be able to find a greater F. In fact, these 4 factorions are the only factorions known to humanity. How can this be?

A more fruitful avenue of research may be the search for $F_{②}$ —factorions "of the second kind," which are formed by the *product* of the factorial values for each of their digits. Additionally, there are hypothetical $F_{③}$—factorions "of the third kind" formed by grouping digits. For example, a factorion of the third kind might have the form

$$abcdef = (ab)! + c! + d! + (ef)!$$

where each letter represents a digit. (Any groupings of digits are allowed for factorions of the third kind.)

Near-factorions $F^{\times}$ are n-digit numbers that are the sum of factorial values for n-1 of their digits. (For example, a number of the form $abc = a! + c!$ would be a $F^{\times}$.) Do they exist? To date, Dr. Googol is unaware of the existence of $F_{②}$, $F_{③}$, or $F^{\times}$, and he would be interested in hearing from readers who can find any.

Parenthetically, he should point out that Herve Bronninan from Princeton University has recently found some magnificent factorions in other bases, most notably 519,326,767, which in base 13 is written as 8.3.7.9.0.12.5.11 and is equal to 8! + 3! + 7! + 9! + 0! + 12! + 5! + 11! (You can interpret this base 13 number as $8 \times 13^7 + 3 \times 13^6 + 7 \times 13^5 + 9 \times 13^4 + 0 \times 13^3 + 12 \times 13^2 + 5 \times 13^1 + 11 \times 13^0$. Some write this number as $83790C5B_{13}$.)

This chapter also discussed the narcissistic number, **153**. **153** is special for other reasons:

⊙ $153 = 1! + 2! + 3! + 4! + 5$

⊙ When the cubes of the digits of any 3-digit number that is a multiple of 3 are added, and the digits of the resulting number are cubed and added, and the process continued, the final result is 153. For instance, start with 369, and you get the sequence 369, 972, 1080, 513, 153.

⊙ 153 is the seventeenth triangular number.

⊙ St. Augustine, the famous Christian theologian, thought that 153 was a mystical number and that 153 saints would rise from the dead in the eschaton. How is that? St. Augustine interpreted the Bible using numbers. For example, he was fascinated by a New Testament event (John 21:11) where the Apostles caught 153 fish from the sea of Tiberias. Seven disciples hauled in the fish, using nets. St. Augustine reasoned that these 7 were saints. Why 7 saints? Since there are 7 gifts from the Holy Ghost that enable people to obey the 10 Commandments, he thought the disciples must therefore be saints. Moreover, 10 + 7 = 17, and if we add together the numbers 1 through 17, we get a total of 153. The hidden meaning of all this is that 153 saints will rise from the dead after the world has come to an end.

As one searches for larger and larger narcissistic numbers, will they eventually run out, as in the case of the lonely factorions? If they are proved to die out in one number system, does this mean they are finite in another? (*News flash*: Martin Gardner wrote to Dr. Googol recently and indicated that the number of narcissistic numbers has been proved finite. They can't have more than 58 digits in our standard base 10 number system.)

Finally, Kevin S. Brown writes that he knows of only three occurrences of $n! + 1 = m^2$, namely $25 = 4! + 1 = 5^2$, $121 = 5! + 1 = 11^2$, and $5041 = 7! + 1 = 71^2$. We do not know if there are any others. Perhaps these "Brown numbers" will be as lonely as the factorions. The prolific mathematician Paul Erdös long ago conjectured that there are only 3 such numbers, and he offered a cash prize for a proof of this!

For various proofs relating to factorions, see my book *Keys to Infinity*.

⊞ See [www.oup-usa.org/sc/0195133420] for program code to search for factorions.

Chapter 89

The abcdefghij Problem

Using the program code at [www.oup-usa.org/sc/0195133420] we can compute values for the variables that satisfy the equation $(ab)c = def \times ghij$. Here are some possibilities: ($a = 4$, $b = 8$, $c = 3$, $d = 1$, $e = 9$, $f = 2$, $g = 0$, $h = 5$, $i = 7$, $j = 6$); ($a = 4$, $b = 8$, $c = 3$, $d = 5$, $e = 7$, $f = 6$, $g = 0$, $h = 1$, $i = 9$, $j = 2$); ($a = 8$, $b = 4$, $c = 3$, $d = 5$, $e = 7$, $f = 6$, $g = 1$, $h = 0$, $i = 2$, $j = 9$). Dr. Googol does not know if there are any solutions to a related problem: $(ab)^c = def \times ghij$.

Here's a much tougher challenge from mathematician Kevin Brown. The number 588,107,520 is expressible in the form $(X^2 - 1)(Y^2 - 1)$ (where X, Y are integers) in 5 distinct ways, and Kevin asks if anyone knows a 6-way-expressible number. So far, no 6-way-expressible number has been found, although such a number has not been proved impossible. Regarding 5-way numbers, Dean Hickerson and Fred Helenius both independently found 5 more, so as of now the complete list of 5-way expressible numbers is 588,107,520; 67,270,694,400; 546,939,993,600; 2,128,050,512,640; 37,400,697,734,400; and 5,566,067,918,611,200. Dr. Googol does not know if there are infinitely many such numbers, or even if there are any more beyond this list.

Dr. Googol leaves you with a final unsolved problem. For positive integers x, y, what are the solutions to equations of the form $axy + bx + cy = d$ where a, b, c, and d are integers?

Chapter 90

Grenade Stacking

As discussed in Laurent Beeckman's article in the May 1994 *American Mathematical Monthly*, if we allow *any* set of k consecutive squares (not necessarily beginning with 1), there are solutions for $k = 1, 2, 11, 23, 24, 33, 47, \ldots$ For each of these we have infinitely many sequences of k consecutive squares whose sum is a square. For example, with

$k = 24$ we not only have the previous sequence, $1^2 + 2^2 + 3^3 + 4^4 + \ldots + 24^2 = 70^2$, but we also have $9^2 + 10^2 + 11^2 + \ldots + 32^2 = 106^2$, $20^2 + 21^2 + 22^2 + \ldots + 43^2 = 158^2$, etc. Can you find others? In general, it seems that the sum of the 24 squares beginning with m^2 is a square for $m = 1, 9, 20, 25, 44, 76, \ldots$

Chapter 91

The 450-Pound Problem

If you select 2 random numbers, what is the chance they will be coprime? The answer is $6/\pi^2$. This is also the probability that a randomly selected integer is "square free" (not divisible by a square! Now that you have this "secret" knowledge, perhaps you can make some money gambling with your friends. Have them pick numbers at random from a pile of 2,000 cards with the numbers 1 through 200. None of you looks at the cards. Can you profit from your knowledge of the odds that the number is square free? (See Chapter 51's "For Further Reading" for more information on coprime numbers.)

Chapter 92

The Hunt for Primes in Pi

Mathematicians are aware of pi-primes, $\pi^\frown$, for $k = 1, 2, 6$, and 38, which correspond to the primes

$\pi^\frown(k) = 3, 31, 314159, 31415926535897932384626433832795028841, \ldots$?

Does anyone know the next $\pi^\frown$ in this sequence? Dr. Googol believes that there are infinitely many primes of the form $\pi^\frown(k)$ but that neither humans nor any lifeforms in the vast universe will ever know the next prime beyond $\pi^\frown(38)$. It is simply too large for our computers to find.

Martin Gardner in his book *Gardner's Whys and Wherefores* notes that several researchers have searched for "piback primes." Symbolized as $\pi^\frown$, these are primes in the first n digits of o running backwards. We would expect them to be more numerous than $\pi^\frown$, because all pibacks end in 3 (the first digit of π), one of the four numbers a prime must end with; the others are 1, 7, and 9. By contrast $\pi^\frown$ numbers can end in any number, which means only 40% of the numbers have a chance to be prime. Seven $\pi^\frown$ numbers have been found: 3, 13, 51413, 951413, 2951413, and 53562951413, and 979853562951413.

If you can find any *e*-primes, write to Dr. Googol.

Chapter 94

Perfect, Amicable, and Sublime Numbers

As Dr. Googol told Monica, the first 4 perfect numbers, 6, 28, 496, and 8,128, were known to the late Greeks and Nicomachus, a disciple of Pythagoras. Perfect numbers are indeed difficult to find.

The first 10 perfect numbers are

1. $2M_2 = 6$
2. $2^2M_3 = 28$
3. $2^4M_5 = 496$
4. $2^6M_7 = 8128$
5. $2^{12}M_{13} = 33550336$
6. $2^{16}M_{17} = 8589869056$ (Discovered in 1588 by Cataldi)
7. $2^{18}M_{19} = 137438691328$ (Discovered in 1588 by Cataldi)
8. $2^{30}M_{31} = 2305843008139952128$ (Discovered in 1772 by Euler)
9. $2^{60}M_{61}$ (Discovered in 1883 by Pervusin)
10. $2^{88}M_{89}$ (Discovered in 1911 by Powers)

The thirtieth perfect number, $2^{216090}M_{216091}$, was found using a Cray supercomputer in 1985 (see Table F94.1).

To understand this list of the first 10 perfect numbers, first note that perfect numbers can be expressed as $2^X(2^{X+1}-1)$ for special values of X. Euclid proved that this rule was sufficient for producing a perfect number, and Euler, 2,000 years later, proved that all even perfect numbers have this form, if 2^N-1 is a prime number. (In this notation, N is $X+1$.) Such numbers are called Mersenne prime numbers M_N after their inventor Marin Mersenne (1588–1648). For example, $127 = 2^7 - 1$ is the seventh Mersene number, denoted by M_7, and it is also prime and the source of the fourth perfect number, 2^6M_7. (Mersenne prime numbers are a special subclass of Mersenne numbers generated by 2^N-1).

Note that Table F94.1 rapidly becomes obsolete as more prime numbers are discovered at a rate of about 1 per year by computer searches such as GIMPS (see "Further Exploring" for Chapters 56 and 76).

Like many of the best mathematicians centuries ago, Marin Mersenne was a theologian. In addition, Father Mersenne was a philosopher, music theorist, and mathematician. He was a friend of Descartes, with whom he studied at a Jesuit college. Mersenne discovered several prime numbers of the form 2^N-1, but he underestimated the future of computing power by stating that all eternity would not be sufficient to decide if a 15- or 20-digit number were prime. Unfortunately the prime number values for N that make 2^N-1 a prime number form no regular sequence. For example, the number is prime when $N = 2,3,5,7,13,17,19,\ldots$. Notice that when N is equal to the prime number 11, $M_{11} = 2,047$ which is not prime because $2,047 = 23 \times 89$.

In 1814, P. Barlow in *A New Mathematical and Philosophical Dictionary* wrote that the eighth perfect number was "probably the greatest perfect number that ever will be

		$2^{N-1}(2^N-1)$	
Number	N	Discovered (year,	human)
1–4	2,3,5,7	in or before the middle ages	
5	13	in or before 1461	
6–7	17,19	1588	Cataldi
8	31	1750	Euler
9	61	1883	Pervouchine
10	89	1911	Powers
11	107	1914	Powers
12	127	1876	Lucas
13–17	521,607,1279,2203,2281	1952	Robinson
18	3217	1957	Riesel
19–20	4253,4423	1961	Hurwitz & Selfridge
21–23	9689,9941,11213	1963	Gillies
24	19937	1971	Tuckerman
25	21701	1978	Noll & Nickel
26	23209	1979	Noll
27	44497	1979	Slowinski & Nelson
28	86243	1982	Slowinski
29	110503	1988	Colquitt & Welsh
30	132049	1983	Slowinski
31	216091	1985	Slowinski
32?	756839	1992	Slowinski & Gage
33?	859433	1993	Slowinski

Table F94.1 Several Perfect Numbers.

discovered for they are merely curious without being useful, and it is not likely that any person will attempt to find one beyond it." Barlow placed such a limit on human knowledge because, before computers, the discovery of Mersenne primes depended on laborious human computations. M_{31} or $2^{31}-1 = 2,147,483,647$ is quite large, even though Euler in 1772 was able to ascertain that it is a prime number.

With the electronic computer, Barlow's limit on humanity's knowledge was rendered invalid. Because of their special form, Mersene numbers are easier to test for primality then other numbers, and therefore all the recent record-breaking primes have been Mersenne numbers—and have automatically led to a new perfect number.

There is a bizarre and puzzling relationship between cubes and perfect numbers. Every even perfect number, except 6, is the sum of the cubes for consecutive odd numbers. For example:

$$28 = 1^3 + 3^3$$
$$496 = 1^3 + 3^3 + 5^3 + 7^3$$
$$8,128 = 1^3 + 3^3 + 5^3 + 7^3 + 9^3 + 11^3 + 13^3 + 15^3$$

▣ [www.oup-usa.org/sc/0195133420] contains a BASIC program listing for computing perfect numbers.

Odd perfect numbers are even more fascinating then even ones for the sole reason that no one knows if odd perfect numbers exist. They may remain forever shrouded in mystery. On the other hand, mathematicians have cataloged a long list of what we *do* know about odd perfect numbers; for example, computer searches as far as 10300 have not found an odd perfect number. Mathematician Albert H. Beiler says, "If an odd perfect number is ever found, it will have to have met more stringent qualifications than exist in a legal contract, and some almost as confusing." Here are just a few:

An odd perfect number

⊙ must leave a remainder of 1 when divided by 12 or a remainder of 9 when divided by 36.

⊙ It must have at least 6 different prime divisors.

⊙ If it is not divisible by 3, it must have at least 9 different prime divisors.

⊙ If it is less than 10^{9118}, it is divisible by the 6th power of some prime.

Author and mathematician David Wells comments, "Researchers, without having produced any odd perfects, have discovered a great deal about them, if it makes sense to say that you know a great deal about something that may not exist." Throughout both ancient and modern history, the feverish hunt for perfect numbers became a religion. The mystical significance of perfect numbers reached a feverish peak around the 17th century. Peter Bungus, for example, was among a growing number of 17th-century mathematicians who combined numbers and religion. In his alchemic book titled *Numerorum Mysteria*, he listed 24 numbers said to be perfect, of which Mersenne later stated that only 8 were correct. Mersenne went on to add 3 more perfect numbers, for $N = 67$, 127, and 257, in an equation that can be used for *even* perfect numbers $(2^{N-1})(2^N - 1)$, but it took a walloping 303 years before mathematicians could check Mersenne's statement to find errors in it. 67 and 257 should not be admitted, and perfect numbers corresponding to $N = 89$ and 107, for which the Mersenne numbers are prime, should be added to the list.

How could Mersenne, back in the 17th century, have conjectured about the existence of such large perfect numbers? After centuries of debate, no one has an answer. Could he have discovered some theorem not yet rediscovered? Recall that empirical methods of his time could hardly have been used to compute these large numbers. (The Mersenne number for $N = 257$ has 78 digits.)

Zealous attempts at perfection are not limited to Peter Bungus and Mersenne. Even in the 1900s there have been startling attempts to find the Holy Grail of huge perfect numbers. For example, on March 27, 1936, newspapers around the world trumpeted Dr. S. I. Krieger's discovery of a 155-digit perfect number $(2^{256}(2^{257} - 1))$. He thought he had proved that $2^{257} - 1$ is prime. The Associated Press release, appearing in the *New York Herald Tribune*, was as follows:

Unfortunately for Dr. Krieger, a few years earlier the number $2^{257} - 1$ had been found to be composite (nonprime). Editors of Mathematical journals therefore wrote letters to the *New York Herald Tribune* complaining that it had sacrificed accuracy for sensationalism in reporting the Krieger story.

PERFECTION IS CLAIMED FOR 155-DIGIT NUMBER
Man Labors 5 Years to Prove Problem Dating from Euclid

New York Herald Tribune, March 27, 1936

Chicago, March 26 (AP).—Dr. Samule I. Krieger laid down his pencil and paper today and asserted he has solved a problem that had baffled mathematicians since Euclid's day—finding a perfect number of more than nineteen digits.

A perfect number is one that is equal to the sum of its divisors, he explained. For example, 28 is the sum of 1, 2, 4, 7, and 14, all of which may be divided into it. Dr. Krieger's perfect number contains 155 digits. Here it is:

26, 815, 615, 859, 885, 194, 199, 148, 049, 996, 411, 692, 254, 958, 731, 641, 184, 786, 755, 447, 122, 887, 443, 528, 060, 146, 978, 161, 514, 511, 280, 138, 383, 284, 395, 055, 028, 465, 118, 831, 722, 842, 125, 059, 853, 682, 308, 859, 384, 882, 528, 256.

Its formula is 2 to the 513th power minus 2 to the 256th power. The doctor said it took him seventeen hours to work it out and five years to prove it correct.

El Madshriti, an Arab of the 11th century, experimented with the erotic effects of amicable numbers by giving a beautiful woman the smaller number 220 to eat in the form of a cookie, and himself eating the larger 284! I am not sure whether his mathematical approach to winning the woman's heart was successful, but this method may be of interest to all modern dating services. Imagine restaurants of the future branding the numbers into 2 pieces of filet mignon for 2 prospective marriage candidates. Perhaps amicable-number tattoos will one day be used for mathematical displays of public affection.

Our Arab friend, El Madshriti, was not the last to make use of amicable numbers to unite the sexes. In the 14th century, the Arab scholar Ibn Khaldun said in reference to amicable numbers:

Persons who occupy themselves with talismans assure that these numbers have a particular influence in establishing union and friendship between two individuals. One prepares a horoscope theme for each individual. On each, one inscribes one of the numbers just indicated, but gives the *strongest number* to the person whose friendship one wishes to gain. There results a bond so close between the two persons that they cannot be separated.

When Ibn Khaldun used the term *strongest number*, he was not certain whether to use the larger of the 2 amicable numbers or the one that had the most divisors.

Since antiquity, Arabs have been interested in different ways of finding amicable numbers. One personal favorite is taken from the Arabian mathematician-astronomer Thabet ben Korrah (A.D. 950). Select any power of 2, such as 2^x, and form the numbers

$$a = 3 \times 2^x - 1$$
$$b = 3 \times 2^{x-1} - 1$$
$$c = 9 \times 2^{2x-1} - 1$$

If these are all primes, then $2^x ab$ and $2^x c$ are amicable. When x is 2, this gives the numbers 220 and 284. ([www.oup-usa.org/sc/0195133420] contains a BASIC program listing for computing amicable numbers.)

The number 672 is one of many *multiply perfect numbers*—numbers such that the sum of *all* their divisors is an exact *multiple* of the number. For example, 120 is a triple perfect number because its divisors $1 + 2 + 3 + 4 + 5 + 6 + 8 + 10 + 12 + 15 + 20 + 24 + 30 + 40 + 60 + 120$ add up to 360, which is 3×120. Similarly, 672 is a triple perfect number.

There have been several recent attempts to explain the mysterious title of Hugo von Hoffmannsthal's tale *The Story of the 672nd Night*. Hugo von Hoffmannsthal (1874–1929), was an Austrian poet, dramatist, and essayist, best known for writing libretti for Richard Strauss's operas. One explanation for his title is the fact that 672 is a multiply perfect number, but literary scholars are not certain that this is Hoffmannsthal's reason for using 672. (Some scholars suspect that the 672 in *The Story of the 672nd Night* is connected with the tale *1001 Arabian Nights*.)

As already mentioned, an even perfect number has the form $2^{N-1}(2^N - 1)$. Harry J. Smith of Saratoga, California, wrote a program using Borland C++ to compute a perfect number if given the exponent of a Mersenne prime. For a large perfect number ($N = 859,433$, see Table F94.1), his result is an output file 530,462 bytes long.

No one knows if perfect numbers eventually die out as one sifts through the landscape of numbers. The mathematical landscape is out there, waiting to be searched. The Pythagoreans could find only 4 perfect numbers, and we can find over 30. Will humanity ever discover more than 40 perfect numbers? There is a limit on humans' mathematical knowledge arising not only from our limited brains but also from our limited computers. In a strange way, the "total" of mathematical knowledge is godlike— unknowable and infinite. As we gain more mathematical knowledge, we grow closer to this god, but can never truly reach him. All around us we catch glimpses of a hidden harmony in the works of humans and nature. From the Great Pyramid of Cheops to patterns in plants, we see evidence of design by precise geometrical laws. Nobly, we continue to search for the connections underlying all that is beautiful and functional.

5,775 and 5,776 are 2 consecutive abundant numbers. Is it possible to find 3 consecutive abundant numbers? It was not until 1975 that the smallest triplet of consecutive abundant numbers was discovered (by Laurent Hodges and Michael Reid):

$$171,078,830 = 2 \times 5 \times 13 \times 23 \times 1973$$
$$171,078,831 = 3^3 \times 7 \times 11 \times 19 \times 61 \times 71$$
$$171,078,832 = 2^4 \times 21 \times 344,917$$

Chapter 96

Cards, Frogs, and Fractal Sequences

There are many definitions of fractal, or self-similar, sequences; the one that seems to fit some of the sequences in this chapter is given by Benoit Mandelbrot in his *The Fractal Geometry of Nature:* "An unbounded set S is self-similar with respect to the ratio r, when the set $r(S)$ is congruent to S." Let me give some examples. Consider a sequence of integers x_1, x_2, x_3, x_4, x_5, . . . as in the previous example, 1, 1, 2, 1, 3, . . . This sequence is self-similar with respect to the ratio 2, because x_2, x_4, x_6, . . . is identical to x_1, x_2, x_3,. . . . Of course, we can generalize and say a sequence is self-similar with respect to the ratio r (r an integer greater than 1) if there is some integer d, $1 \le d \le r$, for which x_d, $x_{(r+d)}$, $x_{(2 \times r+d)}$, $x_{(3 \times r+d)}$, $x_{(4 \times r+d)}$, . . . is identical to x_1, x_2, x_3, x_4, x_5,. . . . For instance, with $r = 4$ we would have every fourth entry of the sequence, and starting with x_1 (and $d = 1$), x_1, x_5, x_9, x_{13}, . . . is the same as x_1, x_2, x_3, x_4, x_5,. . . . Or starting with x_2, we find x_2, x_6, x_{10}, x_{14}, . . . is the same as x_1, x_2, x_3, x_4, x_5,. . . .

In this chapter, I also consider fractal-like sequences that consist of any string that contains copies of itself, even if the string doesn't quite conform to the above rules. For example, consider the letter string:

a, b, a, c, b, a, d, c, b, e, a, d, c, f, b, e, a, d, g, c, f, b, e . . .

If you delete the first occurrence of each letter, you'll see that the remaining string is the same as the original.

a̶,̶ b̶,̶ a, e̶,̶ b, a, d̶,̶ c, b, e̶,̶ a, d, c, f̶,̶ b, e, a, d, g̶,̶ c, f, b, e . . .

I refer to this type of sequence as fractal-like because, like most fractals, it has "parts that resemble the whole."

To arrive at a traditional definition of signature sequence, let θ be an irrational number; $S(\theta) = \{c + d\theta : c, d, \in \mathbb{N}\}$ and let $c_n(\theta) + d_n(\theta)(\theta)$ be the sequence obtained by arranging elements of $S(\theta)$ in increasing order. A sequence x is said to be a signature sequence if there exists a positive irrational number θ such that $x = \{c_n(\theta)\}$, and x is called the signature of θ. The signature of an irrational number is considered a fractal sequence according to various literature (for example, in C. Kimberling's paper in the reference section).

Fractal signature sequences: Here are the first few terms for some miscellaneous fractal signature sequences computed by David E. Shippee of Littleton, Colorado.

Number Signature Sequence

Number	Signature Sequence
$0.55000000 = 11/20$	1 1 2 1 2 1 3 2 1 3 2 1 4 3 2 1 4 3 2 1 5 4 3 2 1 5 4 3 2 1 6 5 4 3 2
$0.707106781 = \sqrt{1/2}$	1 1 2 1 2 3 1 2 3 1 4 2 3 1 4 2 5 3 1 4 2 5 3 1 6 4 2 5 3 1 6 4 2 7 5
$1.0498756 = \sqrt{101} - 9$	1 2 1 3 2 1 4 3 2 1 5 4 3 2 1 6 5 4 3 2 1 7 6 5 4 3 2 1 8 7 6 5 4 3 2
$1.10000000 = 1 + 1/10$	1 2 1 3 2 1 4 3 2 1 5 4 3 2 1 6 5 4 3 2 1 7 6 5 4 3 2 1 8 7 6 5 4 3 2
$1.41421356 = \sqrt{2}$	1 2 1 3 2 1 4 3 2 5 1 4 3 6 2 5 1 4 7 3 6 2 5 8 1 4 7 3 6 9 2 5 8 1 4
$1.50000000 = 1 + 1/2$	1 2 1 3 2 4 1 3 5 2 4 1 6 3 5 2 7 4 1 6 3 8 5 2 7 4 1 9 6 3 8 5 2 10 7
$1.73205081 = \sqrt{3}$	1 2 1 3 2 4 1 3 5 2 4 6 1 3 5 7 2 4 6 1 8 3 5 7 2 9 4 6 1 8 3 10 5 7 2
$2.23606798 = \sqrt{5}$	1 2 3 1 4 2 5 3 1 6 4 2 7 5 3 1 8 6 4 2 9 7 5 3 1 10 8 6 4 2 11 9 7 5 3
$2.71828183 = e$	1 2 3 1 4 2 5 3 6 1 4 7 2 5 8 3 6 9 1 4 7 10 2 5 8 11 3 6 9 11 2 4 7 10
$3.10000000 = \pi$ to 1 decimal	1 2 3 4 1 5 2 6 3 7 4 1 8 5 2 9 6 3 10 7 4 1 11 8 5 2 12 9 6 3 13 10 7 4
$3.14000000 = \pi$ to 2 decimals	1 2 3 4 1 5 2 6 3 7 4 1 8 5 2 9 6 3 10 7 4 1 11 8 5 2 12 9 6 3 13 10 7 4
$3.14100000 = \pi$ to 3 decimals	1 2 3 4 1 5 2 6 3 7 4 1 8 5 2 9 6 3 10 7 4 1 11 8 5 2 12 9 6 3 13 10 7 4
$3.14160000 = \pi$ to 4 decimals	1 2 3 4 1 5 2 6 3 7 4 1 8 5 2 9 6 3 10 7 4 1 11 8 5 2 12 9 6 3 13 10 7 4
$3.14159265 = \pi$ to 8 decimals	1 2 3 4 1 5 2 6 3 7 4 1 8 5 2 9 6 3 10 7 4 1 11 8 5 2 12 9 6 3 13 10 7 4
$7.07106781 = \sqrt{50}$	1 2 3 4 5 6 7 8 1 9 2 10 3 11 4 12 5 13 6 14 7 15 8 1 16 9 2 17 10 3 18
$10.0498756 = \sqrt{101}$	1 2 3 4 5 6 7 8 9 10 11 1 12 2 13 3 14 4 15 5 16 6 17 7 18 8 19 9 20 10

As far as Dr. Googol can tell, all sequences are fractal. Irrational numbers appear to yield unique signatures, but rational numbers do not. For example, examine the signature sequence for 1.5 (1, 2, 1, 3, 2, 4, 1, 3, 5, 2 . . .). This could just as easily be 1, 2, 1, 3, 2, 1, 4, 3, 2, 5 . . . because 4 + 1 × 1.5 = 1 + 3 × 1.5, so the 4, 1 in the first sequence could just as easily be the 1, 4 in the second sequence. David E. Shippee included sequences for 3.1, 3.14, 3.141, and 3.1416 to see how the sequences might converge. Their signatures are all identical. (He used an upper limit of 30 for i and j, giving 900 entries in the sequence.) It seems that one must have many entries to see a distinction; i.e., the sequences converge slowly.

Batrachions: Let us now consider how fast the frog approaches its 0.5 destination at infinity. For example, can you find a value of n beyond which the value of $a(n)/n$ is so tiny that it is forever within 0.05 from the value 1/2? (In other words, $|a(n)/n-1/2|$ < 0.05. The bars indicate the absolute value.)

A difficult problem? John Conway, the prolific British mathematician, offered **$10,000** to the person who could find the first value of n such that the frog's path is always less than 0.55 for higher values of n. A month after Conway made the offer, Colin Mallows of AT&T solved the $10,000 question: $n = 1,489$. Figure F96.1 shows this value on a plot for $0 < n < 10,000$. (For a variety of minor technical reasons, a less accurate number is published in Schroeder's book.) As Dr. Googol dictates this, no one on the planet has found a value for the smallest n such that $a(n)/n$ is always within 0.001 of the value 1/2, that is, $(|a(n)/n-1/2| < 0.001)$. (No one even knows if such a value exists.)

Looking at Figure F96.1, we can see that the frog "hits the pond" periodically. In fact, $a(n)/n$ "hits" 0.5 at values corresponding to powers of 2, for example, at 2^k, $k = 1$, 2, 3, . . . Does each hump reach its maximum at a value of n halfway between the 2^k and 2^{k+1} end points?

Tal Kubo from the Mathematics Department at Harvard University is one of the world's leading experts on this batrachion. He notes that the sequence is subtly connected with a range of seemingly unrelated topics in mathematics: variants of Pascal's triangle, the Gaussian distribution, combinatorial operations on finite sets, and Catalan

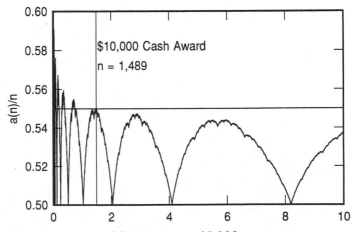

F96.1 Batrachion $a(n)/n$ **for** $0 < n < 10,000$.

ε	Last n such that $|a(n)/n - 1/2| > \epsilon$
1/20	1489 (found by Mallows in 1988)
1/30	758765
1/40	6083008742 (found by Mallows in 1988)
1/50	809308036481621
1/60	1684539346496977501739
1/70	55738373698123373661810220400
1/80	1508884187519093848482894842861205283
1/90	127565909103887972767169084026274554426122918035
1/100	88266080011270776195815899395505310219430599069671270070225

Table F96.1 The Infinite Frog.

numbers. Tal Kubo and Ravi Vakil have developed algorithms to compute the behavior of the batrachion as it nears infinity. Indeed, they have found that the frog tires rather slowly! For example, the frog's jumps are not always less than 0.52 until it has jumped 809,308,036,481,621 times!

Table F96.1 lists the values for different frog jump heights. These values were found by Tal Kubo and Ravi Vakil using a Mathematica program running on a Sun 4 computer.

Colin Mallows, the statistician who conducted the first in-depth study of this class of curve, notes that no finite amount of computations will suffice to prove that the regularities we see in the curve persist indefinitely. He does note that the difference between successive values is either 0 or 1. Is this true indefinitely?

For a variety of novel ways to visualize these sequences, see my book *Keys to Infinity*. Interestingly, it is not clear how one hump in the batrachion is generated from the previous hump. As Mallows has pointed out, $a(100)$, which is located in the sixth hump, is computed as $a(a(99)) + a(100 - a(99)) = a(56) + a(44) = 31 + 26 = 57$. This shows that a point in hump 6 is generated from two points in hump 5 that are far apart.

Various authors, such as Manfred Schroeder, have discussed how mathematical waveforms sound when converted to time waveforms and played as an audio signal. For example, Weierstrass curves (which are continuous but quite jagged) are a rich mine of paradoxes. They're produced by $w(t) = \Sigma_{k=1}^{\infty} A^k \cos B^k t$ where $AB > 1 + 3\pi/2$. If they are recorded on audio tape and replayed at twice the recording speed, the human ear will unexpectedly hear a sound with a lower pitch. Other fractal waveforms do not change pitch at all when the tape speed is changed. It is rumored (but Dr. Googol has not confirmed) that the first batrachion described in this chapter produces a windy, crying sound when converted to an audio waveform. He would be interested in hearing from readers who have conducted such audio experiments on any of the Batrachions. For other musical mappings of number sequences and genetic sequences to sound, see my book *Mazes for the Mind: Computers and the Unexpected*.

```
—  26 27 28 29 — 33 34 35 36 37 38 — 44 45 46 47 — — —
—  25 — — 30 31 32 — — — — 39 42 43 — — 48 — — —
—  24 — — — — — — — — 40 41 — — — 49 50 — —
—  23 —                                   51 — —
21 22 —                                   52 53 54
20 19 —                                   57 56 55
—  18 17          Doughnut Puzzle         58 — —
—  — 16              Solution             59 — —
13 14 15                                  60 61 —
12 — —                                    — 62 63
11 10 9                                   — 65 64
—  — 8                                    — 66 —
—  — 7 — 3 2 — — — — — — — — — — — — — 67 68
—  — 6 5 4 1 — — — — — — — — — — — — — — 69
—  — — — — — — — — — — — — — — — — — — 70
```

Table F98.1 Doughnut Loop Solution.

Chapter 98

Doughnut Loops

Dr. Googol believes the solution in Table F98.1 is the best solution for the doughnut puzzle. Can you find equally long or longer solutions? The maximal path length seems to be 70. In the schematic illustration of the path, the first position of the sequence is marked 1, the second 2, and so on, and the last is marked 70. Assuming the upper left corner to be (1,1) and the lower right (20,15), then this sequence starts at (6,14) and ends at (20,15). The first few numbers on the path are 6 — 34 — 37 — 25 — 15 — 70 — 26 — 20 — 43 — 60 — 9 — 54 —. . . . Since the 54 is the twelfth number in this sequence, its position (1,10) is marked 12 in the solution diagram.

Chapter 99

Everything You Wanted to Know about Triangles but Were Afraid to Ask

Pythagorean triangles with integral sides have been the subject of a huge amount of mathematical inquiry. For example, Albert Beiler, author of *Recreations in the Theory of Numbers*, has been interested in Pythagorean triangles with large consecutive leg values. These triangles are as rare as diamonds for small legs. Triangle 3-4-5 is the first of these exotic gems. The next such one is 21-20-29. The tenth such triangle is quite large: 27304197-27304196-38613965.

You can compute these "praying triangle" leg lengths using the BASIC program listing at [www.oup-usa.org/sc/0195133420]. The recipe is as follows. Start with 1 and multiply by a constant $D = (\sqrt{2} + 1)^2 = 5.828427125\dots$. Truncate the result to an integer value and multiply again by D. Continue this process for as long as you like, creating a list of integers: 1, 5, 29.... To produce the leg-length values for praying triangles, pick 1 of these integers, square it, divide by 2, and then take the square root. The 2 leg lengths are produced by rounding up and rounding down the result.

Now let's discuss "divine triangles." In 1643, French mathematician Pierre de Fermat wrote a letter to his colleague Mersenne asking for a Pythagorean triangle the sum of whose legs and whose hypotenuse were squares. In other words, if the sides are labeled X, Y, and Z, this requires

$$X + Y = a^2$$
$$Z = b^2$$
$$X^2 + Y^2 = Z^2 = b^4$$

It is difficult to believe that the *smallest* 3 numbers satisfying these conditions are $X = 4,565,486,027,761$, $Y = 1,061,652,293,520$, and $Z = 4,687,298,610,289$. Dr. Googol has called triangles of this rare type divine triangles because only a god could imagine another solution to this problem. Why? It turns out that the second triangle would be so large that if its numbers were represented as feet, the triangle's legs would project from Earth to beyond the Sun!

If the ancient Greek mathematician Pythagoras had been told that a race of beings could compute the values for the sides of the second divine triangle, surely he would have believed such beings were gods. Yet today we can compute such a triangle. We have become Pythagoras's gods. We have become gods through computers and mathematics.

Dr. Googol and Mr. Clinton also discussed the interesting general problem of finding Pythagorean triangles with integer values for the sides. A related but fiendishly more difficult task involves searching for solutions to the "integer brick problem." Here one must find the dimensions of a 3-dimensional brick such that the distance between any 2 vertices is an integer. In other words, you must find integer values for a, b, and c (which represent the lengths of the brick's edges) that produce integer values for the various diagonals of each side: d, e, and f. In addition, the 3-dimensional diagonal g spanning the brick must also be an integer. This means that the following equations must have an integer solution:

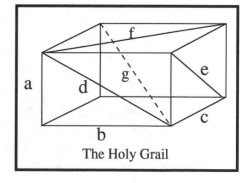

The Holy Grail

$$a^2 + b^2 = d^2$$
$$a^2 + c^2 = e^2$$
$$b^2 + c^2 = f^2$$
$$a^2 + b^2 + c^2 = g^2$$

No solution has been found. However, mathematicians haven't been able to prove that no solution exists. Many solutions have been found with only 1 noninteger side.

Chapter 105

Alien Ice Cream

Wasn't this a killer problem? You can make your own Alien Ice Cream game by changing the instructions but using the same illustration. To solve the problem, go up the stairway at right connecting the ground floor with the second floor. Go through the door. Go out the window and down the ladder. Go up to the third floor using the fire escape stairs. Go down the ladder between the third floor and second floor. Go up the spiral staircase. Go up the ladder to the roof.

The numbers in Figure F105.1 should help guide you.

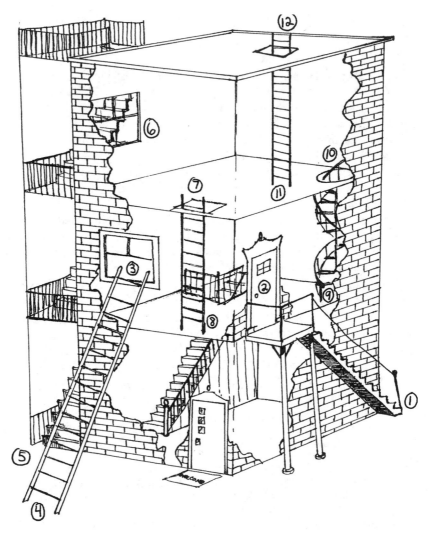

F105.1 Alien Ice Cream. Follow the numbers.

Chapter 106

The Huascarán Box

For the first problem, turn on the red finger for 10 seconds. Turn off the red finger and turn on the green finger. Quickly open the box. If the fan is continually spinning, then the green finger is the one. If the fan is spinning but slows down, then it is connected to the red finger. Otherwise, it is the yellow finger. (Physicist Dick Hess of Rancho Palos Verdes, California, proposed a similar problem in the 1998 *Pi Mu Epsilon Journal*, vol. 10, no. 8, p. 660.)

For the second problem, turn on the red switch and pour some paprika into the hole above the fan. Next, turn off the red switch and wait a while. Next, turn on the green and blue switches. Then, as before, switch off the green and immediately open the box and look. Dr. Googol's colleague Jim McLean points out that you now have 4 possibilities:

1. Fan is turning steadily—blue switch controls.

2. Fan is slowing down and stopping—green switch controls.

3. Fan is stopped, Peruvian paprika is strewn about—red switch controls.

4. Fan is stopped, Peruvian paprika is in a small pile—golden switch controls (no fan has ever been on).

Chapter 107

The Intergalactic Zoo

To be certain that he has 2 animals of the same species, the alien must drop 4 animals— 1 more than the number of different species. To be certain he has a male-female pair of the same species, he must drop 12 animals—1 more than the total number of animal pairs. Didn't get these answers? Try writing each animal's species and gender on separate scraps of paper. Then put all the papers in a box and withdraw them, 1 at a time, without looking. Now that you see how it's done, can you think of other "animal and alien" puzzles?

Incidentally, various authors render the quote at the beginning of this chapter in several flavors. (**"A mathematician is a blind man in a dark room looking for a black cat which isn't there."**) Instead of *mathematician*, some books use *philosopher*. Some authors attribute it to "anonymous" rather than Darwin. Dr. Googol wonders about its true source. Another interesting version floating around the Internet is "A theologian is like a blind man in a dark room searching for a black cat which isn't there—and finding it!"

Chapter 108

The Lobsterman from Lima

No, the lobster does not weigh 15 pounds. One good way to have students work on this problem is to visualize a balance scale. The lobster is on the left side. On the right side are a 10-pound weight and half a lobster. The scale is perfectly balanced. Stop and draw the scale now. Now look at the right side of your balance. Notice that the 10-pound weight is in essence taking the place of half the lobster. That means another 10-pound weight could take the place of the lobster-half. By looking at the drawing, you can see that the lobster weighs 20 pounds. If you are a teacher, you could have your class try to figure this out with algebra, but more important, try to show your class the value of visualization in problem solving. There's nothing quite like drawing a diagram to illustrate a problem before you attempt to solve it.

Now for a real killer question:

> **if the lobster weighs 10 pounds plus twice its own weight, how much does it weigh?**

Can you solve this without resorting to a pencil and paper? Do you see any possible problems with this?

Chapter 109

The Incan Tablets

The second pair completes the set because this pair completes every possible pair of the 4 symbols. Perhaps there are other equally valid solutions?

Chapter 110

Chinchilla Overdrive

Hello. The relevant equation is $L + 10 = 5L - 2$. The answer is 3.

Chapter 111

Peruvian Laser Battle

Figure F111.1 shows a solution. Are there other solutions?

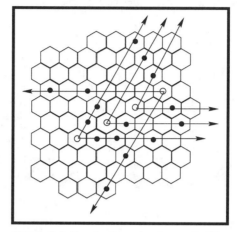

F111.1 Solution to Peruvian Laser Battle.

Chapter 112

The Emerald Gambit

Figure F112.1 shows one solution. Can you find others?

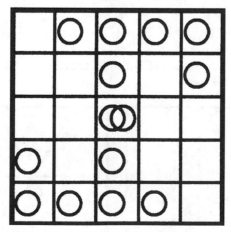

F112.1 One solution to the Emerald Gambit.

Chapter 113

Wise Viracocha

Figures 113.1 and 113.2 show solutions to the puzzles. Can you find others? Try to design other Viracocha puzzles using other coin shapes—for example, triangular, pentagonal, and hexagonal.

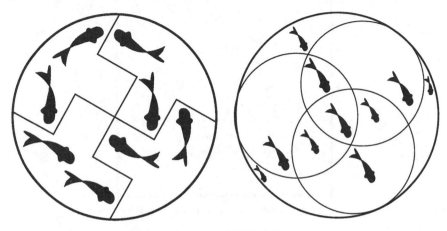

F113.1 Solution to Viracocha's coin.   **F113.2 Solution to Viracocha's pizza.**

Chapter 114

Zoologic

In Figure F114.1, Mr. Gila walks along 47 paths, or 4,700 feet. The path he chooses hits these enclosures in sequence: 18, 20, 19, 17, 18, 20, 21, 13, 14, 10, 9, 5, 6, 10, 11, 7, 6, 2, 3, 7, 8, 12, 11, 15, 14, 22, 23, 15, 16, 12, 8, 4, 3, 2, 1, 5, 9, 13, 21, 22, 23, 24, 16, 28, 25, 26, 27, and 28. As you can see, in several instances he must travel a path twice. Can you find a shorter route?

If Mr. Gila places the 19 panes of glass in the manner shown in Figure F114.2, he will have 10 enclosures of equal size.

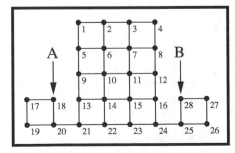

F114.1 Mr. Gila's walk.

F114.2 10 enclosures of equal size.

Chapter 115

Andromeda incident

In Figure F115.1, the 3 saucers have taken up new positions, as indicated by the arrows, and still no 2 saucers are in a straight line. Are there any other solutions?

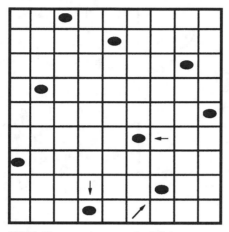

F115.1 New arrangement of flying saucers.

Chapter 116

Yin or Yang

The puzzle is actually based on an ancient problem. Figure F116.1 is the only solution of which Dr. Googol is aware. To satisfy yourself that the pieces are in fact the same size and shape, you can draw this pattern on a piece of paper, cut out the pieces, and superimpose them on one another.

It's also possible for the children to divide the yin and the yang into 4 pieces with the same area but *different* shapes by a single extra cut. Can you figure out how?

F116.1 The chocolate/vanilla cake.

Chapter 117

A Knotty Challenge at Tacna

To solve this knotty problem, consider that there are two possible crossings at each intersection point. This means that there are 2 × 2 × 2 = 8 possible sets of crossings. Of

all these possibilities, only 2 create a knot. (Test this for yourself using a loop of string.) This means that the probability of having a knot is ¼. Don't bet on it happening!

Figure F117.1 shows another possible rope configuration. What are the odds that it forms a knot? Does the probability of knot formation increase with increasing numbers of intersection points? What does this say about "Murphy's Law"—that ropes and strings and electrical cords always seem to get tangled when thrown in a jumble in your garage?

F117.1 Another rope configuration. What are the odds that it forms a knot?

Chapter 118

An Incident at Chavín de Huántar

To decode the "keys to the universe," you must substitute an English letter for each symbol. Rest in peace.

Chapter 119

An Odd Symmetry

You fool! There are no identical positive integers you can put in the mailboxes that will make this work beyond the second row, ⬛+⬛ = ⬛ × ⬛. And the only solution for this row is 2 + 2 = 2 × 2. This problem is so much fun because the solutions drop from infinity to 1 to 0 so quickly.

For example, consider the third line, ⬛+⬛+⬛ = ⬛ × ⬛ × ⬛. Mathematically speaking, we are trying to find values for a in the equation $a + a + a = a \times a \times a$. This is equivalent to $3a = a^3$, which is equivalent to $a^2 = 3$, which has no integer solutions. By induction, we are trying to solve $a^{n-1} = n$. One simple way to determine if this can have integer solutions for higher values of n is to make a graph of $y = a^{n-1}$ and a graph of $y = n$ (which is just a straight line) and see where the 2 lines intersect. As mathematician Dan Winarski points out, after $n = 2$, a^{n-1} is greater than n for all integers greater than or equal to 2. Thus, there are no more integer solutions.

Here is a related problem, developed by Dr. Googol's friend Craig Becker. Are there many solutions to the growing pyramid below? To solve this, use any positive integers that you like.

$$a = a$$
$$a + b = a \times b$$
$$a + b + c = a \times b \times c$$
$$a + b + c + d = a \times b \times c \times d$$
$$a + b + c + d + e = a \times b \times c \times d \times e$$
$$a + b + c + d + e + f = a \times b \times c \times d \times e \times f$$
$$\ldots \text{etc.} \ldots$$

As David Shippee points out, each row has at least 1 solution. For a row with n terms on each side of the equals sign, 1 solution involves the following sequence: $(n-2)$ "1s", 2, n. For example, here is a list of solutions for $n = 2$ to 5:

a	b	c	d	e
2	2			
1	2	3		
1	1	2	4	
1	1	1	2	5

Dr. Googol does not know if there are lots of other solutions, or if there are solutions in which each variable has a different value.

Chapter 120

The Monolith at Madre de Dios

One possible solution is to assign values to the symbols as follows: $4 = ♎$, $3 = ♏$, $2 = ♐$, and $1 = ♑$. In each row, the number assigned to the rightmost symbol is equal to the number assigned to the first symbol, plus the second, minus the third, minus the fourth. Therefore, 1 solution for the missing symbol is ♑.

Chapter 121

Amazon Dissection

Below is 1 possible solution. (Cut the paper, or draw a line, so that all the symbols that fall in the gray squares are on 1 side). Can you find other solutions?

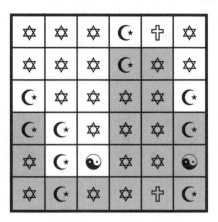

Chapter 122

3 Weird Problems with 3

For problem 1, you can define an arithmetic series as follows: $a_1 = 1$, $a_2 = 2$, $a_3 = 3$, $a_n = a_{n-3} + a_{n-2} + a_{n-1}$ for $n \geq 4$. The sum of each row in the original problem is the sum of its digits. This means that the sum of each row is the sum of the previous 3 rows' sums. One can use this information to write a BASIC program to compute the sum for the thirtieth row: 45,152,016.

For problem 2, it appears that no new atomic species will develop in row 30 that are not already present in row 8. Joseph Zbiciak predicts what species we see in row 30: The species "3" is on the end of every line. Therefore it will be in row 30. The species "31" and the species "331" are both imbedded in a row previous to row 30. Therefore they will be in row 30, because the "middle parts" of each row are duplicated down the list, not modified. The species "1" only shows up every third row. It happens to occur on rows such that (row number) mod 3 = 1. Because 30 mod 3 = 0, the species "1" will *not* occur in row 30. Hence, we have the three species "3", "31", "331" occurring in row 30.

For problem 3, an exact solution is not known. It appears that this algorithm forces 1s out in front all of the time and keeps appending 3s on the end of the row. Hence, you'll see a proliferation of species such as "3331", "33331," "333331," etc. Zbiciak predicts that in row 30, you will have all the species from "3," "31," "331," "3331," "33331," etc., up to "333333333333333333333333331."

Chapter 123

Zen Archery

Problem 1: 8 + 19 + 31 + 41 + 101 = 200. We believe there are roughly 27 distinct solutions. What method did you use to solve this? Must the 8 be present in every solution?

Problem 2: 10 + 19 + 71 = 100.

Chapter 124

Treadmills and Gears

The treadmill does work. (I believe the treadmill would lock if the figure-8 were replaced by a Möbius strip.) The gear train will not lock.

Chapter 125

Anchovy Marriage Test

Monica responded first by throwing her pizza at Dr. Googol. Then she gave him the answers. 9 + 9 + 999 = 1,000, and 646 + 20 = 666 (the small stroke is applied to the first + symbol). A less creative solution is 6 + 6 + 20 ≠ 666. For the last problem, (666/6) × (66-65) = 111. Are there other solutions to these problems? If you were Monica, would you marry Dr. Googol?

Further Reading

PREFACE

Schroeder, M. (1986) *Number Theory in Science and Communication.* 2nd enlarged. Springer: New York

CHAPTER 1. ATTACK OF THE AMATEURS

Cole, K. C. (1998) Beating the pros to the punch. *Los Angeles Times,* Wednesday, March 11, A1: 1.

Mauldin, R. D. (1997) A generalization of Fermat's last theorem: The Beal conjecture and prize problem. *Notices of the American Mathematical Society.* 44 (December): 1436.

Peterson, I. (1998) Picking off more pieces of pi. *Science News.* October 17, 154(16): 255. Also see: http://www.cecm.sfu.ca/projects/pihex/announce5t.html.

Holden, C. (1998) Making *e* easy. *Science,* November 20, 282(5393): 1409.

Peterson, I. (1999) Pi by the billions. *Science News* 156(16): 255.

Stewart, I. (1999) Most-perfect magic squares. *Scientific American.* November, 281(5): 122–123.

CHAPTER 2. WHY DON'T WE USE ROMAN NUMERALS ANYMORE?

Achenbach, J. (1966) *Why Things Are and Why Things Aren't.* Ballantine Books: New York.

CHAPTER 4. THE ULTIMATE BIBLE CODE

Gardner, M. (1998) A quarter-century of recreational mathematics. *Scientific American.* August, 279(2): 68–75.

CHAPTER 5. HOW MUCH BLOOD?

Paulos, John. (1990) *Innumeracy.* Vintage: New York.

CHAPTER 10. NUMBERS BEYOND IMAGINATION

Aho, A. Hopcroft, J. and Ullman. (1974) *Data Structures and Algorithms.* Addison-Wesley: Massachusetts.

Berezin, A. (1987) Super super large numbers. *Journal of Recreational Mathematics.* 19(2): 142–143. This paper discusses the mathematical and philosophical implications of the "superfactorial" function defined by the symbol \$, where $N\$ = N!^{N!^{N!}} \ldots$ (the term $N!$ is repeated $N!$ times).

Clawson, C. (1997) *Mathematical Mysteries.* Plenum: New York.

Davis, P. (1961) *The Lore of Large Numbers.* Random House: New York.

Ellis, K. (1978) Is God a number? In *Number Power In Nature, Art, and Everyday Life.* St. Martin's Press: New York.

Hunter, J., and Madachy, J. (1968) *Mathematical Diversions.* Van Nostrand: New York.

Pickover, C. (1990) Results of the very-large-number contest, *Journal of Recreational Mathematics.* 22(3) 166–169.

Wells, D. (1987) *The Penguin Dictionary of Curious and Unusual Numbers.* Penguin: New York.

CHAPTER 15. QUINCUNX

Dowlatshahi, A. (1979) *Persian Designs and Motifs.* Dover: New York.
Gombrich, E. (1979) *The Sense of Order: A Study in the Psychology of Decorative Art.* Cornell University Press: New York.
Lehner, E. (1969) *Symbols, Signs and Signets.* Dover: New York.
Lockwood, E. and MacMillan, R. (1978) *Geometric Symmetry.* Cambridge: New York.
Pickover, C. (1990) Pentagonal symmetry in historic ornament. In *Five-Fold Symmetry.* I. Hargittai, ed. VCH: New York.
Pickover, C. (1996) Five. *Journal of Recreational Mathematics.* 27(4): 292–296.

CHAPTER 16. JERUSALEM OVERDRIVE

Denes, J., and Keedwell, A. (1974) *Latin Squares and Their Application.* Academic Press: New York.
Gardner, M. (1992) *Fractal Music, Hypercards, and More. . . .* Freeman: NY.
Pickover, C. (2001) *The Zen of Magic Squares, Circles, and Stars.* Princeton University Press: Princeton, New Jersey.

CHAPTER 17. THE PIPES OF PAPUA

Allouche, J.-P., and Shallit, J. (1999) The ubiquitous Prouhet-Thue-Morse sequence. In *Sequences and Their Applications: Proceedings of SETA 1998.* Springer; New York. 1–16.

CHAPTER 18. THE FRACTAL SOCIETY

Pickover, C. (1993) Fractal fantasies. *BYTE.* March: 256.

CHAPTER 19. THE TRIANGLE CYCLE

Pickover, C. (1993) Fractal triangle puzzle. *Journal of Recreational Mathematics.* 25(1): 66.

CHAPTER 20. IQ-BLOCK

Pickover, C. (1992) IQ-Block. *Journal of Recreational Mathematics.* 24(4): 308.
Pickover, C. (1994) IQ-Block. *Journal of Recreational Mathematics.* 26(1): 67–68. This features Joseph Lemire's solutions with holes.

CHAPTER 24. INTERVIEW WITH A NUMBER

Pickover, C. (1995) Vampire numbers. *Theta.* Spring, 9(1): 11–13.
Pickover, C. (1995) Interview with a number. *Discover.* June, 16(6): 136.
Roushe, F. W., and Rogers, D. G. (1997/98) Tame vampires. *Mathematical Spectrum.* 30: 37–39.

CHAPTER 26. SATANIC CYCLES

Pickover, C. (1993) Bicycles from hell. *Theta.* Spring 7(1): 3–4.

CHAPTER 27. PERSISTENCE

Sloane, N. (1973) The persistence of a number. *Journal of Recreational Mathematics.* 6: 97–98.
Guy, R. (1981) *Unsolved Problems in Number Theory.* Springer: New York.

CHAPTER 29. WHY WAS THE FIRST WOMAN MATHEMATICIAN MURDERED?

Deakin, M. (1994) Hypatia and her mathematics. *American Mathematical Monthly*. March, 101(3): 234–243.

Encyclopaedia Brittannica. See http://www.eb.com

Pappas, T. (1997) *Mathematical Scandals*. Wide World Publishing/Tetra: San Carlos, California.

CHAPTER 30. WHAT IF WE RECEIVE MESSAGES FROM THE STARS?

Freudenthal, H. (1960) *Lincos, Design of a Language for Cosmic Intercourse*. North-Holland Publishing: Amsterdam.

CHAPTER 32. EINSTEIN, RAMANUJAN, HAWKING

Berndt, B., and Rankin, R. (1995) *Ramanujan: Letters and Commentary*. American Mathematical Society: Providence, Rhode Island.

Gindikir, S. (1998) Ramanujan the phenomenon. *Quantum*. March/April, 8: 4–9.

Kanigel, R. (1991) *The Man Who Knew Infinity: A Life of the Genius Ramanujan*. Charles Scribner's Sons: New York.

CHAPTER 33. A RANKING OF THE 8 MOST INFLUENTIAL FEMALE MATHEMATICIANS

Encyclopedia Britannica. See http://www.eb.com

O'Connor, John J., and Edmund F. Robertson, "Female mathematicians," http://www-history.mcs.st-and.ac.uk/history/Indexes/Women.html

Smith, S. (1995) *Agnessi to Zeno*. Key Curriculum Press: New York.

CHAPTER 34. A RANKING OF THE 5 SADDEST MATHEMATICAL SCANDALS

Pappas, T. (1997) *Mathematical Scandals*. Wide World Publishing/Tetra: San Carlos, California.

CHAPTER 35. THE 10 MOST IMPORTANT UNSOLVED MATHEMATICAL PROBLEMS

Chudnovsky, D. V., and Chudnovsky, G. V. (1991) Classical constants and functions: Computations and continued fraction expansions. In *Number Theory—New York Seminar 1989–1990*. D. V. Chudnovsky, G. V. Chudnovsky, H. Cohn and M. B. Nathanson, eds. Springer Verlag: New York. 14–74.

Mackenzie, D. (2000) Fermat's Last Theorem's first cousin. *Science*. 287(5454): 792–793.

Smale, S. (1997) Mathematics problems for the next century. *Mathematical Intelligencer*. 20(2): 7–15.

CHAPTER 36. A RANKING OF THE 10 MOST INFLUENTIAL MATHEMATICIANS WHO EVER LIVED

O'Connor, John J., and Edmund F. Robertson, "Biographies of mathematicians," http://www-history.mcs.st-and.ac.uk/history/BiogIndex.html

Singh, S. (1997) *Fermat's Last Theorem*. Fourth Estate: London, England.

CHAPTER 38. A RANKING OF THE 10 MOST INFLUENTIAL MATHEMATICIANS ALIVE TODAY

Encyclopaedia Britannica. See http://www.eb.com
Singh, S. (1997) *Fermat's Last Theorem*. Fourth Estate: London. England.

CHAPTER 39. A RANKING OF THE 10 MOST INTERESTING NUMBERS

Begley, S. (1998) Science finds God. *Newsweek*. July 20, 132(3): 46–52.

CHAPTER 44. THE 15 MOST FAMOUS TRANSCENDENTAL NUMBERS

Pickover, C. (1993) The fifteen most famous transcendental numbers. *Journal of Recreational Mathematics*. 25(1): 12.

CHAPTER 45. WHAT IS NUMERICAL OBSESSIVE-COMPULSIVE DISORDER?

Rapoport, J. (1989) *The Boy Who Couldn't Stop Washing*. Signet: New York.

CHAPTER 46. WHO IS THE NUMBER KING?

Hoffman, P. (1998) *The Man Who Loved Only Numbers*. Hyperion: New York.
Pickover, C. (1998) *Strange Brains and Genius: The Secret Lives of Eccentric Scientists and Madmen*. Quill: New York.
Holden, C. (2000) Analyzing the Erdös star cluster. *Science*. February 4th, 287(5454): 799.

CHAPTER 49. HAILSTONE NUMBERS

Crandall, R. (1978) On the "3x + 1" problem. *Mathematics of Computation*. 32: 1281–1292.
Garner, L. (1981) On the Collatz 3n + 1 problem. *Proceedings of the American Mathematics Society*. 82: 19–22.
Hayes, B. (1984) Computer recreations: On the ups and downs of hailstone numbers. *Scientific American*. January, 250: 10–16.
Legarias, J. (1985) The 3x + 1 problem and its generalizations. *American Mathematics Monthly*. January, 3–23.
Wagon, S. (1985) The Collatz problem. *Mathematical Intelligencer*. 7: 72–76.
Pickover, C. (1989) Hailstone (3n + 1) number graphs. *Journal of Recreational Mathematics*. 21(2): 112–115.

CHAPTER 50. THE SPRING OF KHOSROW CARPET

Burns, A. (1997) Persian recursion. *Mathematics Magazine*. 70(3): 196–199.

CHAPTER 51. THE OMEGA PRISM

Pickover, C. (1996) A note on visualizing the Omega Prism. *The Visual Computer*. 12(9): 451–454.
Schroeder, M. (1986) *Number Theory in Science and Communication*. (2nd enlarged ed.) Springer: New York
Cohen, H. (1993) *A Course in Computational Algebraic Number Theory*. Springer: Berlin.

CHAPTER 52. THE INCREDIBLE HUNT FOR DOUBLE
SMOOTHLY UNDULATING INTEGERS

Shirriff, K. (1994) Comments on double smoothly undulating integers. *Journal of Recreational Mathematics.* 26(2): 103–104.

Ashbacher, C. (1994) Smoothly undulating integers in more than one base. *Journal of Recreational Mathematics.* 26(2): 105–106.

Robinson, D. F. (1994) There are no double smoothly undulating integers in both decimal and binary representation. *Journal of Recreational Mathematics.* 26(2): 102–103.

Schwartz, B. (1994) More on multiple smoothly undulating integers. *Journal of Recreational Mathematics.* 26(2): 108–109.

Pickover, C. (1990) Is there a double smoothly undulating integer? *Journal of Recreational Mathematics.* 22(1): 77–78.

Trigg, C. (1982–83) Palindromic octagonal numerals. *Journal of Recreational Mathematics.* 15(1): 41–46.

CHAPTER 53. ALIEN SNOW: A TOUR OF CHECKERBOARD
WORLDS

Brown, D. (1987) Competition of cellular automata rules. *Complex Systems.* 1: 169–180.

Conway, J., Berlekamp, E., and Guy, R. (1982) *Winning Ways for Your Mathematical Plays.* Academic Press: New York.

Dewdney, A. (1988) *The Armchair Universe.* Freeman: New York.

Levy, S. (1985) The portable universe: Getting to the heart of the matter with cellular automata. *Whole Earth Review Magazine.* Winter, 42–48.

Maeder, D. (1987) The free energy concept in cellular automaton models of solid-solid phase transitions. *Complex Systems.* 1: 131–144.

Peterson, I. (1987) Forest fires, barnacles, and trickling oil. *Science News.* 132: 220–221.

Pickover, C. (1989) Mathematics and beauty VIII: Tesselation automata derived from a single defect. *Computers and Mathematics with Applications.* 17: 361–336.

Poundstone, W. (1985) *The Recursive Universe.* William Morrow: New York.

Schrandt, R., and S. Ulam. (1970) On recursively defined geometrical objects and patterns of growth. In *Essays on Cellular Automata.* A. Burks, ed. University of Illinois Press: Chicago.

Toffoli, T., and Margolus, N. (1987) *Cellular Automata Machines: A New Environment for Modeling.* MIT Press: Cambridge, Massachusetts.

Wolfram, S. (1983) Statistical mechanics of cellular automata. *Review of Modern Physics.* 55: 601–644.

CHAPTER 54. BEAUTY, SYMMETRY, AND PASCAL'S TRIANGLE

Jansson, L. (1973) Spaces, functions, polygons, and Pascal's triangle. *Mathematics Teacher.* 66: 71–77.

Usiskin, Z. (1973). Perfect square patterns in the Pascal triangle. *Mathematics Magazine.* September–October, 203–208.

Bidwell, J. (1973) Pascal's triangle revisited. *Mathematics Teacher.* 66: 448–452.

Gardner, M. (1977) Pascal's triangle. In *Mathematical Carnival.* Vintage Books: New York.

Spencer, D. (1982) *Computers in Number Theory.* Computer Science Press: Rockville, Maryland.

Mandelbrot, B. (1983) *The Fractal Geometry of Nature.* Freeman: New York.

Zhiqing, L. (1985) Pascal's pyramid. *Mathematical Spectrum.* 17(1):1–3.

Wolfram, S. (1984) Geometry of binomial coefficients. *American Mathematics Monthly.* 91: 566–571.

Gordon, J., Goldman, A. and Maps, J. (1986) Superconducting-normal phase boundary of a fractal network in a magnetic field. *Physical Review Letters.* 56: 2280–2283.

Holter, N., Lakhtakia, A., Varadan, V., Vasundara, V., and Messier, R. (1986) On a new class of planar fractals: The Pascal-Sierpinski gaskets. *Journal of Physics A: Mathematics General.* 19: 1753–1759.

Dudley, U. (1987) An infinite triangular array. *Mathematics Magazine.* 61(5): 316–317.

Bondarenko, B. (1990) *Patterns in Pascal's Triangle.* A bibliography lists 406 papers with topics relating to Pascal's triangle. For more information, write Professor B. A. Bondarenko, Institute of Cybernetics, Academy of Science, Uzbekistan, Ul.F. Hodgaeva 34, Tashkent - 143, 700143 Uzbekistan).

Lakhtakia, A., Vasundara, V., Messier, R., and Varadan, V. (1988) Fractal sequences derived from the self-similar extensions of the Sierpinski gasket. *Journal of Physics A: Mathematics General.* 21: 1925–1928.

Edwards, A. (1988) Pascal's triangle—and Bernoulli's and Vieta's. *Mathematical Spectrum.* 33–37.

Pickover, C. (1990) On the aesthetics of Sierpinski gaskets formed from large Pascal's triangles. *Leonardo.* 23(4): 411–417.

Pickover, C. (1993) On computer graphics and the aesthetics of Sierpinski gaskets formed from large Pascal's triangles. In *The Visual Mind: Art and Mathematics.* MIT Press: Cambridge, Massachusetts.

Pickover, C. (1995) Pascal's beast. *Journal of Recreational Mathematics.* 27(2): 81–82.

Micolich, A. and Jonas, D. (1999) Fractal expressionism. *Physics World.* October, 12(10): 25–28.

CHAPTER 55. AUDIOACTIVE DECAY

Hilgemeir, M. (1986) Die Gleichniszahlen-Reihe. *Bild der Wissenschaft.* 12: 194–195.

Pickover, C. (1987) DNA vectorgrams: Representation of cancer gene sequences as movements along a 2-D cellular lattice. *IBM Journal of Research and Development.* 31: 111–119.

Pickover, C., and Khorasani, E. (1991) Visualization of the Gleichniszahlen-Reihe, an unusual number theory sequence. *Mathematical Spectrum.* 23(4): 113–115.

Lakhtakia, A., and Pickover, C. (1993) Observations on the Gleichniszahlen-Reihe, an unusual number theory sequence. *Journal of Recreational Mathematics.* 25(3): 202–205.

Hilgemeir, M. (1997) One metaphor fits all: A fractal voyage with Conway's audioactive decay. In *Fractal Horizons: The Future Use of Fractals.* C. Pickover, ed. St. Martin's Press: New York.

CHAPTER 56. DR. GOOGOL'S PRIME PLAID

Hellemans, A. (1999) Internet security code is cracked. *Science.* September 3, 285: 1472–1473.

Schroeder, M. (1986) *Number Theory in Science and Communication.* Springer: New York.

Spencer, D. (1982) *Computers in Number Theory.* Computer Science Press: Rockville, Maryland.

Stein, M., Ulam, S., and Wells, M. (1964) A visual display of some properties of the distribution of primes. *Mathematics Monthly.* 71(5): 516–520.

CHAPTER 57. SAIPPUAKAUPPIAS

Calandra, M. (1985–86) Integers which are palindromic in both decimal and binary nota-tion. *Journal of Recreational Mathematics.* 18(1): 47.

Gardner, M. (1979) *Mathematical Circus.* Knopf: New York. 242–252.

Lakhtakia, A., and Pickover, C. (1990) Some observations on palindromic numbers. *Journal of Recreational Math.* 22(1): 55–60.

Pilpel, S. (1985–86) Some more double palindromic integers. *Journal of Recreational Mathematics.* 18(3): 174–176.

Peretti, A. (1985) Query 386C. *Notices of the American Mathematics Society.* May.

Schmidt, H. (1988) Palindromes: Density and divisibility. *Mathematics Magazine.* 61(5): 297–299.

Schwartz, B. (1986) Counting palindromes. *Journal of Recreational Mathematics.* 18(3): 177–179.

CHAPTER 58. EMORDNILAP NUMBERS

Ellis, K. (1987) *Number Power.* St. Martin's Press: New York. (122–123).

Gruenberg, F. (1984) Computer recreations. *Scientific American.* April: 19–26.

Bendat, J., and Piersol, R. (1966) *Measurement and Analysis of Random Data.* John Wiley and Sons: New York. For information on power spectra.

Gardner, M. (1979) *Mathematical Circus.* Knopf: New York. 242–252.

Kröber, G. (1998) Structure generation by palindromization. *Computers & Graphics.* 22(2/3): 307–317.

Pickover, C. (1991) Reversed numbers and palindromes. *Journal of Recreational Mathematics.* 26(4): 243–247.

Richardson, R., and Shannon, C. (1996) Palindrome pictures. *Computers & Graphics.* 20(4): 597–603.

Trigg, C. (1972) More on palindromes by reversal-addition. *Mathematics Magazine.* 45: 184–186

Trigg, C. (1973) Versum sequences in the binary system. *Pacific Journal of Mathematics.* 47: 263–275.

CHAPTER 59. THE DUDLEY TRIANGLE

Dudley, U. (1987) An infinite triangular array. *Mathematics Magazine.* 61(5): 316–317.

Pickover, C., and Khorasani, E. (1992) Infinite triangular arrays. *Journal of Recreational Mathematics.* 24(2): 104–110.

CHAPTER 60. MOZART NUMBERS

Hartson, J. (1988) *Drunken Goldfish and Other Irrelevant Scientific Research.* Sterling: New York.

CHAPTER 61. HYPERSPACE PRISONS

Heinrich, B. (1991) The ways of coprophiles. *Science.* 254(5033): 878–879.

Hanski, I., and Cambefort, Y. (1991) *Dung Beetle Ecology.* Princeton University Press: Princeton, New Jersey.

CHAPTER 62. TRIANGULAR NUMBERS

Belier, A. (1966) *Recreations in the Theory of Numbers.* Dover: New York.

Guy, R. (1994) Every number is expressible as the sum of how many polygonal numbers? *American Mathematics Monthly.* February, 101(2): 169–172.

Kordemsky, B. (1972) *The Moscow Puzzles*. Dover: New York.

Wells, D. (1987) *The Penguin Dictionary of Curious and Interesting Numbers*. Penguin: New York. Many of the interesting triangular number formulas come from this book.

CHAPTER 63. HEXAGONAL CATS

Wells, D. (1986) *The Penguin Dictionary of Curious and Interesting Numbers*. Penguin Books: New York.

Pickover, C. (1994) Undulating undecamorphic and undulating pseudofareymorphic integers. *Journal of Recreational Mathematics*. 26(2): 110–116.

Trigg, C. (1987) Hexamorphic numbers. *Journal of Recreational Mathematics*. 19(1): 42–55.

CHAPTER 65. A LOW-CALORIE TREAT

Pickover, C., and Angelo, M. (1995) On the existence of cakemorphic integers. *Journal of Recreational Mathematics*. 27(1): 13–16.

Trigg, C. (1980–81) A matter of morphic nomenclature. *Journal of Recreational Mathematics*. 13(1): 48–49.

Trigg, C. (1987) Hexamorphic numbers. *Journal of Recreational Mathematics*. 19(1): 42–55.

Wells, D. (1986) *The Penguin Dictionary of Curious and Interesting Numbers*. Penguin Books: New York.

CHAPTER 66. THE HUNT FOR ELUSIVE SQUARIONS

Spencer, D. (1982) *Computers in Number Theory*. Computer Science Press: Rockville, Maryland.

Pickover, C., and Gursky, M. (1991) Pair square numbers. *Journal of Recreational Mathematics*. 23(4): 312–314.

CHAPTER 67. KATYDID SEQUENCES

Guy, R. (1983) Don't try to solve these problems! *American Mathematics Monthly*. January, 90(1): 35.

CHAPTER 68. PENTAGONAL PIE

Chen, S. (1989) *The IBM Programmer's Challenge*. Tab Books: Blue Ridge, Pennsylvania. Contains a computer program for calculating Catalan numbers.

CHAPTER 69. AN *A*?

Schroeder, M. (1991) *Fractals, Chaos, Power Laws*. Freeman: New York.

CHAPTER 70. HUMBLE BITS

Schroeder, M. (1991) *Fractals, Chaos, Power Laws*. Freeman: New York.

Mandelbrot, B. (1982) *The Fractal Geometry of Nature*. Freeman: New York.

Pickover, C., and Lakhtakia, A. (1989) Diophantine equation graphs for $x^2y = c$. *Journal of Recreational Mathematics*. 21(3): 167–170.

Pickover, C. (1992) Intricate patterns from logical operators. *Theta*. Spring, 6(1): 11–14.

Szyszkowicz, M. (1991) Patterns generated by logical operators. *Computers and Graphics*. 15(2): 299–300.

CHAPTER 71. MR. FIBONACCI'S NEIGHBORHOOD

Ashbacher, C. (1989) Repfigit numbers. *Journal of Recreational Mathematics*. 21(4): 310–311.

Keith, M. (1987) Repfigit numbers. *Journal of Recreational Mathematics*. 19(1): 41–42.

Wagon, S. (1985) The Collatz problem. *Mathematical Intelligencer.* 7: 72–76.

Peterson, I.(1999) Fibonacci at random. *Science News.* 155(24):376–377.

Pickover, C. (1990) All known replicating Fibonacci-digits less than one billion, *Journal of Recreational Mathematics.* 22(3): 176–178.

Keith, M. (1994) All repfigit numbers less than 100 billion. *Journal of Recreational Mathematics.* 26(3): 181–184.

Heleen, B. (1994) Finding repfigits—a new approach. *Journal of Recreational Mathematics.* 26(3): 184–187.

Robinson, N. (1994) All known replicating Fibonacci digits less than one thousand billion. *Journal of Recreational Mathematics.* 26(3): 188–191.

Shirriff, K. (1994) Computing replicating Fibonacci digits. *Journal of Recreational Mathematics.* 26(3): 188–192.

Esche, H. (1994) Non-decimal replicating Fibonacci digits. *Journal of Recreational Mathematics.* 26(3): 193–195.

CHAPTER 72. APOCALYPTIC NUMBERS

Ashbacher, C. (1995) Apocalyptic primes. *Journal of Recreational Mathematics.* 28(3): 237–238.

Kierstead, F. (1995) Commentary on apocalyptic primes. *Journal of Recreational Mathematics.* 28(3): 238.

Pickover, C. (1993) Apocalypse numbers. *Math Spectrum.* 26(1): 10–11.

CHAPTER 77. ⊎-NUMBERS FROM LOS ALAMOS

Cooper, N. (1989) *From Cardinals to Chaos.* Cambridge University Press: New York. Topics: Stan Ulam, iteration, strange attractors, Monte Carlo methods, the human brain, random number generators, number theory, and genetics.

Guy, R. (1981) *Unsolved Problems in Number Theory.* Springer: New York.

Recamoan, B. (1973) Questions on a sequence of Ulam. *American Mathematics Monthly.* 80: 919–920.

CHAPTER 78. CREATOR NUMBERS ♌

Guy, R. (1981) *Unsolved Problems in Number Theory.* Springer: New York.

Pickover, C., and Shirriff, K. (1992) The terrible twos problem. *Theta: A Journal of Mathematics.* Autumn, 6(2): 3–7.

CHAPTER 79. PRINCETON NUMBERS

Robbins, D. (1991) The story of 1, 2, 7, 42, 429, 7436, . . . *Mathematical Intelligencer.* 13(2): 12–18.

CHAPTER 82. APOCALYPTIC POWERS

Pickover, C. (1994) Apocalpytic and double apocalyptic powers. *Mathematics and Computer Education.* Spring: 211–212.

Rickert, J. (1998) Apocalyptic powers. *Journal of Recreational Mathematics.* 29(2): 102–106.

CHAPTER 85. THE ALIENS FROM *INDEPENDENCE DAY*

Pennington, J. (1957) The red and white cows. *American Mathematics Monthly.* March, 64: 197–198.

Schroeder, M. (1991) *Fractals, Chaos, Power Laws.* Freeman: New York. Discusses similar sequences including what Schroeder has called the rabbit sequence: 1 0 1 1 0 1 0 1. . . .

CHAPTER 87. UNDULATION IN MONACO

Pickover, C. (1996) The undulation of the monks. *Mathematical Spectrum.* 28(2): 31–32.

CHAPTER 92. THE HUNT FOR PRIMES IN PI

Gardner, M. (1999) *Gardner's Whys and Wherefores.* Prometheus Books: Amherst, New York.

CHAPTER 94. PERFECT, ABUNDANT, AMICABLE, AND SUBLIME NUMBERS

Belier, A. (1966) *Recreations in the Theory of Numbers.* Dover: New York.

Brown, H., and Fleigel, H. (1995) Almost perfect numbers. *Journal of Recreational Mathematics.* 27(4): 255–261.

Hodges, L., and Reid, M. (1995) Three consecutive abundant numbers. *Journal of Recreational Mathematics.* 27(2): 156.

Spencer, D. (1982) *Computers in Number Theory.* Computer Science Press: Rockville, Maryland.

CHAPTER 96. CARDS, FROGS, AND FRACTAL SEQUENCES

Conway, J. (1988) Some crazy sequences. Videotaped talk, AT&T Bell Labs, July 15.

Hofstadter, D. (1980) *Gödel, Escher, Bach.* Vintage Books: New York.

Kimberling, C., (1995) Numeration systems and fractal sequences. *Acta Arithmetica.* 73: 103–117.

Kimberling, C., and Shultz, H. (1999) Card sorting by dispersions and fractal sequences. *Ars Combinatoria.* Forthcoming.

Kimberling, C. (1997) Fractal sequences and interspersions. *Ars Combinatoria.* 45: 157–168.

Schroeder, M. (1991) *Fractals, Chaos, Power Laws.* Freeman: New York.

Mallows, C. (1991) Conway's challenge sequence. *American Mathematics Monthly.* January: 5–20.

Pickover, C. (1995) The crying of fractal batrachion 1,489. *Computers and Graphics.* 19(4): 611–615.

CHAPTER 97. FRACTAL CHECKERS

Pickover, C. (1993) Recursive worlds. *Dr. Dobb's Software Journal.* September, 18(9): 18–26.

CHAPTER 99. EVERYTHING YOU WANTED TO KNOW ABOUT TRIANGLES BUT WERE AFRAID TO ASK

Belier, A. (1966) *Recreations in the Theory of Numbers.* Dover: New York.

CHAPTER 100. CAVERN GENESIS AS A SELF-ORGANIZING SYSTEM

Fournier, A., Fussell, D., and Carpenter, L. (1982) Computer rendering of stochastic models. *Communications of the ACM.* 25: 371–384.

Mandelbrot, B. (1983) *The Fractal Geometry of Nature.* Freeman: New York.

Voss, R. (1985) Random fractal forgeries. In *Fundamental Algorithms for Computer Graphics.* R. Earnshaw, ed. Springer: New York.

Musgrave, F., Kolb, C., and Mace, R. (1989) The synthesis and rendering of eroded fractal terrain. *Computer Graphics* (ACM-SIGGRAPH). 23(3): 41–48.

Pickover, C. (1985) *Keys to Infinity.* Wiley: New York.

Cahill, T. (1991) Charting the splendors of Lechuguilla cave. *National Geographic.* March 179(3): 34–59. Contains breathtaking photographs of real caverns.

Jackson, D. (1982) *Underground Worlds.* Time-Life Books: Alexandria, Virginia.

Pickover, C. (1997) *The Loom of God: Mathematical Tapestries at the Edge of Time.* Plenum: New York.

Pickover, C. (1994) *Mazes for the Mind: Computers and the Unexpected.* St. Martin's Press: New York.

Pickover, C. (1993) *Computers and the Imagination.* St. Martin's Press: New York.

Pickover, C. (1992) *Computers, Pattern, Chaos, and Beauty.* St. Martin's Press: New York.

Pickover, C. (1994) *Chaos in Wonderland: Visual Adventures in a Fractal World.* St. Martin's Press: New York.

Pickover, C. (1998) Cavern genesis as a self-organizing system. *Leonardo.* 31(3):206–224.

CHAPTER 101. MAGIC SQUARES, TESSERACTS, AND OTHER ODDITIES

Boardman, S. (1996) Theta problem page. *Theta.* 10(2): 25.

Gardner, M. (1998) Magic squares cornered. *Nature.* September, 395(6699): 216–217. Contains information on the latest breakthrough research with magic squares.

Kurchan, R. (1991) An all pandigital magic square (question posed by Rudolf Ondrejka). *Journal of Recreational Mathematics.* 23(1): 69–78.

Hendricks, J. (1990) The magic tesseracts of Order 3 complete. *Journal of Recreational Mathematics.* 22(1): 16–26.

Hendricks, J. (1962) The five- and six-dimensional magic hypercubes of Order 3. *Canadian Matehmatical Bulletin.* May, 5(2): 171–189.

Hendricks, J. (1995) Magic tesseract. In *The Pattern Book: Fractals, Art, and Nature.* C. Pickover, ed. World Scientific: River Edge, New Jersey.

CHAPTER 102. FABERGÉ EGGS SYNTHESIS

Audsley, W., and Audsley, G. (1978) *Designs and Patterns from Historic Ornament.* Dover: New York.

Dowlatshahi, A. (1979) *Persian Designs and Motifs.* Dover: New York.

Hayes, B. (1986) On the bathtub algorithm for dot-matrix holograms. *Computer Language.* 3: 21–25.

Pickover, C. (1987) A recipe for self-decorating eggs. *Computer Language.* 4(11): 55–58.

Pickover, C. (1998) Fabergé eggs via residue analysis. *Leonardo.* 31(4): 280.

CHAPTER 103. BEAUTY AND GAUSSIAN RATIONAL NUMBERS

Ford, L. R. (1938) Fractions. *American Mathematical Monthly.* 45: 586–601.

Pickover, C. (1995) *Keys to Infinity.* Wiley: New York.

Pickover, C. (1997) A note on geometric representations of Gaussian rational numbers. *Visual Computer.* 13:127–130.

Pickover, C. (1995) Fractal milkshakes and infinite archery. *Leonardo.* 28(4): 333–334.

CHAPTER 104. A BRIEF HISTORY OF SMITH NUMBERS

Dudley, U. (1994) Smith numbers. *Mathematics Magazine.* February, 67(1): 62–65. This is the best written history of Smith numbers and their role in mathematical research.

Wilansky, A. (1982) Smith numbers. *Two-Year College Mathematics Journal.* 13: 21.

Oltikar, S., Sham, E., and Wayland, K. (1983) Construction of Smith numbers. *Mathematics Magazine.* 56: 36–37.

About the Author

You already know about Dr. Googol from the introductory section describing his life. Perhaps you may also have some interest in Cliff Pickover's background.

Clifford A. Pickover received his Ph.D. from Yale University's Department of Molecular Biophysics and Biochemistry. He graduated first in his class from Franklin and Marshall College, after completing the four-year undergraduate program in three years. His many books have been translated into Italian, German, Japanese, Chinese, Korean, Portuguese, and Polish. He is author of the popular books *The Girl Who Gave Birth to Rabbits* (Prometheus, 2000), *Cryptorunes: Codes and Secret Writing* (Pomegranate, 2000), *The Zen of Magic Squares, Circles and Stars* (Princeton University Press, 2001), *Surfing Through Hyperspace* (Oxford University Press, 1999), *The Science of Aliens* (Basic Books, 1998), *Time: A Traveler's Guide* (Oxford University Press, 1998), *Strange Brains and Genius: The Secret Lives of Eccentric Scientists and Madmen* (Plenum, 1998), *The Alien IQ Test* (Basic Books, 1997), *The Loom of God* (Plenum, 1997), *Black Holes— A Traveler's Guide* (Wiley, 1996), and *Keys to Infinity* (Wiley, 1995). He is also author of numerous other highly-acclaimed books including *Chaos in Wonderland: Visual Adventures in a Fractal World* (1994), *Mazes for the Mind: Computers and the Unexpected* (1992), *Computers and the Imagination* (1991) and *Computers, Pattern, Chaos, and Beauty* (1990), all published by St. Martin's Press—as well as the author of over 200 articles concerning topics in science, art, and mathematics. He is also coauthor, with Piers Anthony, of *Spider Legs*, a science-fiction novel recently listed as Barnes and Noble's second-bestselling science-fiction title.

Pickover is currently an associate editor for the scientific journals *Computers and Graphics* and *Theta Mathematics Journal* and is an editorial board member for Odyssey, *Idealistic Studies, Leonardo,* and *YLEM*. He has been a guest editor for several others.

Pickover is editor of the books *Chaos and Fractals: A Computer Graphical Journey* (Elsevier, 1998), *The Pattern Book: Fractals, Art, and Nature* (World Scientific, 1995), *Visions of the Future: Art, Technology, and Computing in the Next Century* (St. Martin's Press, 1993), *Future Health* (St. Martin's Press, 1995), *Fractal Horizons* (St. Martin's Press, 1996), and *Visualizing Biological Information* (World Scientific, 1995), and coeditor of the books *Spiral Symmetry* (World Scientific, 1992) and *Frontiers in Scientific Visualization* (Wiley, 1994). His primary interest is finding new ways to continually expand creativity by melding art, science, mathematics, and other seemingly disparate areas of human endeavor.

The *Los Angeles Times* recently proclaimed, "Pickover has published nearly a book a year in which he stretches the limits of computers, art and thought." Pickover received first prize in the Institute of Physics "Beauty of Physics Photographic Competition." His computer graphics have been featured on the cover of many popular magazines, and his research has recently received considerable attention by the press—including CNN's *Science and Technology Week*, the Discovery Channel, *Science News*, the *Washington Post*, *Wired*, and the *Christian Science Monitor*—and also in international exhibitions and museums. *Omni* magazine recently described him as "van Leeuwenhoek's twentieth-century equivalent." *Scientific American* has several times featured his graphic work, calling it "strange and beautiful, stunningly realistic." *Wired* magazine wrote, "Bucky Fuller thought big, Arthur C. Clarke thinks big, but Cliff Pickover outdoes them both." Among his many patents, Pickover has received U.S. Patent 5,095,302 for a 3-D computer mouse, 5,564,004 for strange computer icons, and 5,682,486 for black-hole transporter interfaces to computers.

Dr. Pickover is currently a Research Staff Member at the IBM T. J. Watson Research Center, where he has received twenty invention achievement awards, three research division awards, and five external honor awards. For many years, Dr. Pickover was also lead columnist for the "Brain-Boggler" column in *Discover* magazine, and he is currently the "Brain-Strain" columnist in *Odyssey*.

Dr. Pickover's hobbies include practicing Ch'ang-Shih Tai-Chi Ch'uan and Shaolin Kung Fu, raising golden and green severums (large Amazonian fish), and playing the piano (mostly jazz). He is also a member of The SETI League, a group of signal-processing enthusiasts who systematically search the sky for intelligent extraterrestrial life. His web site has received over 300,000 visits: http://www.pickover.com. He can be reached at P.O. Box 549, Millwood, New York 10546–0549 USA.

index